CONTEMPORARY MATHEMATICS

305

Quantum Computation and Information

AMS Special Session
Quantum Computation and Information
January 19–21, 2000
Washington, D.C.

Samuel J. Lomonaco, Jr.
Howard E. Brandt
Editors

American Mathematical Society
Providence, Rhode Island

Editorial Board

Dennis DeTurck, managing editor

Andreas Blass   Andy R. Magid   Michael Vogelius

This volume contains the proceedings of an AMS Special Session on Quantum Computation and Information held in Washington, D.C., on January 19–21, 2000.

2000 *Mathematics Subject Classification.* Primary 81P68, 94-02;
Secondary 68Q05, 81-01, 81R99, 94A60, 94A99.

Library of Congress Cataloging-in-Publication Data
AMS Special Session Quantum Computation and Information (2000 : Washington, D.C.)
 Quantum computation and information : AMS Special Session Quantum Computation and Information, Washington, D.C., January 19–21, 2000 / Samuel J. Lomonaco, Jr., Howard E. Brandt, editors.
   p. cm. — (Contemporary mathematics ; v. 305)
 Includes bibliographic references.
 ISBN 0-8218-2140-7 (alk. paper)
 1. Quantum computers—Congresses. I. Lomonaco, Samuel J. II. Brandt, Howard E. III. Title. IV. Contemporary mathematics (American Mathematical Society) ; v. 305.

QA76.889.A47 2000
004.1′4–dc21   2002026239

Copying and reprinting. Material in this book may be reproduced by any means for educational and scientific purposes without fee or permission with the exception of reproduction by services that collect fees for delivery of documents and provided that the customary acknowledgment of the source is given. This consent does not extend to other kinds of copying for general distribution, for advertising or promotional purposes, or for resale. Requests for permission for commercial use of material should be addressed to the Acquisitions Department, American Mathematical Society, 201 Charles Street, Providence, Rhode Island 02904-2294, USA. Requests can also be made by e-mail to reprint-permission@ams.org.

 Excluded from these provisions is material in articles for which the author holds copyright. In such cases, requests for permission to use or reprint should be addressed directly to the author(s). (Copyright ownership is indicated in the notice in the lower right-hand corner of the first page of each article.)

© 2002 by the American Mathematical Society. All rights reserved.
The American Mathematical Society retains all rights
except those granted to the United States Government.
Printed in the United States of America.

∞ The paper used in this book is acid-free and falls within the guidelines
established to ensure permanence and durability.
Visit the AMS home page at http://www.ams.org/

10 9 8 7 6 5 4 3 2 1   07 06 05 04 03 02

Contents

Preface	v
Gilles Brassard Awarded Pot de Vin Prize	vii
List of Participants	viii

Space searches with a quantum robot
  PAUL BENIOFF — 1

Perturbation theory and numerical modeling of quantum logic operations with a large number of qubits
  G. P. BERMAN, G. D. DOOLEN, D. I. KAMENEV, G. V. LÓPEZ, AND V. I. TSIFRINOVICH — 13

Inconclusive rate with a positive operator valued measure
  HOWARD E. BRANDT — 43

Quantum amplitude amplification and estimation
  GILLES BRASSARD, PETER HØYER, MICHELE MOSCA, AND ALAIN TAPP — 53

Manipulating the entanglement of one copy of a two-particle pure entangled state
  LUCIEN HARDY — 75

Geometric algebra in quantum information processing
  TIMOTHY F. HAVEL AND CHRIS J. L. DORAN — 81

Quantum computing and the Jones polynomial
  LOUIS H. KAUFFMAN — 101

Quantum hidden subgroup algorithms: A mathematical perspective
  SAMUEL J. LOMONACO, JR. AND LOUIS H. KAUFFMAN — 139

Improved two-party and multi-party purification protocols
  ELITZA N. MANEVA AND JOHN A. SMOLIN — 203

Quantum games and quantum algorithms
  DAVID A. MEYER — 213

A proof that measured data and equations of quantum mechanics can be linked only by guesswork
  JOHN M. MYERS AND F. HADI MADJID — 221

Quantum computation by geometrical means
    JIANNIS PACHOS   245

Pauli exchange and quantum error correction
    MARY BETH RUSKAI   251

Relative entropy in quantum information theory
    BENJAMIN SCHUMACHER AND MICHAEL D. WESTMORELAND   265

An unentangled Gleason's theorem
    NOLAN R. WALLACH   291

Entangled chains
    WILLIAM K. WOOTTERS   299

Preface

This book is a collection of papers given by invited speakers at the AMS Special Session on Quantum Computation and Information held at the Annual Meeting of the American Mathematical Society in Washington, DC, January 19-21, 2000. This AMS Special Session was held in conjunction with the AMS Short Course on Quantum Computation, January 17-18, 2000, which has been published separately as the AMS PSAPM Volume 58 entitled "Quantum Computation: A Grand Mathematical Challenge for the Twenty-First Century and the Millennium."

This Special Session, together with its accompanying Short Course, was the first time at an AMS meeting that the new and emerging discipline of quantum information science was formally introduced to the AMS community. It is hoped that this event, together with the two books recording this occasion, will act as a catalyst to encourage members of the mathematical community to take advantage of the many mathematical research opportunities arising from the Grand Challenge of Quantum Information Science.

This book was partially supported by the Defense Advanced Research Projects Agency (DARPA) and Air Force Materiel Command USAF under agreement number F30602-01-0522, by Army Research Office (ARO) Grant #P-38804-PH-QC, by the National Institute of Standards and Technology (NIST), and by L-O-O-P Fund No. 2000WADC.

The editors thank UMBC and ARL for support in this endeavor. Thanks are also due to Jekkin Shah for spending many hours transforming the papers contributed to this volume into AMS LaTeX. Moreover, the editors thank the staff of the American Mathematical Society for their editorial support.

Finally, thanks are due to all the AMS Session participants whose efforts and contributions made all this possible.

Samuel J. Lomonaco, Jr.		Howard E. Brandt
Univ. of Maryland Baltimore County		Army Research Lab
Baltimore, MD 21250	and	Adelphi, MD 20783
Lomonaco@umbc.edu		hbrandt@arl.army.mil
www.csee.umbc.edu/~lomonaco		

June, 2002

Gilles Brassard Awarded Pot de Vin Prize

The Millennium 2000 Pot de Vin Prize was awarded to Gilles Brassard for his many outstanding contributions to quantum computation and quantum information science. This award was presented at the AMS Special Session on Quantum Computation and Information held at the Annual Meeting of the American Mathematical Society in Washington, DC, in January 2000.

Description of the Pot de Vin Prize

The Pot de Vin Prize is an award coveted by Nobel Laureates and Fields Medalists. It was established to bribe (FR, pot de vin) major contributors to the field of quantum computation and quantum information into giving an invited talk. All Nobel Laureates and Fields Medalists are automatically disqualified from receiving the prize. The Pot de Vin Prize is sponsored by the L-O-O-P Fund.

List of Participants

The First Special Session on Quantum Computation and Information
Annual Meeting of the American Mathematical Society
Washington, DC
January 19-21, 2000

Paul Benioff
Charles H. Bennett
Gennady P. Berman
Howard E. Brandt
Gilles Brassard
Christopher A. Fuchs
Daniel Gottesman
Lov K. Grover
Lucien Hardy
Timothy F. Havel
Louis H. Kauffman
Alexei Y. Kitaev
Samuel J. Lomonaco

Gianfranco F. Mascari
David A. Meyer
John M. Myers
Jiannis Pachos
Mary Beth Ruskai
Benjamin W. Schumacher
Peter W. Shor
John A. Smolin
Nolan R. Wallach
Michael D. Westmoreland
Umesh V. Vazirani
William K. Wootters
Wojceich H. Zurek

Space Searches with a Quantum Robot

Paul Benioff

ABSTRACT. Quantum robots are described as mobile quantum computers and ancillary systems that move in and interact with arbitrary environments. Their dynamics is given as tasks which consist of sequences of alternating computation and action phases. A task example is considered in which a quantum robot searches a space region to find the location of a system. The possibility that the search can be more efficient than a classical search is examined by considering use of Grover's Algorithm to process the search results. For reversible searches this is problematic for two reasons. One is the removal of entanglements generated by the search process. The other is that even if the entanglement problem can be avoided, the search process in 2 dimensional space regions is no more efficient than a classical search. However quantum searches of higher dimensional space regions are more efficient than classical searches. Reasons why quantum robots are interesting independent of these results are briefly summarized.

1. Introduction

Quantum computers have been the subject of much study, mainly because of computations that can be done more efficiently than on classical computers. Well known examples include Shor's and Grover's algorithms, [1, 2]. Quantum robots have also been recently described as mobile systems, with an on board quantum computer and ancillary systems, that move in and interact with environments of quantum systems [3]. Dynamics of quantum robots are described as tasks consisting of alternating computation and action phases.

Quantum robots can be used to carry out many types of tasks. These range from simple ones such as searching a region of space to determine the unknown location of a system to complex tasks such as carrying out physical experiments. The fact that the spatial searches are similar to the data base searches which are efficiently implemented using Grover's Algorithm [2] suggests that similar results might hold for use of Grover's Algorithm to process results of a quantum search of a spatial region. This is an example of the possible applicability of Grover's algorithm to various physical measurements [6]. If this can be done for the search task, then

---

2000 *Mathematics Subject Classification.* Primary 81P68, 93C85; Secondary 94A99.

This work is supported by the U.S. Department of Energy, Nuclear Physics Division, under contract W-31-109-ENG-38.

one would have an example of a task that can be carried out more efficiently by a quantum robot than by any classical robot.

This possibility is analyzed here for a search of a 2 dimensional space region by a quantum robot to locate a system. It is seen that for reversible searches with time independent unitary dynamics, there are two problems preventing the efficient use of Grover's Algorithm. One is that it appears impossible to remove entanglements generated during the search process. The other problem is that the action in the search task, which is the equivalent of the "oracle function", which is assumed in Grover's Algorithm to be evaluated on any argument in one step, takes many steps to evaluate. The result is that even if the entanglement problem is ignored, quantum searches of 2 dimensional space regions are no more efficient than classical searches. However quantum searches of higher dimensional space regions are more efficient than classical searches.

The plan of this paper is to give, in the next section, a brief description of quantum robots and a summary of how they are different from quantum computers. This is followed, in Section 2, by a description of the dynamics of tasks as sequences of alternating computation and action phases. An explicit description of the dynamics is given as a Feynman sum over computation-action phase paths. An example of a quantum robot searching a space area to determine the unknown location of a system is then described, Section 3. The next section is concerned with the use of Grover's algorithm to process the search results. A very brief summary of Grover's Algorithm, in Subsection 4.1, is followed by a description of the problems encountered, Subsection 4.2. The paper finishes with a discussion of why quantum robots are interesting independent of these results.

2. Quantum Robots

2.1. Comparison with Quantum Computers.
Quantum robots are similar to quantum computers in that an important component is an on board quantum computer. Other systems such as a memory system m, an output system o, and a control qubit c are also present in quantum robots [3]. A relatively minor difference is that quantum robots are mobile whereas quantum computers are stationary relative to the environment. For quantum Turing machine models of quantum computers the head moves but the quantum registers are stationary. For networks of quantum gates the qubit systems move but the gates are stationary. This is shown in physical models of interacting qubits. Examples include ion trap models [4] or nuclear magnetic resonance models [5]. In these cases the ion traps and the liquid of active molecules are stationary.

As is the case for quantum computers the effects of the environment on the component systems of the quantum robot need to be minimized. Methods to achieve this include the possible use of shielding or quantum error correction codes [8]. However other than this the dynamical properties of the environment are completely arbitrary.

This is quite different than the case for quantum computers. To see this assume for quantum Turing machines that the quantum registers are the environment of the head. Similarly for gate networks the moving qubits may be considered as the environment of the network of quantum gates. Here the dynamics of these systems is quite restricted in that the states of the registers or moving qubits can change only during interaction with the head or the gate systems. Also the types of changes that can occur are limited to those appropriate for the specific computation being carried out.

This shows the main difference between quantum robots and quantum computers, namely, that for quantum computers the states of the qubits must not change spontaneously in the absence of interactions related to the computation. No such dynamical restrictions apply to the environment of quantum robots. The states of environmental systems may change spontaneously whether the quantum robot is or is not interacting with them.

Another aspect relates to the requirement that the quantum robot cannot be a multistate system with one or a very few number of degrees of freedom, but must include a quantum computer. One reason is the need for computations as part of the implementation of any task (see below). Another is that the quantum robot may need to be able to respond to a large number of different environmental states. Also a large repertoire of different (task dependent) responses to the same environmental state must be available. If the total number N of needed responses is large then the only physically reasonable approach is to make the number of degrees of freedom of the quantum robot proportional to $\log N$. This is satisfied by including a quantum computer on board.

2.2. Task Dynamics. The dynamics of quantum robots are described as tasks consisting of alternating computation and action phases [**3**]. The purpose of each computation phase is to determine what is to be done in the next action phase. The computation may depend on the states of o, m, and the local environment as input. The goal is to put o in one of a set $\mathcal{B}$ of basis states each of which specifies an action. During the computation the quantum robot does not move. Interactions with the environment, if any, are limited to local entanglement interactions of the type occurring in the measurement process (premeasurements in the sense of Peres [**7**]).

The purpose of an action phase is to carry out the action determined by the previous computation phase. The action is determined by the state of o and may include local premeasurements of the environment state. Activities during this phase include motion of the quantum robot and local changes in the state of the environment. The action is independent of the state of the on board quantum computer and the m system. If o is in a state in $\mathcal{B}$, then the state does not change during the action.

The purpose of the control qubit c is to regulate which type of phase is active. The computation [action] phase is active only if c is in state $|1\rangle$, $[|0\rangle]$. Thus the last step of the computation [action] phase is the change $|1\rangle \to |0\rangle$, $[|0\rangle \to |1\rangle]$.

The overall system evolution is described here using a discrete space time lattice. In this case a unitary elementary step operator Γ gives the overall system evolution during an elementary time step Δ. Since Γ has, in general, nonzero matrix elements between environmental degrees of freedom and quantum robot degrees of freedom, it describes the evolution of the environment and the quantum robot as well as interactions between the environment and the quantum robot.

It is useful to decompose Γ into two terms based on the states of the control qubit. If P_0^c and P_1^c are projection operators on the respective control qubit states $|0\rangle$ and $|1\rangle$, then

(2.1) $$\Gamma = \Gamma(P_0^c + P_1^c) = \Gamma_a + \Gamma_c.$$

Here Γ_a and Γ_c are step operators for the action and computation phases. Interactions among environmental degrees of freedom as well as degrees of freedom of the quantum robot other than those taking part in the task dynamics, if any, are also included in both operators.

Some of the conditions described for the computation and action phases are reflected in properties that the operators Γ_a and Γ_c must satisfy. In particular, if P_x^{QR}, P_d^o are projection operators for finding the quantum robot at each lattice position x ($x = x_1, x_2, \cdots, x_d$ in d-dimensional space) and the output system in any state $|d\rangle$ in $\mathcal{B}$, then one has

$$\begin{aligned}\Gamma_c P_x^{QR} &= P_x^{QR} \Gamma_c \\ \Gamma_a P_d^o &= P_d^o \Gamma_a\end{aligned} \tag{2.2}$$

These commutation relations express the requirements that the position of the quantum robot does not change during the computation phase and, except for possible entanglements with environmental states, the state of the output system is not changed during the action phase. These entanglements would occur if o was in a linear superposition of $\mathcal{B}$ states each of which resulted in different environment states during the action. Another property of Γ_a is that it is the identity operator on the space of states for the on board quantum computer and memory system degrees of freedom. Note also that Γ_a and Γ_c do not commute.

If Ψ_0 is the overall system state at time 0 then the state at time $n\Delta$ is given by $\Psi_n = (\Gamma_a + \Gamma_c)^n \Psi_0$. The amplitude for finding the quantum robot and environment in a state $|w', j\rangle$ is given by

$$\Psi_n(w', j) = \sum_{w,i} \langle w', j|(\Gamma_a + \Gamma_c)^n|w, i\rangle \Psi_0(w, i). \tag{2.3}$$

Here $|w\rangle, |w'\rangle$ denote the states of all environmental and quantum robot systems except the control qubit in some suitable basis, and $i, j = 0, 1$ refer to the states of c.

All the information about the dynamics of the system is given in the matrix elements $\langle w', j|(\Gamma_a + \Gamma_c)^n|w, i\rangle$. For each w, w', n, i, j the matrix element can be expanded in a Feynman sum over phase paths [3, 9]. One first expands $(\Gamma_a + \Gamma_c)^n$ as a sum of products of Γ_a and Γ_c:

$$(\Gamma_a + \Gamma_c)^n = \sum_{v_1=a,c} \sum_{t=1}^{n} \sum_{h_1, h_2, \cdots, h_t = 1}^{\delta(\sum, n)} (P_0^c + P_1^c)(\Gamma_{v_t})^{h_t} (\Gamma_{v_{t-1}})^{h_{t-1}}, \\ \cdots, (\Gamma_{v_2})^{h_2} (\Gamma_{v_1})^{h_1}. \tag{2.4}$$

In this expansion the number of phases is given by the value of t which ranges from $t = 1$ corresponding to one phase with n steps to $t = n$ corresponding to n alternating phases each of 1 step. The duration of the ℓth phase is given by the value of h_ℓ for $\ell = 1, 2, \cdots, t$. The requirement that the total number of steps equals n, or $h_1 + h_2 +, \cdots, + h_t = n$, is indicated by the upper limit $\delta(\sum, n)$ on the h sum. The alternation of phases is shown by v where $v_{m+1} = a$ (or c) if $v_m = c$ (or a). The factor $P_0^c + P_1^c$ expresses the fact that the tth phase may not be completed.

Expansion in a complete set of states between each of the phase operators $(\Gamma_{v_\ell})^{h_\ell}$ gives the desired path sum:

$$\langle w',j|(\Gamma_a+\Gamma_c)^n|w,i\rangle = \sum_{t=1}^{n} \sum_{p_2,\cdots,p_t} \sum_{h_1,h_2,\cdots,h_t=1}^{\delta(\sum,n)} \langle w',j|(\Gamma_{v_t})^{h_t}|p_t\rangle$$
$$\cdots,\langle p_3|(\Gamma_{v(2)})^{h_2}|p_2\rangle\langle p_2|(\Gamma_i)^{h_1}|w,i\rangle$$
(2.5)

Here the sum is over all paths p of states of length $t+1$ with beginning and endpoints given by the states $|w\rangle$ and $|w'\rangle$. That is $|p_1\rangle = |w\rangle, |p_{t+1}\rangle = |w'\rangle$. The states of the control qubit have been suppressed as they correspond to the values of v. Note that $v(1) = i$.

Each term in this large sum gives the amplitude for finding t alternating phases in the first n steps where the ℓth phase begins with all systems (except for c) in state $|p_\ell\rangle$ and ends after h_ℓ steps in state $|p_{\ell+1}\rangle$. The sums express the dispersion in the duration or number of steps in each phase (h sums), in the number of phases (t sum), and in the initial and terminal states for each phase (p sums).

3. An Example of Quantum Searching

Quantum robots are well suited for carrying out search tasks. As a simple example consider a search task where a quantum robot searches a large square area R of $N \times N$ sites to locate a system s. To keep things simple s is assumed to be motionless and located at just one unknown site. The goal of the search is to determine the location of s in R.

The quantum robot consists of an on board quantum computer, memory and output systems, and a control qubit. The on board computer is assumed here to be a quantum Turing machine consisting of a head moving on a cyclic lattice of $O(\log N)$ qubits. $O(-)$ denotes of the order of. The memory system also is a cyclic lattice which is taken here to have about the same number of qubits and to lie adjacent to the computation lattice. A schematic representation of the quantum robot located at a corner (the origin) of R is shown in the figure.

One method of carrying out the search is to let the coordinates X, Y with $0 \le X, Y \le N-1$ of each point of R define a search path. If the memory is initially in state $|X,Y\rangle_m$ the quantum robot, starting from the location $0,0$ moves X sites in the x direction, then Y sites in the y direction and looks for s at its location. After recording the presence or absence of s at the site and further processing, if any, the quantum robot returns along the path to the origin.

A more detailed description starts with the qubits in the memory, m, and computation lattice, $\mathcal{L}$, in the state $|X,Y\rangle_m|0\rangle_\mathcal{L}$ the output system o in state $|dn\rangle_o$ and a computation phase active (c in state $|1\rangle_c$). After copying the m state onto $\mathcal{L}$ to give the state $|X,Y\rangle_m|X,Y\rangle_\mathcal{L}$, the computation phase checks to see if $X = 0$ or $X > 0$. If $X > 0$ the computation phase continues by subtracting 1 from $|X,Y\rangle_\mathcal{L}$ to give $|X-1,Y\rangle_\mathcal{L}$. It ends by changing the o state to $|+x\rangle_o$ and the c state to $|0\rangle_c$.

The action phase consists of one step (one iteration of Γ_a) in which the quantum robot moves one lattice site in the $+x$ direction and the c state is converted back to $|1\rangle_c$. The process is repeated until the state with $X = 0$ is reached on $\mathcal{L}$. Then the above process is repeated for Y (the o state now becomes $|+y\rangle_o$ to denote one step motion in the $+y$ direction) until $Y = 0$ is reached in the state of $\mathcal{L}$.

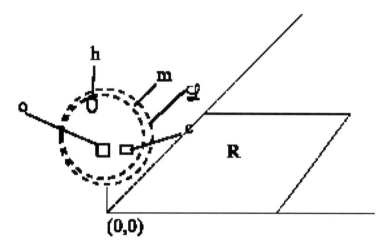

FIGURE 1. A Schematic Model of a Quantum Robot at the Origin of R. Both the memory system lattice (m) and Quantum Turing machine lattice $\mathcal{L}$ are shown with the head h that moves on the lattices. The control qubit (c) and output system (o) are also shown. The quantum robot is greatly magnified relative to R to show details.

At this point the presence or absence of s at the location X, Y of the quantum robot is recorded during a computation phase and, after further processing, if any, the quantum robot returns along the same path. This is done by interleaving motion of the quantum robot in the $-y$ and $-x$ directions, with corresponding o states $|-y\rangle_o, |-x\rangle_o$, with adding 1 to the y, and then x components of the $\mathcal{L}$ state with checking if the values Y and then X are reached. This is done by stepwise comparison with the state of m which remains unchanged.

When the state of $\mathcal{L}$ is the same as the state of m the quantum robot has returned to the starting point at the origin of R. A computation phase changes the o state to $|dn\rangle_o$ and transfers motion to some ballast system. As has been noted this is necessary to preserve reversibility and the corresponding unitarity of the dynamics [10].

Examples of ballast motion consist of repetitions of adding 1 to a large lattice of M qubits or emitting a particle which moves away from R. In the first case with a finite number 2^M of ballast states, the quantum robot remains in the final state of the search degrees of freedom for a finite time only before the search process is undone. This does not occur for the second case with an infinite number of ballast states.

4. Grover's Algorithm and the Quantum Search

Before applying Grover's Algorithm to process the results of the search, it is useful to understand what it does and how it works. A very brief summary, that follows Grover [2] and Chen et al [11], is given next.

4.1. Grover's Algorithm.

Suppose one has a data base B of N elements and a function f that takes the value 0 on all elements except one, ω, on which f has value 1. It is assumed that ω is completely unknown and that a procedure is available for obtaining the value of f on any element of the data base in 1 step. Let each x in B correspond to a unique length n binary string and $|x\rangle_B$ be the corresponding n-qubit state.

Let the initial state for the search be given by

$$\phi = \frac{1}{\sqrt{N}} \sum_{x \epsilon B} |x\rangle \tag{4.1}$$

where the sum is over all N elements x in B. This corresponds to a coherent sum over all product $|0\rangle, |1\rangle$ states of n qubits in a quantum computer if $N = 2^n$. This state is easily constructed from the constant 0 state $|\underline{0}\rangle = \otimes_{j=1}^{n} |0\rangle_j$ by applying the operator $(1/\sqrt{2})(\sigma_z + \sigma_x)$ to each qubit. Here σ_x, σ_z are the Pauli matrices. This is referred to as the Walsh-Hadamard transformation W. Thus $\phi = W|\underline{0}\rangle$.

Define the unitary operator Q by $Q = -I_\phi I_\omega$ where $I_\phi = 1 - 2P_\phi$ and $I_\omega = 1 - 2P_\omega$. Both P_ϕ and P_ω are projection operators on the states ϕ and $|\omega\rangle$. Let $|\alpha\rangle = (1/\sqrt{N-1}) \sum_{x \neq \omega} |x\rangle$ be the coherent sum over all states $|x\rangle$ with x in B and different from ω. Since $|\alpha\rangle$ and $|\omega\rangle$ are orthonormal they form a binary basis for a 2 dimensional Hilbert space.

One can expand ϕ in this basis:

$$\phi = \sqrt{\frac{N-1}{N}} |\alpha\rangle + \frac{1}{\sqrt{N}} |\omega\rangle. \tag{4.2}$$

In the same basis Q has the representation

$$Q = \begin{pmatrix} \cos\theta & \sin\theta \\ -\sin\theta & \cos\theta \end{pmatrix} \tag{4.3}$$

where $\cos\theta = 1 - 2/N$ and $\sin\theta = 2\sqrt{N-1}/N$.

This shows that Q acting on ϕ corresponds to a rotation by θ, and m iterations of Q correspond to a rotation by $m\theta$. So, carrying out m iterations of Q on ϕ where $m\theta \approx \pi/2$, rotates ϕ from a state that is almost orthogonal to $|\omega\rangle$ to a state that is almost parallel to $|\omega\rangle$. Measurement of this final state gives with high probability, the value of ω.

Grover's algorithm derives its efficiency from the fact that this rotation is achieved with $m \sim \sqrt{N}$ whereas classically $\sim N$ steps are needed to find ω with high probability. The iteration of Q must be stopped at the right value of m because additional iterations will continue to rotate ϕ.

Efficient implementation of this algorithm on a quantum computer, corresponds to iteration of Q on each component of ϕ. This requires that it is possible to determine, in a small number of steps, if $x = \omega$ or $x \neq \omega$. This is often described in terms of an unknown or "oracle" function f on B, where $f(x) = 0[=1]$ if $x \neq \omega[x = \omega]$, that can be evaluated in one step on any x in B. I_ϕ is implementable in $O(n)$ steps as $\phi = W|\underline{0}\rangle$ and W is the product of n single qubit operators.

Grover [**2, 6**] first introduced the algorithm for searching an unstructured data base of $N = 2^n$ elements for a single element (f has value 1 on just one element). Since then the algorithm has been much studied under various generalizations. These include searches for several elements (f has value 1 on several elements) [**11, 12, 13**], searches in which N is arbitrary [**12**], and searches in which the initial amplitude distribution of the component states is arbitrary [**14**]. It has also been shown that the algorithm is optimally efficient [**15**]. Further development is described in other work [**16, 17**]. However, as has been recently emphasized [**18**], all these searches depend on the fact that the evolving state is and remains a coherent superposition of components corresponding to elements of the data base.

4.2. Problems with the Use of Grover's Algorithm. The description in Section 3 of the quantum search was for the initial memory state $|X,Y\rangle_m$. However if the memory system lattice is in the initial state

$$\psi_m = (1/N) \sum_{X,Y=0}^{N-1} |X,Y\rangle_m, \tag{4.4}$$

then the description also applies to each component state $|X,Y\rangle$.

As was the case for the initial state ϕ, the state ψ_m can be efficiently prepared from the state $|0\rangle_m$ using the Walsh-Hadamard transformation. $(1/\sqrt{2})(\sigma_z + \sigma_x)$ on each qubit of m. For this initial memory state all N^2 searches are carried out coherently. Since the path lengths range from 0 to $2N$, the quantum robot can search all sites of R and return to the origin in $O(N \log N)$ steps. Since this is less than the number of steps, $O(N^2 \log N)$, required by a classical robot, the question arises if Grover's algorithm can be used to process the final memory state to determine the location of s. If this is possible, the overall search and processing should require $O(N \log N)$ steps which is less than that required by a classical robot.

It is worth a digression at this point to see that Grover's Algorithm is not applicable to the usual method of recording the presence or absence of s at a site. To see this assume that s is at site X_0, Y_0 and that an extra qubit, r of the memory is set aside to record the presence or absence of s. If r is initially in state $|0\rangle_r$ and is changed to state $|1\rangle_r$ only in the presence of s, then the initial memory state $\phi_I = (1/N)\sum_{X,Y=0}^{N-1}|X,Y\rangle_m|0\rangle_r$ is changed to the final memory state

$$\phi_f = (1/N)(\sum_{X,Y \neq X_0 Y_0} |X,Y\rangle_m|0\rangle_r + |X_0,Y_0\rangle_m|1\rangle_r). \tag{4.5}$$

after the quantum robot has returned to the origin of R.

The idea then would be to use Grover's algorithm [**2**] by carrying out N iterations of a unitary operator U to amplify the component state $|X_0,Y_0\rangle_m|1\rangle_r$ at the expense of the other components. Following Grover and others [**2, 11**], define U by $U = -I_{\phi_i}I_{|1\rangle_r}$ where $I_{\phi_i} = 1 - 2P_{\phi_i}$ and $I_{|1\rangle_r} = 1 - 2P_{|1\rangle_r}$. Here P_{ϕ_i} and $P_{|1\rangle_r}$ are projection operators on the memory state ϕ_i and the record state $|1\rangle_r$. $P_{|1\rangle_r}$ is the identity on other memory degrees of freedom.

It is clear that for this case iterations of U can be carried out efficiently. However, the problem is that the initial state ϕ_i contains a component, $|X_0,Y_0\rangle_m|0\rangle_r$, not present in the final state ϕ_f, Eq. 4.5. Also the state ϕ_i does not contain the state $|X_0,Y_0\rangle_m|1\rangle_r$. In this case U does not have a two dimensional representation in the basis pair

$$\frac{1}{\sqrt{N^2-1}} \sum_{X,Y \neq X_0 Y_0} |X,Y\rangle_m|0\rangle_r \; ; \; |X_0 Y_0\rangle_m|1\rangle_r \tag{4.6}$$

obtained from Eq. 4.5. As a result U cannot be represented as a rotation that, under iteration, rotates the desired component to be almost parallel to the initial state.

This problem can be avoided by changing the method of recording the presence or absence of s and not using an extra qubit r. Here, following Grover [2, 6], the sign of the component corresponding to the location of s is changed. In this case the initial memory state, Eq. 4.4, becomes the final memory state

$$\psi_f = (1/N)\left(\sum_{X,Y \neq X_0 Y_0} |X,Y\rangle_m - |X_0, Y_0\rangle_m\right). \tag{4.7}$$

after return of the quantum robot. In this case U is defined as $U = -I_{\psi_m} I_{X_0 Y_0}$ where $I_{\psi_m} = 1 - 2P_m$ and $I_{X_0 Y_0} = 1 - 2P_{X_0 Y_0}$. Here P_m and $P_{X_0 Y_0}$ are projection operators on the memory states ψ_m and $|X_0 Y_0\rangle_m$.

The problem here is that the only way to determine the action of $I_{X_0 Y_0}$ is by repeating the search part of the process. This is not efficient as it requires $O(N \log N)$ steps. In the language of much of the work on Grover's algorithm this corresponds to the fact that it requires $O(N \log N)$ steps to determine the value of the oracle function instead of just one step as is usually assumed. In this case the advantage of quantum over classical searching is lost for 2 dimensional regions as use of Grover's Algorithm would require $O(N)$ searches each requiring $O(N \log N)$ steps.

This suggests that a method be considered in which the Grover iterations are done prior to return when the quantum robot is at the path endpoint. At this point the component memory states are entangled with the quantum robot position states as the overall state has the form $(1/N)\sum_{X,Y} |X,Y\rangle_m |X,Y\rangle_{QR}$ where $|X,Y\rangle_{QR}$ is the quantum robot position state for the site X, Y.

In this case $I_{X_0 Y_0}$ can be efficiently carried out on each initial component memory state $|X,Y\rangle$ by a local observation to see if s is or is not at the site X, Y. Also the action of $U = -I_{\psi_m} I_{X_0 Y_0}$ on each component memory state is given by

$$U|X,Y\rangle_m = -I_m |X,Y\rangle_m = \frac{2}{\sqrt{N}} \psi_m - |X,Y\rangle_m \tag{4.8}$$

if $|X,Y\rangle_m \neq |X_0, Y_0\rangle_m$ and

$$U|X_0, Y_0\rangle_m = I_m |X_0, Y_0\rangle_m = -\frac{2}{\sqrt{N}} \psi_m + |X_0, Y_0\rangle_m \tag{4.9}$$

if $|X,Y\rangle_m = |X_0, Y_0\rangle_m$. Here, as before, $\psi_m = (1/N)\sum_{X,Y=0}^{N-1} |X,Y\rangle_m$.

Here the problem is that there is no efficient way to carry out more than one iteration of U. As noted above the first iteration can be done efficiently. However additional iterations require that the action of $I_{X_0 Y_0}$ on memory component states $|X', Y'\rangle$ be evaluated for arbitrary values of X', Y' while the quantum robot remains at site X, Y. This cannot be done efficiently as the quantum robot has no way of knowing whether s is or is not at these different locations. To know this the quantum robot must go to the site X', Y' to see if s is there. This is inefficient as such a trip requires $O(N \log N)$ steps.(Actions are efficient if they require $O(\log N)$ steps or less. Low powers of $\log N$ are also acceptable.)

One sees from this that implementation of Grover's Algorithm using either of these methods requires $O(N)$ iterations of U (as R has N^2 sites) where each iteration requires $O(N \log N)$ steps. The resulting number of steps required, $O(N^2 \log N)$, is the same as that needed by a classical robot. So quantum searches of 2 dimensional space regions combined with Grover's Algorithm are no more efficient than classical searches.

It is of interest to note that quantum searches of higher dimensional space regions combined with Grover's Algorithm are more efficient than classical searches. To see this assume a search of a d dimensional cube of N^d sites with the memory in the initial state

$$\Psi = \frac{1}{N^{d/2}} \sum_{X_1,\cdots X_d=0}^{N-1} |X_1, X_2, \cdots X_d\rangle_m \tag{4.10}$$

Carrying out Grover's Algorithm requires $O(N^{d/2})$ iterations of U where, as before, each iteration of U requires $O(N \log N)$ steps. This follows from the fact that the number of dimensions appears as a multiplicative factor for the number of steps. Also $O(dN \log N) = O(N \log N)$. So the overall process requires $O(N^{(d/2)+1} \log N)$ steps. For $d > 2$ this is more efficient than a classical search requiring $O(N^d \log N)$ steps.

The discussion so far has ignored the entanglement problems. These problems, which apply to all the above cases, result from the fact that the task evolution, starting from the initial unentangled product state $\psi_m |0\rangle_{\mathcal{L}} |0, 0\rangle_{QR} \cdots$, generates entanglements between the position states $|X', Y'\rangle_{QR}$ of the quantum robot and the components $|X, Y\rangle_m$ of the memory state ψ_m. In order for Grover's Algorithm to work it is necessary to remove this entanglement at the end of each search cycle or iteration of U so that the final memory state is ψ_f.[1]

This entanglement occurs because the unitary dynamics is reversible and the number of steps needed to complete the search task is different for different component states of ψ_m. Here the number ranges from $O(1)$ for the path $|0, 0\rangle_m$ to $O(2N)$ for the path $|N-1, N-1\rangle_m$. This means that the various components of the quantum robot complete an iteration of U at different times. This is independent of whether the Algorithm is completed after or prior to return.

Because of the reversibility each component cannot simply stop and wait until the longest search component is complete. It must instead embark on motion of irrelevant or ballistic degrees of freedom. This means that the memory components $|X, Y\rangle_m$ exchange entanglement with the quantum robot position states for entanglement with states of ballistic degrees of freedom. Since use of Grover's Algorithm requires the removal of this entanglement [18], the question arises whether it is possible to insert delays into each of the memory components that are computed, for example, after the quantum robot returns to the origin of R at the end of each cycle. If this works then there would be some time or step number at which the entanglement is removed and the original product structure of the initial state recovered, with ψ_f replacing ψ_m.

This use of delays to remove the entanglements reversibly requires that no memory of the magnitude of the delay be left in the delay degrees of freedom. Otherwise one ends up with entanglement with the delay degrees of freedom. Also determining the magnitude of each delay is not trivial as it depends not only on the lengths of each of the paths but on the number of steps in the computation phases used to determine motion along the paths. This includes the dependence of the number of steps required to subtract 1 from a number M on the value of M (through the number of "carry 1" operations needed [20]).

---

[1]The entanglement referred to here is different from that of the memory state qubits. The latter is generated during iteration of the Grover operator and is necessary for successful operation of Grover's Algorithm on multiqubit states [19].

Based on these considerations it is seems doubtful that one can use Grover's Algorithm to efficiently process the results of a quantum search of a space region R. Even if the entanglement problem were solvable, the above results show that, for 2 dimensional space regions, use of Grover's Algorithm is no more efficient than a classical search. For higher dimensional searches the Algorithm is more efficient. Note that this conclusion is independent of the details of the quantum robot. It applies to any quantum system such as a mobile head that contains sufficient information on board to tell it where to go, what to do on arrival at the endpoint, and how to return to the origin.

5. Discussion

In spite of these pessimistic results, quantum robots are interesting objects of study. For instance they may be useful test beds for study of control of quantum systems [21, 22] as the dependence of the task dynamics on the local environmental state is, for some tasks, similar to a feedback loop.

Quantum robots and the associated task dynamics also make clear what is and is not being done in any task. This is shown by the quantum search task in that the quantum robot does no monitoring or control of its behavior. It (or the on board quantum computer) has no knowledge of where it is in R at any point or even if it is in R. For each component memory state $|X,Y\rangle_m$ there are X computation phases with the output system o in state $|+x\rangle_o$ and Y phases with o in state $|+y\rangle_o$. These phases are interspersed with X and Y action phases during which anything can happen. For example the quantum robot might move outside R or it might not move at all. Of course for these cases it is unlikely that the quantum robot would return to the origin at the end of the task.

This illustrates a valuable aspect of the description of the task dynamics of quantum robots as sequences of alternating computation and action phases. This is that, for the search task examples described here, it makes very clear the lack of awareness and control the quantum robot has over what has happened in the action phases and what it is doing. This argument applies to the computation phases also. For example the "subtract 1" steps could carry out an arbitrary change to the memory state and the task would continue. In this case the task would no longer be a search task but would be something else.

These considerations are also part of foundational reasons why quantum robots and quantum computers are interesting. If quantum mechanics (or some extension such as quantum field theory) is assumed to be universally applicable, then all systems involved in the validation of quantum mechanics are quantum systems. This includes the systems that make theoretical computations (which includes quantum computers) and the systems that carry out experiments (which includes quantum robots). Thus, in some sense quantum mechanics must describe its own validation, to the maximum extent possible. Exploration of this and the questions of self consistency and possible incompleteness that may occur make this an interesting path of inquiry.

In addition quantum robots, and to some extent quantum computers, are natural systems for investigating several questions. In particular what physical properties must a quantum system have such that

- It is aware of its environment?
- It has significant characteristics of intelligence?

- It changes states of some quantum systems so that the new states can be interpreted as text having meaning to the system generating the text [23]?

It is hoped to examine these and related questions in future work.

References

[1] P. Shor, in *Proceedings, 35th Annual Symposium on the Foundations of Computer Science*, S. Goldwasser (Ed), IEEE Computer Society Press, Los Alamitos, CA, 1994, pp 124-134; SIAM J. Computing, **26** (1997), 1481-1509.

[2] L.K.Grover, in *Proceedings of 28th Annual ACM Symposium on Theory of Computing* ACM Press New York 1996, pp. 212-219; Phys. Rev. Letters, **79** (1997), 325-328; G. Brassard, Science **275** (1997), 627-628.

[3] P. Benioff, Phys. Rev. A **58** (1998), 893-904; *Quantum Robots* in, Feynman and Computation, Exploring the Limits of Computers, Anthony Hey, Ed, Perseus Books, Reading, MA. 1998; Los Alamos Archives Preprint Quant-ph/9807032.

[4] J. I. Cirac and P. Zoller, Phys. Rev. Letters **74** (1995), 4091-4094; C. Monroe, D. M. Meekhof, B. E. .King, W. M. Itano, and D. J. Wineland, Phys Rev. Letters **75** (1995), 4714-4717; P. Domokos, J. M. Raimond, M Brune, and S. Haroche, Phys. Rev. A **52** (1995), 3554-3559; C. Monroe, D. Leibfried, B. E. King, D. M. Meekhof, W. M. Itano, and D. J. Wineland, Phys. Rev. A **55** (1997), R2489-2491; J. F. Poyatos, J. I. Cirac, and P. Zoller, Phys. Rev. Letters, **81** (1998), 1322-1325.

[5] N. A. Gershenfeld and I. L. Chuang, Science **275** (1997), 350-356; D. G. Cory, A. F. Fahmy, and T. F. Havel, Proc. Nat. Acad. Sci. **94** (1997) 1634-1639.

[6] L. K. Grover, Phys. Rev. Letters, **80** (1998), 4329-4332.

[7] A. Peres, Phys. Lett. **101A** (1984), 249-250.

[8] R. Laflamme, C. Miquel, J. P. Paz, and W. Zurek, Phys. Rev. Letters **77** (1996), 198-201; D. P. DiVincenzo and P. W. Shor, Phys. Rev. Letters **77** (1996), 3260-3263; E. M. Raines, R. H. Hardin. P. W. Shor, and N. J. A. Sloane, Phys. Rev. Letters **79** (1997), 953-954; E. Knill, R. Laflamme, and W. Zurek, Science **279** (1998), 342-346; D. Cory, M. Price, W. Maas, E. Knill, R. Laflamme, W. Zurek, T. Havel, and S. Somaroo, Phys. Rev. Letters **81** (1998), 2152-2155.

[9] H. E. Brandt, Progr. Quantum Electronics, **22** (1998), 257-370.

[10] P. A. Benioff, Int. Jour. Theoret. Physics **21** (1982), 177-202.

[11] G. Chen, S. A. Fulling, and M. O. Scully, Los Alamos Archives, Quant-ph/9909040.

[12] M. Boyer, G. Brassard, P. Hoyer, and A. Tapp, Fortschritte Der Physik, **46** (1998), 493-506.

[13] L. K. Grover, Los Alamos Archives, Quant-ph/9912001.

[14] E. Biham, O. Biham, D. Biron, M. Grassl, and D. Lidar, Phys. Rev. A, **60** (1999), 2742-2745.

[15] C. Zalka, Phys. Rev. A **60** (1999), 2746-2751.

[16] E. Farhi and S. Guttmann, Los Alamos Archives, Quant-ph/9711035.

[17] G. L. Long, Y.S. Li, W. L. Zhang, and L. Niu, Physics Letters A, **262** (1999), 27-34.

[18] J. P. Barnes and W. S. Warren, Phys. Rev. A **60** (1999),4363-4374.

[19] S. Lloyd, Phys. Rev. A **61** (2000) 010301(R).

[20] P. Benioff, Phys. Rev. Letters **78** (1997), 590-593; Phys. Rev. B **55** (1997) 9482-9494.

[21] L. Viola, S. Lloyd, and E. Knill, Phys. Rev. Letters **83** (1999), 4888-4891; S. Lloyd and J. E. Slotine, Los Alamos Archives Quant-ph/9905064.

[22] W. S. Warren, H. Rabitz, and M Dahleh, Science **259** (1993), 1581- 1589; V. Ramakrishna and H. Rabitz, Phys. Rev. A **51** (1995), 960-966; Phys. Rev. A **54** (1996), 1715-1716; M. Demiralp and H. Rabitz, Phys. Rev. A, **57** (1998), 2420-2425.

[23] P. Benioff, Phys. Rev. A **59** (1999), 4223-4237.

(Paul Benioff) PHYSICS DIVISION, ARGONNE NATIONAL LABORATORY, ARGONNE ILLINOIS, 60439

E-mail address, Paul Benioff: **pbenioff@anl.gov**

Perturbation Theory and Numerical Modeling of Quantum Logic Operations with a Large Number of Qubits

G. P. Berman, G. D. Doolen, D. I. Kamenev, G. V. López, and V. I. Tsifrinovich

ABSTRACT. The perturbation theory is developed based on small parameters which naturally appear in solid state quantum computation. We report the simulations of the dynamics of quantum logic operations with a large number of qubits (up to 1000). A nuclear spin chain is considered in which selective excitations of spins are provided by having a uniform gradient of the external magnetic field. Quantum logic operations are utilized by applying resonant electromagnetic pulses. The spins interact with their nearest neighbors. We simulate the creation of the long-distance entanglement between remote qubits in the spin chain. Our method enables us to minimize unwanted non-resonant effects in a controlled way. The method we use cannot simulate complicated quantum logic (a quantum computer is required to do this), but it can be useful to test the experimental performance of simple quantum logic operations. We show that: (a) the probability distribution of unwanted states has a "band" structure, (b) the directions of spins in typical unwanted states are highly correlated, and (c) many of the unwanted states are high-energy states of a quantum computer (a spin chain). Our approach can be applied to simple quantum logic gates and fragments of quantum algorithms involving a large number of qubits.

Introduction

Recently much progress has been made in single particle technologies. These technologies allow one to manipulate a single electron, a single atom, and a single ion. A distinguishing feature of all these technologies is that they are "quantum"; quantum effects are crucial for the preparation of the initial state, for performing useful operations, and for reading out the final state. One of the future quantum technologies is quantum computation. In a quantum computer, the information is loaded in a register of quantum bits – "qubits". A qubit is a quantum object (generalized spin 1/2) which can occupy two quantum states, $|0\rangle$ and $|1\rangle$, and an arbitrary superposition of these states: $\Psi = C_0|0\rangle + C_1|1\rangle$. (The only constraint is:

2000 *Mathematics Subject Classification.* Primary 81Q15; Secondary 65L15.

This work was supported by the Department of Energy under contract W-7405-ENG-36, by the National Security Agency (NSA), and by the Advanced Research and Development Activity (ARDA).

$|C_0|^2 + |C_1|^2 = 1$.) A quantum computer is remarkably efficient in executing newly invented quantum algorithms (the quantum Fourier transform, Shor's algorithm for prime factorization, Grover's algorithm for data base searching, and others). Using these algorithms, quantum computing promises to solve problems that are intractable on digital computers. The main advantage of quantum computation is the rapid, parallel execution of quantum logic operations. One of the promising directions in quantum computation is solid state quantum computation. (See the review [**1**].) In solid state computers, a qubit can be represented by a single nuclear spin 1/2 [**2, 3**], or a single electron spin 1/2 [**4, 5**], a single Cooper pair [**6, 7**]. This type of quantum computers is quite different from quantum computers in which a qubit is represented by an ensemble of spins 1/2 [**8, 9, 10**].

Crucial mathematical problems must be solved in order to understand the dynamical aspects of quantum computation. One of these problems is the creation of the dynamical theory of quantum computation – the main subject of our paper. The processes of the creation of quantum data bases, the storage and searching of quantum information, the implementation of quantum logic gates, and all the steps involved in quantum computation are dynamical processes. When qubits representing a register in a quantum computer are in superpositional states, they are not eigenstates of the Hamiltonian describing the quantum computer. These superpositions are time-dependent. Understanding their dynamics is very important. To design a working quantum computer, simulations of quantum logic operations and fragments of quantum computation are essential. The results of these simulations will enable engineers to optimize and test quantum computers.

There are two main obstacles to perform useful quantum logic operations with large number of qubits on a digital (classical) computer: (1) the related Hilbert space is extremely large, $D_N = 2^N$, where N is the number of qubits (spins), and (2) even if the initial state of a quantum computer does not involve many basic states (eigenstates), the number of excited eigenstates can rapidly grow during the process of performing quantum logic operations. Generally both of these obstacles exist. At the same time, many useful quantum operations can be simulated on a digital computer even if the number of qubits is quite large, say 1000. How one can do this? One way is to create (and use) a *perturbation theory of quantum computation*. To do this, one should consider a quantum computer as a many-particle quantum system and introduce small parameters. Usually, in *all* physical problems there exist small parameters which allow one to simplify the problem to find approximate solutions. The objective in this approach is to build a solution in a controlled way.

As an example, it is useful to consider an electron in a hydrogen atom interacting with a laser field. The electron has an infinite (formally) number of discrete levels plus a continuous part of its energy spectrum. So, a single electron in a Hydrogen atom has a Hilbert space larger than *all* finite qubit quantum computers. Assume that the electron is populated initially on some energy level(s). The action of the laser field on the electron leads in many cases to a regular and controlled dynamics of the electron. In particular, if the amplitude of the laser field is small enough, the electron interacting with this laser field can be considered as a two-level quantum system interacting with an external time-periodic field. Why is the electron excited to only few energy levels? This happens because the existence of small parameters makes the probability of unwanted events (excitation of the electron to most energy levels including the continuum) negligibly small.

The existence of small parameters allows one to use perturbation approaches when calculating the quantum dynamics of the electron. This strategy can be used to create a perturbation theory of quantum computation.

In this paper we present our results, based on the existence of small parameters, for simulating quantum logic operations with a large number (up to 1000) of qubits. Our perturbation approach does not provide a substitution for computing quantum algorithms on a quantum computer. One will need quantum computer to do the required calculations. Our method will allow one to simulate the required benchmarks, quantum logic test operations, and fragments of quantum algorithms. Our method will enable engineers to optimize a working quantum computer.

In section 1, we formulate our model and the Hamiltonian of a quantum computer based on a one-dimensional nuclear spin chain. Spins interact through nearest-neighbor Ising interactions. All spins are placed in a magnetic field with the uniform gradient. This field gradient enables the selective excitation of spins. Quantum logic operations are provided by applying the required resonant electromagnetic pulses. In section 2, we present the dynamical equations of motion in the interaction representation. As well, we discuss an alternative description based on the transformation to the rotating frame. In the latter case the effective Hamiltonian is independent of time and solution of the problem can be obtained using the eigenstates of this Hamiltonian. In section 3, we derive simplified dynamical equations taking into account only resonant and near-resonant transitions (and neglecting non-resonant transitions). We introduce the small parameters which characterize the probability of generation of unwanted states in the result of near-resonant transitions. We describe a $2\pi k$-method which allows us to minimize the influence of unwanted near-resonant effects which can destroy quantum logic operations. In section 4, we describe a quantum Control-Not gate for remote qubits in a quantum computer with a large number of qubits. Analytical solution for this gate is presented in section 5, for the $2\pi k$ condition. In section 6, we present small parameters of the problem, in explicit form. The equation for the total probability of generation of unwanted states, including the states generated in the result of near-resonant and non-resonant transition, is derived in section 7. In section 8, we present results of numerical simulations of the quantum Control-Not gate for remote qubits in quantum computers with 200 and 1000 qubits when the probability of the near-resonant transitions is relatively large (when the conditions of $2\pi k$ method are not satisfied). We show that: (a) the probability distribution of unwanted states has a "band" structure, (b) the directions of spins in typical unwanted states are highly correlated, and (c) many of the unwanted states are high-energy states of the quantum computer (a spin chain). The total probability of error (including the states generated in the result of near-resonant and non-resonant transitions) is computed numerically in section 9 for small ($N = 10$) and large ($N = 1000$) number of qubits. A range of parameters is found in which the probability of error does not exceed a definite threshold. We test our perturbative approach using exact numerical solution when the number of qubits is small ($N = 10$). In section 10, we show how the problem of quantum computation (even with large number of qubits) can be formulated classically in terms of interacting one-dimensional oscillators. In the conclusion, we summarize our results.

1. Formulation of the model

The mathematical model of a quantum computer used in this paper is based on a one-dimensional Ising nuclear-spin system – a chain of N identical nuclear spins (qubits). Application of Ising spin systems for quantum computations was first suggested in Ref.[11]. Today, these systems are used in liquid nuclear magnetic resonance (NMR) quantum computation with small number of qubits [12]. The register (a 1D chain of N identical nuclear spins) is placed in a magnetic field,

$$\vec{B}(t,z) = \left(b_\perp^{(n)} \cos\left[\nu^{(n)} t + \varphi^{(n)}\right], -b_\perp^{(n)} \sin\left[\nu^{(n)} t + \varphi^{(n)}\right], B^z(z) \right), \quad (1)$$

where $t^{(n)} \leq t \leq t^{(n+1)}$ and $n = 1, ..., M$. In (1), $B^z(z)$ is a slightly non-uniform magnetic field oriented in the positive z-direction. The quantities $b_\perp^{(n)} > 0$, $\nu^{(n)} > 0$ and $\varphi^{(n)}$ are, respectively, the amplitude, the frequency, and the phase of the circular polarized (in the $x-y$ plane) magnetic field. This magnetic field has the form of rectangular pulses of the length (time duration) $\tau^{(n)} = t^{(n+1)} - t^{(n)}$. The total number of pulses which is required to perform a given quantum computation (protocol) is M. Schematically, our quantum computer is shown in Fig. 1.

1.1. The quantum computer Hamiltonian. The quantum Hamiltonian which describes our quantum computer has the form,

$$\hat{H} = \hat{H}_0 + \hat{V} = -\sum_{k=0}^{N-1} \omega_k \hat{I}_k^z - 2J \sum_{k=0}^{N-2} \hat{I}_k^z \hat{I}_{k+1}^z + \hat{V}. \quad (2)$$

(We set $\hbar = 1$.) The operator $\hat{V}$ describes the interaction of spins with pulses of the rf field,

$$\hat{V} = \sum_{n=1}^{M} \hat{V}^{(n)}, \quad (3)$$

and $\hat{V}^{(n)}$ describes the interaction of spins with n-th pulse of the rf field,

$$\hat{V}^{(n)} = -\Theta^{(n)}(t) \frac{\Omega^{(n)}}{2} \sum_{k=0}^{N-1} \left\{ \hat{I}_k^- \exp\left[-i\left(\nu^{(n)} t + \varphi^{(n)}\right)\right] + \hat{I}_k^+ \exp\left[i\left(\nu^{(n)} t + \varphi^{(n)}\right)\right] \right\}, \quad (4)$$

where $\Theta^{(n)}(t)$ equals 1 only during the nth pulse. The operators in (2)-(4) have the following explicit form in the z-representation (the representation in which the operators $\hat{I}_k^z$ are diagonal),

$$\hat{I}_k^z = \frac{1}{2} |0_k\rangle\langle 0_k| - \frac{1}{2} |1_k\rangle\langle 1_k|, \quad (5)$$

$$\hat{I}_k^+ = \hat{I}_k^x + i\hat{I}_k^y = |0_k\rangle\langle 1_k|, \quad \hat{I}_k^- = \hat{I}_k^x - i\hat{I}_k^y = |1_k\rangle\langle 0_k|.$$

We shall use the Dirac notation for the complete set of eigenstates (the stationary states) of the quantum computer described by the Hamiltonian $\hat{H}_0$ in (2). The eigenstates of the spin chain can be described as a combination of 2^N individual states of nuclear spins,

$$|00...00\rangle, \; |00...01\rangle, ..., |11...11\rangle, \quad (6)$$

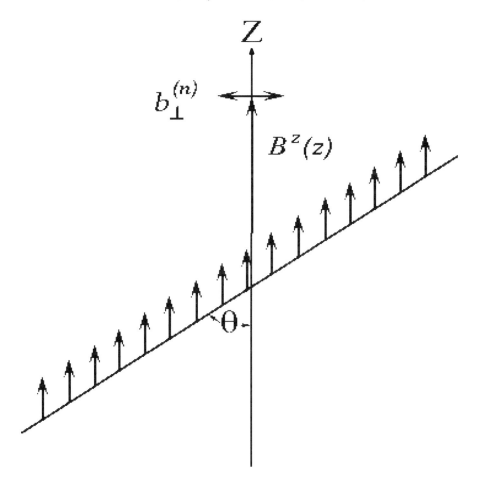

FIGURE 1. Nuclear spin quantum computer (the ground state of nuclear spins). $B^z(z)$ is the slightly non-uniform magnetic field; $b_\perp^{(n)}$ is the amplitude of radio-frequency field of the n-th rf pulse. The chain of spins makes an angle θ with the direction of the magnetic field $B^z(z)$.

where the state $|0\rangle$ corresponds to the direction of a nuclear spin along the direction of the magnetic field, $B^z(z)$, and the state $|1\rangle$ corresponds to the opposite direction. The subscript "k" is omitted.

In (2)-(4), ω_k is the Larmor frequency of the k-th spin (neglecting interactions between spins), $\omega_k = \gamma B^z(z_k)$, and γ is the nuclear gyromagnetic ratio. (For example, for a proton in the field $B^z(z_k) = 10$T, one has the nuclear magnetic resonance (NMR) frequency $f_k = \omega_k/2\pi \approx 430$MHz.) We assume, for definiteness, that the gradient of the magnetic field is positive, $\partial B^z(z)/\partial z > 0$. Suppose that the frequency difference of two neighboring spins is, $\delta f = \delta\omega/2\pi \approx 10$kHz, where $\delta\omega = \omega_{k+1} - \omega_k$. Thus, if the frequency of the edge spin is 430MHz, the frequency of the spin at the other end of the chain of 1000 qubits is ≈ 440MHz. Then, the value of $B^z(z)$ increases by $\Delta B^z = 0.23$T along the spin chain. Taking the distance between the neighboring spins, $a \approx 2$Å, we obtain the characteristic value for the gradient of the magnetic field, $\partial B^z(z)/\partial z \approx 0.23/1000a\cos\theta$, where θ is the angle between the direction of the chain and the z-axis (Fig. 1). Below we will take $\cos\theta = 1/\sqrt{3}$. (This allows one to suppress the dipole interaction between spins in the eigenstates of the Hamiltonian $\hat{H}_0$.) Thus, the gradient of the magnetic field is $\partial B^z(z)/\partial z \approx 2\times 10^6$ T/m. The quantity $\Omega^{(n)} = \gamma b_\perp^{(n)}$ is the Rabi frequency of the n-th pulse.

2. Quantum dynamics of the computer

We discuss below the dynamics of the spin chain described by the Hamiltonian (2)-(4). The quantum state of the quantum computer is described by the wave function $\Psi(t)$. The dynamics of this function is given by the solution of the Schrödinger equation,

$$i\dot{\Psi}(t) = \hat{H}\Psi(t), \tag{7}$$

where dot means derivative over time. This linear equation and an initial condition, $\Psi(0)$, define the state of the system at the time t. In this section we describe two different approaches which allow to solve Eq. (7).

2.1. Equations of motion in the interaction representation. In order to compute the dynamics it is convenient to go over to interaction representation. The wave function, $\Psi_{int}(t)$, in the interaction representation, is connected to the wave function, $\Psi(t)$, in the laboratory system of coordinates by the transformation,

$$\Psi(t) = \hat{U}_0 \Psi_{int}(t), \quad \hat{U}_0 = \exp(-i\hat{H}_0 t), \tag{8}$$

where, $\hat{H}_0$ is defined in (2). We choose the eigenfunctions, $|p\rangle$, and the eigenvalues, E_p, of the Hamiltonian $\hat{H}_0$ as the basis states and expand the wave function $\Psi_{int}(t)$ in a series,

$$\Psi_{int}(t) = \sum_{p=0}^{2^N-1} C_p(t)|p\rangle, \tag{9}$$

where the states $|p\rangle$ satisfy the equation

$$\hat{H}_0|p\rangle = E_p|p\rangle, \quad p = 1,...,2^N. \tag{10}$$

The wave function in the laboratory system of coordinates has the form,

$$\Psi(t) = \sum_p C_p(t)|p\rangle \exp(-iE_p t), \tag{11}$$

Using the Schrödinger equation (7) for the wave function, $\Psi(t)$, we obtain the equations of motion for the amplitudes $C_p(t)$,

$$i\dot{C}_p = \sum_{m=0}^{2^N-1} C_m V_{pm}^{(n)} \exp\left[i(E_p - E_m)t + ir_{pm}\left(\nu^{(n)}t + \varphi^{(n)}\right)\right], \tag{12}$$

which represent the wave function in the interaction representation. Here $r_{pm} = \mp 1$ for $E_p > E_m$ and $E_p < E_m$, respectively, $V_{pm}^{(n)} = -\Omega^{(n)}/2$ for the states $|p\rangle$ and $|m\rangle$ which are connected by a single-spin transition, and $V_{pm}^{(n)} = 0$ for all other states.

2.2. The dynamics in the rotating frame.
Sometimes, to calculate the quantum dynamics generated by the Hamiltonian (2) during the action of the n-th pulse, $t^{(n)} \leq t \leq t^{(n+1)}$, it is convenient to make a transformation to the system of coordinates which rotates with the frequency $\nu^{(n)}$ of the magnetic field, $\vec{b}_\perp$. To do this, one can use the unitary transformation [13],

$$\hat{U}^{(n)} = \exp\left[i\left(\nu^{(n)}t + \varphi^{(n)}\right)\sum_{k=0}^{N-1}\hat{I}_k^z\right]. \tag{13}$$

In the rotating system of coordinates the new wave function, $\Psi_{rot}(t)$ and the new Hamiltonian, $\mathcal{H}^{(n)}$ have the form,

$$\Psi_{rot}(t) = \hat{U}^{(n)+}\Psi(t), \tag{14}$$

$$\hat{\mathcal{H}}^{(n)} = \hat{U}^{(n)+}\hat{H}^{(n)}\hat{U}^{(n)} = -\sum_{k=0}^{N-1}\left[\left(\omega_k - \nu^{(n)}\right)\hat{I}_k^z + \Omega^{(n)}\hat{I}_k^x + 2J\hat{I}_k^z\hat{I}_{k+1}^z\right], \tag{15}$$

where the term $\hat{I}_{N-1}^z\hat{I}_N^z$ must be excluded. In the rotating system of coordinates the effective magnetic field which acts on the k-th spin during the action of the n-th pulse has the components,

$$B_k^{(n)} = \left(\Omega^{(n)}, 0, \omega_k - \nu^{(n)}\right)/\gamma. \tag{16}$$

The advantage of the Hamiltonian (15) is that it is time-independent. In this case one can use the eigenstates of the Hamiltonian $\hat{\mathcal{H}}^{(n)}$ to calculate the dynamics during the nth pulse without solution of the system of differential equations (12). As before, we expand the wave function in the rotating frame, $\Psi_{rot}(t)$, over the eigenstates of the Hamiltonian $\hat{H}_0$,

$$\Psi_{rot}(t) = \sum_{p=0}^{2^N-1} A_p(t)|p\rangle. \tag{17}$$

The wave function in the laboratory system of coordinates is,

$$\Psi(t) = \sum_p A_p(t)|p\rangle \exp\left(-i\chi_p^{(n)}t + i\xi_p^{(n)}\right), \tag{18}$$

where $\chi_p^{(n)} = -\left(\nu^{(n)}/2\right)\sum_{k=0}^{2^N-1}\sigma_k^p$, $\xi_p^{(n)} = \left(\varphi^{(n)}/2\right)\sum_{k=0}^{2^N-1}\sigma_k^p$, $\sigma_k^p = -1$ if the state $|p\rangle$ contains the kth spin in in the state $|1\rangle$, and $\sigma_k^p = 1$ if the state $|p\rangle$ contains the kth spin in the state $|0\rangle$. Below we take $\varphi^{(n)} = \xi_p^{(n)} = 0$ for all n. The Schrödinger equation for the amplitudes $A_p(t)$, has the form,

$$i\dot{A}_p(t) = \left(E_p - \chi_p^{(n)}\right)A_p(t) - \frac{\Omega^{(n)}}{2}\sum_{p'}A_{p'}(t), \tag{19}$$

where the sum is taken over the states $|p'\rangle$ connected by a single-spin transition with the state $|p\rangle$.

Equation (19) can be written in the form $i\dot{A}_p(t) = \mathcal{H}_{pp'}^{(n)}A_{p'}(t)$, where the effective Hamiltonian, $\hat{\mathcal{H}}^{(n)}$, is independent of time. When the number of spins,

N, is not very large ($N < 30$), one can diagonalize the Hamiltonian matrix $\mathcal{H}_{pp'}^{(n)}$ and find its (time-independent) eigenfunctions,

$$\psi_q^{(n)} = \sum_p A_p^{q\,(n)})|p\rangle \qquad (20)$$

and eigenvalues, $e_q^{(n)}$. The amplitudes $A_p^{q\,(n)}$ are the eigenfunctions of the Hamiltonian $\hat{\mathcal{H}}^{(n)}$ in the representation of the Hamiltonian $\hat{H}_0$.

Using the eigenstates of the Hamiltonian $\hat{\mathcal{H}}^{(n)}$ one can calculate the dynamics generated by the nth pulse as,

$$A_p(t_n) = A_p = \sum_{p_0} A_{p_0} \sum_q A_{p_0}^{q\,(n)} A_p^{q\,(n)} \exp\left(-ie_q^{(n)} \tau_n\right), \qquad (21)$$

where $t_n = t_{n-1} + \tau_n$, τ_n is the duration of the pulse, $A_{p_0} = A_p(t_{n-1})$ is the amplitude before the action of the nth pulse, $A_p = A_p(t_n)$ is the amplitude after the action of the nth pulse.

The amplitudes, $A_p(t)$, in the rotating frame are related to the amplitudes, $C_p(t)$, in the interaction representation by (see Eqs. (11) and (18)),

$$A_p(t) = \exp(-i\mathcal{E}_p^{(n)} t) C_p(t), \qquad (22)$$

Where $\mathcal{E}_p^{(n)} = E_p - \chi_p^{(n)}$ are the diagonal elements of the Hamiltonian matrix $\mathcal{H}_{pp'}^{(n)}$. Since the rf pulses in the protocol are different, the effective time-independent Hamiltonians, $\mathcal{H}_{pp'}^{(n)}$, are also different for different n. Hence, before each nth pulse we should make the transformation to the rotating frame, $C_p(t_{n-1}) \to A_p(t_{n-1})$, and after the pulse we must return to the interaction representation $A_p(t_n) \to C_p(t_n)$.

In the rotating frame the most difficult problem is the diagonalization of the sparse symmetric matrices (for each pulse) of the size $2^N \otimes 2^N$. (In each row of this matrix there are only $N+1$ nonzero matrix elements.) This approach can be used directly (without perturbative consideration) only for small quantum computers, with the number of qubits $N < 30$. For our purposes, to model quantum dynamics in a quantum computer with large number of qubits ($N = 1000$), we shall use a perturbation approach based on the existence of small parameters.

3. Resonant and near-resonant transitions

The number of equations in (12) grows exponentially as N increases. Thus, this system of equations cannot be solved directly when N is large. Our intention is to solve Eqs. (12) or (19) approximately, but in a controlled way. This can be done if we make use of small parameters which exist for the system. The explicit expressions for small parameters will be given below, in sections 8 and 9.

3.1. Resonant and non-resonant frequencies. In this section, we describe the main simplification procedure which is based on the separation of resonant and non-resonant interactions. This procedure is commonly used in studying linear and non-linear dynamical systems [14, 15, 16]. The main idea of this approach is the following. When the frequency of the external field, $\nu^{(n)}$, is equal (or close) to the eigenfrequency of the system under consideration, $\omega_{pm} \equiv E_p - E_m$, and when a small parameter exists (the interaction is relatively weak), one can separate the slow (resonant) dynamics from the fast (non-resonant) dynamics. The main contribution to the evolution of the system is associated with resonant effects.

Usually, non-resonant effects are small, and can be considered using perturbation approaches. In quantum computation, the problem of non-resonant effects is more complicated because these effects can accumulate in time creating significant errors in the process of quantum computation. So, in quantum computation, the non-resonant effects must not only be taken into consideration, but they must also be minimized. Below, we describe methods which allow us to minimize the influence of non-resonant effects.

In classical dynamical systems, the separation of resonant and non-resonant effects is very efficient when "action-angle" variables are used. In this case, the "action" is a slow variable and the "angle" (the "phase") is a fast variable. From this point of view, Eqs. (12) are convenient because they are written in the "energy" representation in which the slow (resonant) effects and fast (non-resonant) effects can be naturally separated.

First, we note that as the number of spins, N, increases, the number of eigenstates, $|p\rangle$, increases exponentially (as 2^N), but the number of *resonant frequencies* in our system grows only linearly in N. Indeed, the number of resonant frequencies is $3N - 2$ because only single-spin transitions are allowed by the operator $\hat{V}$ in (3). The definition of the resonant frequency is the following. We introduce the frequency of the single-spin transition: $\omega_{pm} \equiv E_p - E_m$, where E_p and E_m are two eigenstates, $|p\rangle$ and $|m\rangle$, of the Hamiltonian $\hat{H}_0$ with opposite orientations of only the k-th spin, $(k = 0, ..., N-1)$. For example,

$$(23) \qquad |p\rangle = |n_{N-1}n_{N-2}...0_k...n_1 n_0\rangle, \quad |m\rangle = |n_{N-1}n_{N-2}...1_k...n_1 n_0\rangle,$$

where $n_k = 0, 1$. The resonant frequency for transition between pth and mth states, $\nu_{res}^{(n)}$, of the n-th external *rf* pulse is defined as follows:

$$(24) \qquad \nu_{res}^{(n)} = |\omega_{pm}|.$$

The only resonant frequencies of our nuclear spin quantum computer are,

$$(25) \qquad \omega_k \pm J, \qquad (k = 0, \text{ or } k = N - 1),$$

$$\omega_k, \; \omega_k \pm 2J, \qquad (1 \leq k \leq N - 2),$$

where the Larmor frequencies, $\omega_k = \gamma B^z(z_k)$. All resonant frequencies in (25) are assumed to be positive. For end spins with $k = 0$ and $k = N - 1$, the upper and lower signs in (25) correspond to the states $|0\rangle$ or $|1\rangle$ of their only neighboring spins. For inner spins with $1 \leq k \leq N - 2$ the frequency ω_k corresponds to having nearest neighbors whose spins are in opposite directions to each other. The "+" sign corresponds to having the nearest neighbors in their ground state. The "-" sign corresponds to having the nearest neighbors in their excited state.

As an example, consider the resonant frequency of a single-spin transition between the following two eigenstates,

$$|p\rangle = |n_{N-1}n_{N-2}...1_{k-1}0_k 1_{k+1}...n_1 n_0\rangle,$$

$$|m\rangle = |n_{N-1}n_{N-2}...1_{k-1}1_k 1_{k+1}...n_1 n_0\rangle.$$

These two eigenstates differ by a transition of the k-th spin from its ground state, $|0_k\rangle$, to its excited state, $|1_k\rangle$. All other spins in the quantum computer remain in their initial states. According to (2), the eigenvalues of the states $|p\rangle$ and $|m\rangle$ are:

$$(26) \qquad E_p = E_{nres} - \frac{\omega_k}{2} + J, \qquad E_m = E_{nres} + \frac{\omega_k}{2} - J,$$

where E_{nres} is the total energy of all $N-1$ (except the k-th) which did not participate in the transition. It follows from (26) that in this case, the resonance frequency, $\nu_{res}^{(n)}$, of the n-th pulse is,

(27) $$\nu_{res}^{(n)} = |E_p - E_m| = \omega_k - 2J.$$

This frequency belongs to the set of the resonant frequencies introduced in (25). In the following, we assume that our protocol includes only the resonant frequencies from (18), and we will omit the subscript "res".

3.2. Dynamical equations for resonant and near-resonant transitions.
In this sub-section we derive the approximate equations for resonant and near-resonant transitions, in the system. Consider an arbitrary eigenstate of the Hamiltonian $\hat{H}_0$,

(28) $$|n_{N-1}n_{N-2}...n_1n_0\rangle,$$

where, as before, the subscript indicates the position of the spin in a chain, and $n_k = 0, 1$. Assume that one applies to the spin chain a resonant rf pulse of a frequency, $\nu^{(n)}$, from (25). Then, one has two possibilities:
1) The frequency of the pulse, $\nu^{(n)}$, is the resonant frequency of the state (28).
2) The frequency, $\nu^{(n)}$, differs from the closest resonant frequency of the state (28) by the value $\Delta = \pm 2J$ or $\pm 4J$ (near-resonant frequency).

In the first case, one has a resonant transition. In the second case, one has a near-resonant transition. If the small parameters exist, $J/\delta\omega \ll 1$, and $\Omega/\delta\omega \ll 1$, we can neglect all non-resonant transitions related to a flip of k'th non-resonant spin. Below, in section 9, we will write a rigorous condition for $\delta\omega$ which is required in order to neglect all non-resonant transitions.

Thus, considering the dynamics of any state (28) under the action of an rf pulse with the frequency, $\nu^{(n)}$, from (24), (25), we need only to take into account one transition. This transition will be either a resonant one or a near-resonant one. This allows us to reduce the system of equations (12) to the set of only two coupled equations (two-level approximation),

(29) $$i\dot{C}_p = -(\Omega^{(n)}/2)\exp[i(E_p - E_m - \nu^{(n)})t]C_m,$$
$$i\dot{C}_m = -(\Omega^{(n)}/2)\exp[i(E_m - E_p + \nu^{(n)})t]C_p,$$

where $E_p > E_m$, $|p\rangle$ and $|m\rangle$ are any two eigenstates which are connected by a single-spin transition and whose energies differ by $\nu^{(n)}$ or $\nu^{(n)} + \Delta_{pm}^{(n)}$, where $\Delta_{pm}^{(n)} = E_p - E_m - \nu^{(n)}$ is the frequency detuning for the transition between the states $|p\rangle$ and $|m\rangle$.

3.3. Solutions of the dynamical equations in two-level approximation.
The solution of Eqs. (29) for the case when the system is initially in the eigenstate $|m\rangle$, is,

(30)
$$C_m(t_0 + \tau) = [\cos(\lambda_{pm}\tau/2) + i(\Delta_{pm}/\lambda_{pm})\sin(\lambda_{pm}\tau/2)] \times \exp(-i\tau\Delta_{pm}/2),$$
$$C_p(t_0 + \tau) = i(\Omega/\lambda_{pm})\sin(\lambda_{pm}\tau/2) \times \exp(it_0\Delta_{pm} + i\tau\Delta_{pm}/2),$$
$$C_m(t_0) = 1, \quad C_p(t_0) = 0,$$

where $\lambda_{pm} = (\Omega^2 + \Delta_{pm}^2)^{1/2}$, t_0 is the time of the beginning of the pulse; τ is the duration of the pulse. Here and below we omit the upper index "(n)" which indicates the number of the *rf* pulse. If the system is initially in the upper state, $|p\rangle$, ($C_m(t_0) = 0$, $C_p(t_0) = 1$), the solution of (29) is,

(31) $C_p(t_0 + \tau) = [\cos(\lambda_{pm}\tau/2) - i(\Delta_{pm}/\lambda_{pm})\sin(\lambda_{pm}\tau/2)] \times \exp(i\tau\Delta_{pm}/2),$

$C_m(t_0 + \tau) = i(\Omega/\lambda_{pm})\sin(\lambda_{pm}\tau/2) \times \exp(-it_0\Delta_{pm} - i\tau\Delta_{pm}/2),$

$C_p(t_0) = 1, \qquad C_m(t_0) = 0.$

For the resonant transition ($\Delta_{pm} = 0$), the expressions (30) and (31) transform into the well-known equations for the Rabi transitions. For example, we have from (30) for $\Delta_{pm} = 0$,

$$C_m(t_0 + \tau) = \cos(\Omega\tau/2), \qquad C_p(t_0 + \tau) = i\sin(\Omega\tau/2).$$

In particular, for $\Omega\tau = \pi$ (the so-called π-pulse), the above expressions describe the complete transition from the state $|m\rangle$ to the state $|p\rangle$.

For the near-resonant transitions, consider two characteristic parameters in expressions (30) and (31):

$$\epsilon_{pm}^{(1)} = \Omega/\lambda_{pm}, \qquad \epsilon_{pm}^{(2)} = \sin(\lambda_{pm}\tau/2).$$

If either of these two parameters, $\epsilon_{pm}^{(1)}$ or $\epsilon_{pm}^{(2)}$, is zero, the probability of a non-resonant transition vanishes. The second parameter, $\epsilon_{pm}^{(2)}$, is equal zero when,

(32) $$\lambda_{pm}\tau = 2\pi k, \qquad (k = 1, 2, ..),$$

where k is the number of revolutions of the non-resonant (average) spin about the effective field in the rotating frame. Eq. (32) is the condition for the "$2\pi k$"-method to eliminate the near-resonant transitions. (See [**13**], Chapter 22, and [**17, 18**].)

3.4. The $2\pi k$-method. We shall present here the explicit conditions for the "$2\pi k$" rotation from the state $|p\rangle$ to the state $|m\rangle$. For a π-pulse ($\Omega\tau = \pi$), the values of Ω which satisfy the $2\pi k$ condition are, according to Eq. (32),

(33)
$$\Omega_{pm}^{[k]} = |\Delta_{pm}|/\sqrt{4k^2-1} = |\Delta_{pm}|/\sqrt{3}, \ |\Delta_{pm}|/\sqrt{15}, \ |\Delta_{pm}|/\sqrt{35}, \ |\Delta_{pm}|/\sqrt{63}, ...$$

For a $\pi/2$-pulse, the corresponding values of Ω are,

(34) $$\Omega_{pm}^{[2k]} = |\Delta_{pm}|/\sqrt{16k^2-1} = |\Delta_{pm}|/\sqrt{15}, \ |\Delta_{pm}|/\sqrt{63}, ...$$

(If the Rabi frequency, Ω, satisfies the $2\pi k$ condition (34) for a $\pi/2$-pulse, it automatically satisfies the condition (33) for a π-pulse.)

3.5. Resonant and near-resonant transitions in the rotating frame. We consider here the structure of the effective time-independent Hamiltonian matrix $\mathcal{H}_{pp'}$ in the rotating frame. We discuss the two-level approximation in the rotating system of coordinates and demonstrate equivalence of the solution in the rotating frame with the solution (30) (or (31)) in the interaction representation.

Let us discuss the structure of the matrix $\mathcal{H}_{pp'}$. Since the spins in the chain are identical, all nonzero non-diagonal matrix elements are the same and equal to $(-\Omega/2)$. At $\Omega \ll \delta\omega$ the absolute values of the diagonal elements in general case are much larger than the absolute values of the off-diagonal elements. We explain below how the resonance is coded in the structure of the diagonal elements of the Hamiltonian matrix, $\mathcal{H}_{pp'}$.

Suppose that the kth spin in the chain has resonant or near-resonant NMR frequency. The energy separation between pth and mth diagonal elements of the matrix $\mathcal{H}_{pp'}$ related by the flip of resonant (or near-resonant) kth spin is much less than the energy separation between the pth diagonal elements and diagonal elements related to other states, which differ from the state $|p\rangle$ by a flip of a non-resonant k'th ($k' \neq k$) spin. In this case one can neglect the interaction of the pth state with all states except for the state $|m\rangle$, and the Hamiltonian matrix $\mathcal{H}_{pp'}$ breaks up into $2^N/2$, approximately independent 2×2 matrices,

$$\begin{pmatrix} \mathcal{E}_m & V \\ V & \mathcal{E}_{p,} \end{pmatrix} \tag{35}$$

where $E_p = E_m + \Delta_{pm}$, $|\Delta_{pm}| \sim J$ or zero, and $V = -\Omega/2$ is the perturbation amplitude.

Suppose, for example, that $N = 5$ and third spin ($k = 3$) has resonant or near-resonant frequency. (We start enumeration from the right as shown in (23).) Then, the block 2×2 will be organized, for example, by the following states: $|01010\rangle$ and $|00010\rangle$; $|01111\rangle$ and $|00111\rangle$; $|00001\rangle$ and $|01001\rangle$, and so on. In order to find the state $|m\rangle$, which form 2×2 block with a definite state $|p\rangle$, one should flip the resonant spin of the state $|p\rangle$. In other words, positions of $N - 1$ (non-resonant) spins of these states are equivalent, while position of the resonant spin is different.

We now obtain the solution in the two-level approximation. The dynamics is given by Eq. (21). Since we deal only with a single 2×2 block of the matrix $\mathcal{H}_{pp'}$, (but not with the whole matrix), the dynamics in this approximation is generated only by the eigenstates of one block. In order to demonstrate equivalence of two descriptions (in the rotating frame and in the interaction representation) we will choose the same initial condition as in (30). Then, we will make the transformation (22) to the rotating frame, $C_p \to A_p$. After that we shall compute the dynamics by Eq. (21) and return to the interaction representation, $A_p \to C_p$, using Eq. (22). Eventually, we will obtain Eq. (30). The result is the same in the rotating frame and in the interaction representation. Physically these two approaches are equivalent (exactly, but not only in the two-level approximation), but mathematically they are different. In the interaction representation we calculate the dynamics generated by the time-dependent Hamiltonian. There are no stationary states in this case and one should solve the system of differential equations. On the other hand, in the rotating frame the effective Hamiltonian $\hat{\mathcal{H}}$ is time-independent and the wave function evolves in time because it is not the eigenfunction of $\hat{\mathcal{H}}$.

The eigenvalues $e_q^{(0)}$, $e_Q^{(0)}$, and the eigenfunctions of the 2×2 matrix (35) are (we put $\Delta_{pm} = \Delta$ and $\lambda_{pm} = \lambda$),

$$e_q^{(0)} = \mathcal{E}_m + \frac{\Delta}{2} - \frac{\lambda}{2}, \quad \begin{pmatrix} A_m^{q\,(0)} \\ A_p^{q\,(0)} \end{pmatrix} = \frac{1}{\sqrt{(\lambda - \Delta)^2 + \Omega^2}} \begin{pmatrix} \Omega \\ \lambda - \Delta \end{pmatrix}, \tag{36}$$

$$e_Q^{(0)} = \mathcal{E}_m + \frac{\Delta}{2} + \frac{\lambda}{2}, \quad \begin{pmatrix} A_m^{Q\,(0)} \\ A_p^{Q\,(0)} \end{pmatrix} = \frac{1}{\sqrt{(\lambda - \Delta)^2 + \Omega^2}} \begin{pmatrix} -(\lambda - \Delta) \\ \Omega \end{pmatrix}. \tag{37}$$

Suppose that before the nth pulse the system is in the state $|m\rangle$, i.e. the conditions

$$C_m(t_0) = 1, \quad C_p(t_0) = 0,$$

are satisfied. After the transformation, (22), to the rotating frame we obtain

$$A_m(t_0) = \exp(-i\mathcal{E}_m t_0) C_m(t_0) = \exp(-i\mathcal{E}_m t_0), \quad A_p(t_0) = 0.$$

The dynamics is given by Eq. (21), which in our case takes the form:

$$A_m(t) = A_m(t_0)\left[\left(A_m^{q\,(0)}\right)^2 \exp\left(-ie_q^{(0)}\tau\right) + \left(A_m^{Q\,(0)}\right)^2 \exp\left(-ie_Q^{(0)}\tau\right)\right] =$$

$$\frac{\exp\left[-i[\mathcal{E}_m t - (\Delta/2)\tau]\right]}{\Omega^2 + (\lambda - \Delta)^2}\left\{\Omega^2 e^{-i\lambda\tau/2} + (\lambda - \Delta)^2 e^{i\lambda\tau/2}\right\},$$

where $t = t_0 + \tau$. Applying the back transformation,

$$C_m(t) = \exp(i\mathcal{E}_m t) A_m(t),$$

and taking the real and imaginary parts of the expression in curl brackets we obtain the first equation (30). For another amplitude we have,

$$A_p(t) = A_m(t_0)\left[A_m^{q\,(0)} A_p^{q\,(0)} \exp\left(-ie_q^{(0)}\tau\right) + A_m^{Q\,(0)} A_p^{Q\,(0)} \exp\left(-ie_Q^{(0)}\tau\right)\right] =$$

$$i\frac{\Omega}{\lambda} \exp\left\{-i\left[\mathcal{E}_m t - (\Delta/2)\tau\right]\right\} \sin(\lambda\tau/2).$$

Applying the back transformation,

$$C_p(t) = \exp[i(\mathcal{E}_m + \Delta)t] A_p(t),$$

we obtain the second equation in (30).

One may demonstrate the equivalence of our two approaches in a different, more simple, way. The dynamical equations with the 2×2 Hamiltonian (35) are,

$$i\dot{A}_p = \mathcal{E}_p A_p + V A_m,$$

$$i\dot{A}_m = V A_p + \mathcal{E}_m A_m.$$

After the transformation (22) to the interaction representation we obtain,

(38)
$$i\dot{C}_p = V \exp[i(\mathcal{E}_p - \mathcal{E}_m)t] C_m,$$

$$i\dot{C}_m = V \exp[i(\mathcal{E}_m - \mathcal{E}_p)t] C_p,$$

where $\mathcal{E}_p - \mathcal{E}_m = E_p - E_m - \nu = \Delta_{pm}$, $V = -\Omega/2$. One can see that Eqs. (38) are equivalent to Eqs. (29) with the solution given by (30) or (31).

4. A quantum Control-Not gate for remote qubits

The quantum Control-Not (CN_{kl}) gate is a unitary operator which transforms the eigenstate,

(39)
$$|n_{N-1}...n_k........n_l....n_0\rangle,$$

into the state,

(40)
$$|n_{N-1}...n_k........\bar{n}_l....n_0\rangle,$$

where $\bar{n}_l = 1 - n_l$ if $n_k = 1$; and $\bar{n}_l = n_l$ if $n_k = 0$. The k-th and l-th qubits are called the control and the target qubits of the CN_{kl} gate. One can also introduce a modified quantum CN gate which performs the same transformation (39), (40) accompanied by phase shifts which are different for different eigenstates [13]. It is well-known that the quantum CN gate can produce an entangled state of two qubits, which cannot be represented as a product of the wave functions of the individual qubits.

In this section, we shall consider an implementation of the quantum CN gate in the Ising spin chain with the left end spin as the control qubit and the right end spin as the target qubit, i.e. $CN_{N-1,0}$ for a spin chain with large number of qubits, N ($N = 200$ or $N = 1000$). Using this quantum gate we will create entanglement between the end qubits in the spin chain.

Assume that initially a quantum computer is in its ground state,

$$\Psi = |0...0\rangle. \tag{41}$$

Then we apply a $\pi/2$-pulse with the resonant frequency,

$$\nu = \omega_{N-1}. \tag{42}$$

The $\pi/2$-pulse means that the duration of this pulse is $\tau = \pi/2\Omega$. This pulse produces a superpositional state of the $(N-1)$-th (left) qubit,

$$\Psi = (1/\sqrt{2})(|0...0\rangle + i|1...0\rangle). \tag{43}$$

To implement a modified quantum $CN_{N-1,0}$ gate we apply to the spin chain $M = 2N - 3$ π-pulses which transform the state $|10...00\rangle$ to the state $|10...01\rangle$ by the following scheme: $|1000...0\rangle \to |1100...0\rangle \to |1110...0\rangle \to |1010...0\rangle \to |1011...0\rangle \to |1001...0\rangle \to \cdots \to |100...11\rangle \to |100...01\rangle$. All frequencies of our protocol are resonant for these transitions. If we apply the same protocol to the system in the ground state, then with large probability the system will remain in the ground state because these pulses have the detunings from resonant transitions, $\Delta_n = \Delta_{0m}^{(n)} \neq 0$. Here the ground state $|0\rangle = |00...00\rangle$ is related to the state $|m\rangle$ by the flip of the kth spin with the near-resonant NMR frequency. The first π-pulse has the frequency $\omega = \omega_{N-2}$. For the second π-pulse $\omega = \omega_{N-3}$. For the third π-pulse, $\omega = \omega_{N-2} - 2J$, etc. All detunings, Δ_n, in our protocol are the same, $\Delta_n = 2J$, except for the third pulse, where $\Delta_3 = 4J$.

5. An analytic solution with application of a $2\pi k$-method

Assume that all M π-pulses satisfy the $2\pi k$ condition, with the same value of k:

$$\lambda_n^{[k]} \tau_n = 2\pi k, \text{ (and } \Omega_n^{[k]} \tau_n = \pi\text{)}, \tag{44}$$

where $\lambda_n^{[k]} = \sqrt{\left(\Omega_n^{[k]}\right)^2 + \Delta_n^2}$. In this case, we can derive an analytic expression for the wave function, Ψ, after the action of a $\pi/2$-pulse and M π-pulses. This solution has the form,

$$\Psi = C_0|00..0\rangle + C_1|10...1\rangle, \tag{45}$$

where,

$$C_0 = \frac{(-1)^{kM}}{\sqrt{2}} \exp(-i\pi M \sqrt{4k^2 - 1}/2), \qquad C_1 = \frac{(-1)^{N-1}}{\sqrt{2}}, \tag{46}$$

5.1. Large k asymptotics. For $k \gg 1$, we get the same solution for odd and even k:

$$C_0 \approx 1/\sqrt{2}.$$

This result is easy to understand. For a π-pulse, the Rabi frequency is,

$$\Omega_n^{[k]} = |\Delta_n|/\sqrt{4k^2 - 1}. \tag{47}$$

Large values of k correspond to small values of the parameter:

$$\Omega_n^{[k]}/|\Delta_n| \ll 1.$$

As $\Omega_n^{[k]}/|\Delta_n|$ approaches zero, the non-resonant pulse becomes unable to change the quantum state.

5.2. Small k behavior. For small values of k, the non-resonant pulse can change the phase of the initial state. For example, for $k = 1$ we have,

$$C_0 = \exp[i\pi M(1 - \sqrt{3}/2)]/\sqrt{2}. \tag{48}$$

After the first π-pulse, the phase shift is approximately $24°$. This phase shift grows as the number of π-pulses, M, increases. This increasing phase is an effect which can be easily controlled.

6. Small parameters

We shall now introduce the small parameters of the problem. Consider the probability of non-resonant transition. This probability will be small if ε_n is small:

$$\varepsilon_n = (\Omega_n/\lambda_n)^2 \sin^2(\lambda_n \tau_n/2) \ll 1. \tag{49}$$

(It follows from (30) that the expression for $|C_m|^2$ can be written in the form: $|C_m|^2 = 1 - \varepsilon_n$, and $|C_p|^2 = \varepsilon_n$.) In order to minimize the errors caused by the near-resonant transitions, it is reasonable to keep the values of ε_n the same and small. Since the values of the detuning are the same for all pulses, $\Delta_n = \Delta = 2J$ (except for the third pulse, where $\Delta_3 = 4J$), and because ε_n depends only on the ratio $|\Omega_n/\Delta_n|$ we take the values of Ω_n to be the same, $\Omega_n = \Omega$ ($n \neq 3$) and $\Omega_3 = 2\Omega$. In this case ε_n is independent of n ($\varepsilon_n = \varepsilon$). If we take into consideration the change of phase of the generated unwanted state, this change will be small if,

$$\Omega_n/|\Delta_n| \approx \Omega_n/\lambda_n \ll 1. \tag{50}$$

(For the $2\pi k$-method, this condition requires $k \gg 1$.)

Next, we will discuss the probability of the near-resonant transitions, after the action of M pulses, using small parameters ε_n. Analytic expressions for probabilities $|C_0|^2$ and $|C_1|^2$ are:

$$|C_0|^2 = \frac{1}{2} \prod_{n=1}^{M} (1 - \varepsilon_n), \qquad |C_1|^2 = \frac{1}{2}. \tag{51}$$

If all values of ε_n are the same for all pulses, $\varepsilon_n = \varepsilon$, then in the first non-vanishing approximation of the perturbation theory we obtain,

$$|C_0|^2 = \frac{1}{2}(1 - M\varepsilon), \qquad |C_1|^2 = \frac{1}{2}. \tag{52}$$

The decrease of the probability, $|C_0|^2$, is caused by the generation of unwanted states. One can see that the deviation from the value, $|C_0|^2 = 1/2$, grows as the number of π-pulses, M, increases. It means that the small parameter of the problem is $M\varepsilon$ rather than ε. When $M \approx 2N$ is large, even a small deviation from the $2\pi k$ condition can produce large distortions from the desired wave function (45).

7. Non-resonant transitions

In this section we estimate the errors caused by non-resonant transitions and write the formula for the total probability of errors including into consideration both near-resonant and non-resonant transitions. In order to estimate the probability of non-resonant transitions it is convenient to work in the rotating frame, where the effective Hamiltonian is independent of time.

Consider a transition between the states $|m\rangle$ and $|m'\rangle$ related by a flip of a non-resonant k'th spin. The absolute value of the difference between the mth and m'th diagonal elements of the matrix $\mathcal{H}_{pp'}$ is of order or greater than $\delta\omega$, because they belong to different 2×2 blocks (35). Since the absolute values of the matrix elements which connect the different blocks are small, $|V| \ll \delta\omega$, we can use the standard perturbation theory [19]. The wave function, ψ_q, in (20) can be written in the form,

$$\psi_q = \psi_q^{(0)} + \sum_{q'}{}' \frac{v_{qq'}}{e_q^{(0)} - e_{q'}^{(0)}} \psi_{q'}^{(0)}, \tag{53}$$

where superscript "(n)" is omitted, prime in the sum means that the term with $q' = q$ is omitted, ψ_q is the eigenfunction of the Hamiltonian $\mathcal{H}$, the qth eigenstate is related to the mth diagonal element and the q'th eigenstate is related to the m'th diagonal element, $v_{qq'} = 2V\langle\psi_q^{(0)}|I_{k'}^x|\psi_{q'}^{(0)}\rangle$ is the matrix element for transition between the states $\psi_q^{(0)}$ and $\psi_{q'}^{(0)}$, the sum over q' takes into consideration all possible non-resonant one-spin-flip transitions from the state $|m\rangle$. Because the matrix $\mathcal{H}_{pp'}$ is divided into 2^{N-1} relatively independent 2×2 blocks, the energy, $e_q^{(0)}$ ($e_{q'}^{(0)}$), and the wave function, $\psi_q^{(0)}$ ($\psi_{q'}^{(0)}$), in (53) are, respectively, the eigenvalue (36) and the eigenfunction,

$$\psi_q^{(0)} = A_m^{q\,(0)}|m\rangle + A_p^{q\,(0)}|p\rangle, \quad (\psi_{q'}^{(0)} = A_m^{q'\,(0)}|m'\rangle + A_p^{q'\,(0)}|p'\rangle) \tag{54}$$

of the single 2×2 block (35) with all other elements being equal to zero.

When the block (35) is related to the near-resonant transition ($\Delta \sim J$) and when $\Omega \ll J$, the eigenfunctions of this block are,

$$\psi_q^{(0)} \approx [1 - (\Omega^2/32J^2)]|m\rangle + (\Omega/4J)|p\rangle, \tag{55}$$

$$\psi_Q^{(0)} \approx -(\Omega/4J)|m\rangle + [1 - (\Omega^2/32J^2)]|p\rangle.$$

On the other hand, if this block is related to the resonant transition ($\Delta = 0$), we have,

$$\psi_q^{(0)} = (1/\sqrt{2})(|m\rangle + |p\rangle), \quad \psi_Q^{(0)} = (1/\sqrt{2})(|m\rangle - |p\rangle). \tag{56}$$

The probability of the non-resonant transition from the state $|m\rangle$ to the state $|m'\rangle$ connected by a flip of the non-resonant k'th spin is,

$$P_{mm'} = |\langle m'|\psi_q\rangle|^2. \tag{57}$$

Note, that only one term in the sum in (53) contributes to the probability $P_{mm'}$, so that,

$$P_{mm'} \approx \left(\frac{V}{\mathcal{E}_m - \mathcal{E}_{m'}}\right)^2 \approx \left(\frac{V}{|k - k'|\delta\omega}\right)^2, \tag{58}$$

where we put $e_q^{(0)} \approx \mathcal{E}_m$, $e_{q'}^{(0)} \approx \mathcal{E}_{m'}$, $v_{qq'} \approx V$, $|k - k'|$ is the "distance" from the non-resonant k'th spin to the resonant kth spin.

The total probability of generation of all unwanted states by one pulse in the result of non-resonant transitions can be obtained by summation of $P_{mm'}$ over m'. Since each state $|m'\rangle$ differs from the state $|m\rangle$ by the flip of one k' spin, one can replace the summation over the states m' to the summation over the spins $k' \neq k$. For example, the total probability, μ_{N-1} (here the subscript of μ stands for the number of the resonant spin, $k = N - 1$), of generation of unwanted states by the initial $\pi/2$-pulse is,

$$(59) \qquad \mu_{N-1} = \mu \sum_{k'=0}^{N-2} \frac{1}{|N-1-k'|^2}, \qquad \mu = \left(\frac{\Omega}{2\delta\omega}\right)^2.$$

After the initial $\pi/2$-pulse (which creates a superposition of two states with equal probabilities (43) from the ground state) the probability of the correct result is $\mathcal{P}_{\pi/2} = 1 - \mu_{N-1}$ and the probability of error is $P_{\pi/2} = 1 - \mathcal{P}_{\pi/2} = \mu_{N-1}$. After the first π-pulse the probability of error becomes,

$$(60) \qquad P_1 = 1 - \frac{1}{2}(1-\mu_{L-1})(1-\mu_{L-2}-\varepsilon) - \frac{1}{2}(1-\mu_{L-1})(1-\mu_{L-2}).$$

Here we suppose that the values of $\varepsilon = \varepsilon_n$ are the same for all pulses, and $\Omega_n = \Omega$ for $n \neq 3$ and $\Omega_3 = 2\Omega$.

The probability of error in implementation of the whole logic gate by $\pi/2$-pulse and all $2N - 3$ π-pulses is,

$$P = P_{2N-3} = 1 - \frac{1}{2}(1-\mu_{N-1})(1-\mu_{N-2}-\varepsilon)(1-4\mu_{N-2}-\varepsilon)(1-\mu_0-\varepsilon)\prod_{i=1}^{N-3}(1-\mu_i-\varepsilon)^2$$

$$(61) \qquad -\frac{1}{2}(1-\mu_{N-2})(1-4\mu_{N-2})\prod_{i=0}^{N-3}(1-\mu_i)^2,$$

where μ_i can be derived from (59) changing $N - 1$ to i. Two last terms in (61) are connected with two terms in (45). The probability of non-resonant transitions generated by the third pulse is approximately four times larger (terms with the factor 4 in (61)) than other probabilities because the Rabi frequency for this pulse is larger, $\Omega_3 = 2\Omega$ (to keep ε the same). Eq. (61) takes into account all near-resonant transitions, characterized by the parameter ε, and non-resonant transitions, characterized by the parameters μ_i. There is no need to consider the phases of the unwanted states, it is enough to consider only their probabilities.

Eq. (61) is convenient for choosing optimal parameters for application to the scalable solid state quantum computer with a large number of qubits. On the one hand, the number of qubits is a scalable parameter in (61), which can be easily changed and increased. (For example one can put $N = 1000$.) On the other hand, our approach takes into consideration all significant near-resonant and non-resonant processes generating errors in the quantum logic gates. The parameter ε can be minimized (down to the value $\varepsilon = 0$) by the $2\pi k$-method for any number of qubits. However, another parameter, μ, can be significantly decreases only by increasing the gradient of the magnetic field, or $\delta\omega$. Since the number of non-resonant transitions is large, their contribution to the probability of errors quickly increases with N increasing. As a consequence, for given $\delta\omega$ there is a restriction on the number of qubits in our computer, $N < N_{max}$, if we want to keep the errors below some definite threshold (see section 9).

8. Computer simulations for finite ε

Here we report our results on computer simulations of a quantum CN gate for remote qubits in quantum computer with 200 and 1000 qubits. We have developed a numerical code which allows us to study the dynamics of all quantum states with the probabilities no less than 10^{-6} for a spin chain with up to 1000 qubits. All frequencies in this and next sections are dimensionless and measured in units of J. The values of Ω for the results presented in this section are the same for all pulses, $\Omega_n = \Omega$, so that $\varepsilon_n = \varepsilon(\Omega/2J) = \varepsilon$ for $n \neq 3$ and $\varepsilon_3 = \varepsilon(\Omega/4J)$. All probabilities in this section are doubled. The sum of the probabilities of all these states was $2 - O(10^{-6})$ (the normalization condition). In this section we suppose that ε is large, $\varepsilon \gg \mu$ (we suppose $\mu = 0$ and $\delta\omega = \infty$), and one can neglect the non-resonant transitions, since their contribution to the probability of errors is small. The numerical results for the case of finite μ are presented in the next section.

In Fig. 2a, we show the probability of the excited unwanted states after implementation of the $CN_{199,0}$ gate, for $N = 200$ and $\Omega = 0.14$. Because we chose $|\Delta| = 2J$, it follows from (33) that the closest value of the Rabi frequency, Ω, which satisfies the $2\pi k$ condition is:

$$\Omega^{[7]} = \frac{2}{\sqrt{195}} \approx 0.1432.$$

This value differs slightly from the one used in the simulations whose results are presented in Fig. 2a ($\Omega = 0.14$) by 3.2×10^{-3}. At $\Omega = \Omega^{[7]}$ there are no the near-resonant transitions and the probability of errors is of order of μ, i.e. very small. One can see in Fig. 2a that even this small deviation from the $2\pi k$ condition results in generating over 7000 unwanted states during the total protocol. On the horizontal axis in Fig. 2a, the unwanted states are shown in the order of their generation. A total of 7385 unwanted states were generated whose probability $P' \geq 10^{-6}$. (In all figures 2-4 only the states with $P' \geq 10^{-6}$ are taken into account.) The probability distribution of unwanted states clearly contains two "bands". One group of these states has the probability, $P' \sim 10^{-6}$ (the bold "line" near the horizontal axis). The second group of states has the probability $P' \sim 10^{-3}$ (the upper "curve" in Fig. 2a). Fig. 2b shows an enlargement of the upper "band" of the Fig. 2a. One can see some sub-structure in this "band". Fig. 2c shows the sub-structure in the lower "band" of Fig. 2a. Our simulations show the existence of sub-structure in the upper and lower "bands" shown in Fig. 2c. There exists a hierarchy in the structure of the distribution function of unwanted generated states. Figs. 3(b-d) show the typical structure of unwanted states of the spin chain. In Figs. 3(a-d) "0" corresponds to the ground state, $|0\rangle$, of a qubit, and "1" corresponds to the excited state, $|1\rangle$, of a qubit. Fig. 3a shows the ground state of the spin chain. (All qubits are in their ground state, $|0\rangle$.) P' in Fig. 3, is the probability of a state of the whole chain. All unwanted states in Figs 3(b-d) belong to the upper and lower "bands" shown in Fig. 2c, and they have probabilities, $P' \sim 10^{-6}$.

Note, that many unwanted states are the high energy states of the spin chain (many-spin excitations). Typical unwanted states contain highly correlated spin

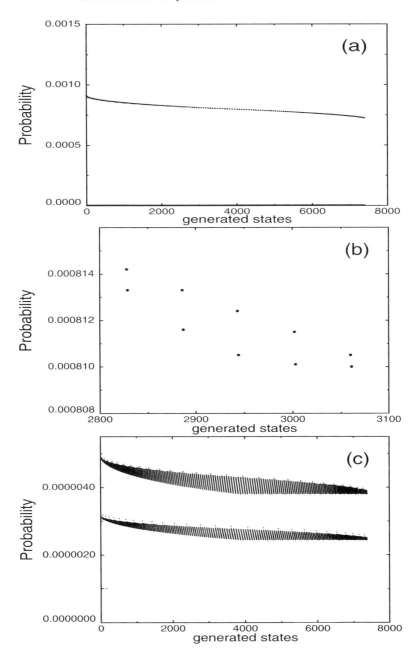

FIGURE 2. Probabilities of unwanted states. The total number of qubits: $N = 200$; $\Omega = 0.14$. (The value of Ω is measured in units of J.) The number of unwanted states with probabilities $|C_n|^2 \geq 10^{-6}$ is 7385. The unwanted states are presented in the order of their generation. (a) Two "bands": the upper band and the lower band; (b) The upper band of Fig. 2a, magnified; (c) The lower band of Fig. 2a, magnified.

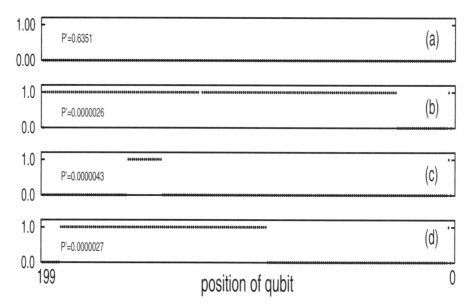

FIGURE 3. (a) The ground state of the spin chain; (b-d) Typical unwanted states with probabilities $P' \geq 10^{-6}$. The horizontal axis shows the position of a qubit in the spin chain of $N = 200$ spins. The vertical axis shows the ground state, $|0\rangle$, or the excited state, $|1\rangle$, of the qubit. Examples of "high energy" (b), "low energy" (c), and "intermediate energy" (d) unwanted states of the whole chain.

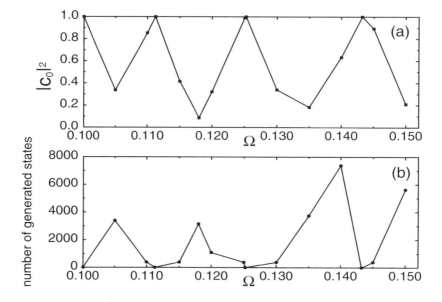

FIGURE 4. (a) Probability, $|C_0|^2$, as a function of Ω. (The value of Ω is measured in units of J.) The total number of qubits, $N = 200$; (b) The total number of unwanted states with probabilities $|C_m|^2 \geq 10^{-6}$.

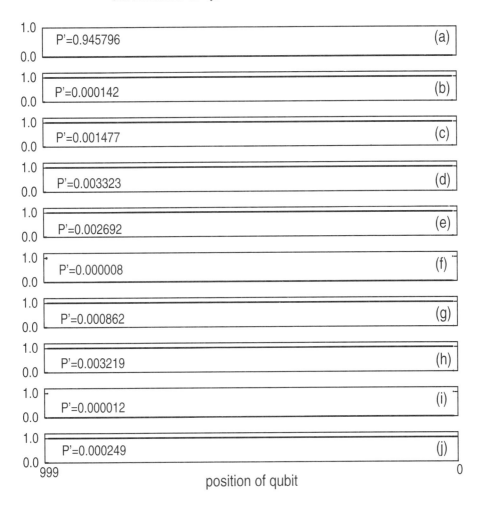

FIGURE 5. (a) The ground state of the chain; (b-j) Examples of unwanted states ($N = 1000$); $\Omega = 0.1$ for all π-pulses except for the π-pulses from the 10-th to the 40-th for which $\Omega = 0.1 + \eta$, where η is a random parameter in the range: $-0.05 < \eta < 0.05$.

excitations. Fig. 4 shows the probability of the ground state, $|C_0|^2$, and the total number of unwanted states (with probability $P' \geq 10^{-6}$) as a function of the Rabi frequency, Ω. (The point $\Omega = 0.14$ corresponds to the results shown in Fig. 2.) The maximum value of $|C_0|^2$ and the minimal total number of unwanted states correspond to values of Ω which satisfy the $2\pi k$ condition for 396 of the total number of π-pulses, 397. (The third π-pulse does not satisfy the $2\pi k$ condition.) One can see from Fig. 4 that application of the $2\pi k$ condition significantly improves the performance of the quantum Control-Not gate. Fig. 5 shows the ground state and the examples of unwanted states for $N = 1000$. In this case, for all π-pulses $\Omega = 0.1$, except for the π-pulses from 10-th to 40-th for which $\Omega = 0.1 + \eta$, where

η is a random number in the range: $-0.05 < \eta < 0.05$. The value $\Omega = 0.1$ corresponds to the $2\pi k$ condition (44) with $k = 10$. One can see from Fig. 5 that the states of the quantum computer are strongly correlated and some of them are highly excited. To avoid these unwanted strongly correlated and highly excited states it will be necessary to apply both a $2\pi k$-method and adequate error correction codes.

9. Numerical results for total error

In the previous section we presented the numerical results when the conditions of the $2\pi k$-method are not satisfied and $\varepsilon \gg \mu$. In this case one can neglect the non-resonant transitions and calculate the error caused by the near-resonant transitions by Eq. (51) or (52). One can minimize these errors by choosing $\varepsilon_n = 0$ for all n ($2\pi k$ condition). We show in this section that even under the conditions of the $2\pi k$-method the error can be relatively large due to non-resonant transitions, and the probability of errors increases when the number of qubits, N, increases.

We present here the results of computer simulations taking into consideration both, near-resonant and non-resonant transitions. We test our approximate formula (61) by exact numerical solution using Eqs. (21) and (22) when the number of qubits is not very large ($N = 10$), so that the total number of states in the Hilbert space is $2^{10} = 1024$. As before, the frequencies are dimensionless and measured in units of J. The values of ε for the results presented in this section are the same for all pulses, $\varepsilon_n = \varepsilon$, so that $\Omega_n = \Omega$ for $n \neq 3$ and $\Omega_3 = 2\Omega$. The norm of the wave function is equal (as usual) to unity and $\delta\omega$ (and μ) is finite.

In Figs. 6(a,b) we compare the total probability, P, of generation of unwanted states, found from the analytical estimate (61), with the result of exact numerical solution for the chain containing $N = 10$ spins. As follows from these figures, there is a very good correspondence between our approximate and exact solutions. From Fig. 6a one can see that at $\varepsilon = 0$ the value of P has tendency to decrease as $const/\delta\omega^2$. In Fig. 6b we plot the probability, P, as a function of Ω. When $\varepsilon \gg \mu$, and $J \ll \delta\omega$, the probability of generation of unwanted states is mostly defined by the value of ε and is almost independent of $\delta\omega$ since $\varepsilon \gg \mu$. The value of $\delta\omega$ fixes the values of the minima in Fig. 6b: for larger $\delta\omega$ the minima in the plot in Fig. 6b become deeper. The values of different minima in Fig. 6b indicate the contribution of non-resonant processes to the probability P. Since the value of Ω in Fig. 6b does not change significantly, the contribution of non-resonant processes to the probability of errors is approximately the same for all Ω. One can see that this contribution is negligible in comparison with the contribution of the near-resonant processes (defined by ε) for all Ω, except for the small regions of Ω, where $\varepsilon = 0$.

We should note that one more condition (except for $\varepsilon, \mu \ll 1$) must be satisfied for Eq. (61) to be valid. The value of the interaction constant, J, should be small in comparison with the difference between the spin frequencies, $J \ll \delta\omega$. In Figs. 7a and 7b we plot the probability, P, as a function of $\delta\omega$ for $\varepsilon = 0$ and $\varepsilon \neq 0$. (Since the frequencies are measured in units of J the value $\delta\omega = 1$ in Figs. 7(a,b) corresponds to $\delta\omega = J$.) One can see that our results are valid only when $J \ll \delta\omega$ ($\delta\omega \gg 1$ in Figs. 7(a,b)), in spite of the fact that the parameter $J/\delta\omega$ does not appear explicitly in (61). From Fig. 7b one can see that the probability of unwanted states, P, for $\varepsilon \gg \mu$ (for large $\delta\omega$) becomes relatively independent of $\delta\omega$.

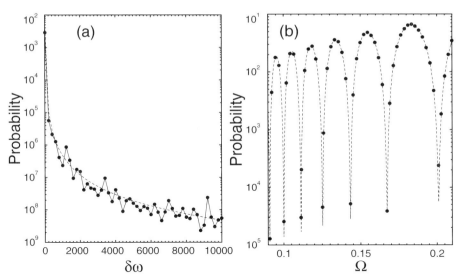

FIGURE 6. The probability, P, of generation of unwanted states. Filled circles are the results of exact numerical solution, dashed lines are the analytic estimates (61). (a) P as a function of $\delta\omega$, $\varepsilon = 0$, $\Omega = \Omega^{[8]}$ from (47). (b) P as a function of Ω, $\delta\omega = 100$. $N = 10$

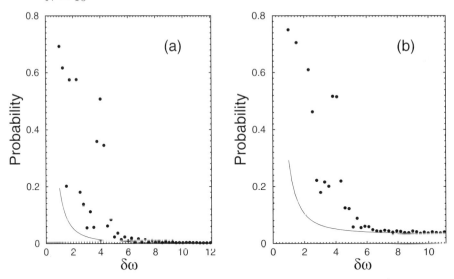

FIGURE 7. The total probability, P, of unwanted states as a function of $\delta\omega$ when the value of $\delta\omega$ is comparable with the value of the interaction constant, J. (a) $\Omega = \Omega^{[8]}$ ($\varepsilon = 0$), (b) $\Omega = 0.15$ ($\varepsilon = 0.0039$). Filled circles are the results of exact numerical solution, solid line is the analytical estimate (61). $N = 10$.

In this case the value of P is defined by the parameter ε which does not depend on $\delta\omega$.

Equation (61) is convenient for choosing the real experimental parameters required for operation of the solid state quantum computer. Suppose that we are able to correct the errors with the probability less than $P_0 = 10^{-5}$. Our perturbation theory allows us to calculate the region of parameters for which the probability of error, P, will be less than P_0. In Figs. 8(a-d) we plot the diagrams obtained using our perturbative approach for $N = 10$. Inside the hatched areas the probability of generation of unwanted states is less than P_0. We should note that building the plots like those in Figs. 8(a-d) using exact solution of the problem requires a significant number of computer time even for relatively small number of qubits ($N = 10$). On the other hand, the number N in our perturbation theory is the parameter which can be increased without any problems.

From Figs. 8(c,d) one can see that the sizes of the hatched areas are small, much smaller than the areas between the neighboring hatched regions. The total size of the hatched regions is mostly defined by the probability, P_0, of tolerant error which can be corrected using additional error correction codes. From comparison of Fig. 8c for $P_0 = 10^{-5}$ and Fig. 8d for $P_0 = 10^{-4}$ one can see that the sizes of the hatched regions increase with P_0 increasing. The coordinate of the point A in Ω is $\Omega = \Omega^{[k]}$, where the values of k are indicated in the figures.

In almost all quantum algorithms the phase of the wave function is important. We numerically compared the phase of the wave function on the boundaries of the hatched regions in Figs. 8(a,b) with the phase in the centers of these regions, (where Ω satisfies $2\pi k$ condition, see Eq. (46), $k = 5$ and $k = 11$) at fixed $\delta\omega$. The deviation in phase is only $\sim 0.15\%$. This is much less that the corresponding change in the probabilities of errors (by several orders).

In Figs. 9(a,b) we plot the same as in Figs. 8(a,b) but for $N = 1000$. One can see that increasing the number of qubits increases the errors and the sizes of the hatched areas decrease considerably. From Figs. 8(a-d) and 9(a-b) one can see that even when the values of Ω satisfy $2\pi k$-method (position of the point A in Ω in Figs. 8(a,b) and extreme left points of the hatched regions in Figs. 8(c,d) and 9(a,b)) the error can be relatively large because of the non-resonant processes. From comparison of Figs. 8(a,b) for $N = 10$ with Figs. 9(a,b) for $N = 1000$ one can see that the minimal value of $\delta\omega$ required to make the errors small increases considerably with N increasing.

We now analyze the probability of errors as a function of $\delta\omega$. When the value of $\delta\omega$ is large enough, the probability of error (and the widths of the hatched areas in Ω) becomes practically independent of $\delta\omega$. This is because at $\delta\omega \gg 1$ and at $\varepsilon \gg \mu$ the error is mostly defined by ε, which is independent of $\delta\omega$. As a consequence, one can, for example, estimate the widths of the hatched areas at $\delta\omega \gg 1$ taking into account only the near-resonant transitions (formula (52)).

In order to test our approximate result we solved the problem exactly using the eigenstates of the matrices $\mathcal{H}^{(n)}$, $n = 1, \ldots M$, for the parameters which correspond to the lower boundary of the hatched regions in the Figs. 8a and 8b (curves AB). In Figs. 10a and 10b we compare the total probability of errors, P, obtained using the exact and approximate solutions. One can see again, as in Figs. 6a and 6b, that there is a good correspondence between the exact and approximate solutions. A similar correspondence can be demonstrated for other points of the parametric space ($\delta\omega$, Ω) when the conditions $\Omega \ll J \ll \delta\omega$ are satisfied.

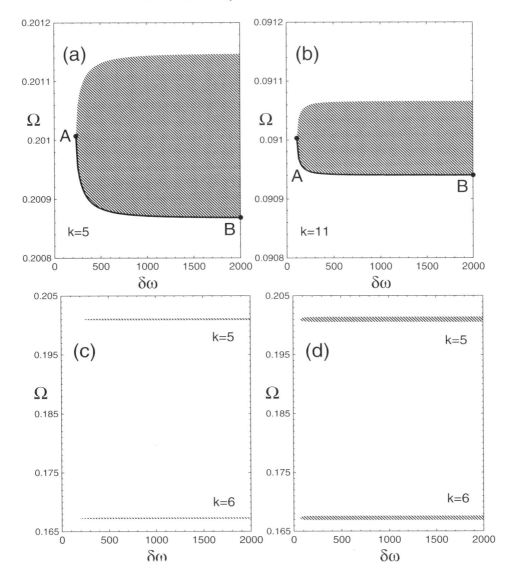

FIGURE 8. The probability of generation of unwanted states, P, at different values of $\delta\omega$ and Ω. In the hatched regions $P < P_0$. The different pictures (a) - (d) indicate the different regions in Ω in different scales. (a), (b), (c) $P_0 = 10^{-5}$; (d) $P_0 = 10^{-4}$. $N = 10$.

10. A classical Hamiltonian for quantum computation

In this section we demonstrate that the process of quantum computation, including creation of the entanglement, the dynamics of quantum controlled operations, and the dynamics of complicated quantum algorithms can be modeled using classical Hamiltonians. That is not surprising because the basic quantum equations (12) are c-number equations and can be formally considered as an effective classical system of equations. The above results show that in some cases,

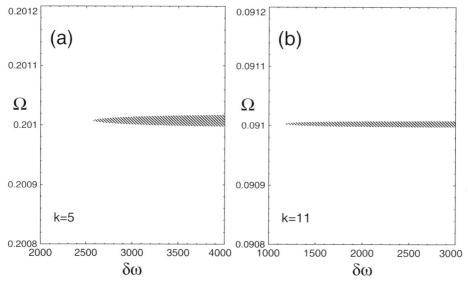

FIGURE 9. (a) The same as in Fig. 8a but for $N = 1000$. (b) The same as in Fig. 8b but for $N = 1000$. The scale in $\delta\omega$ shifted to the larger values in comparison with Figs. 8a, 8b.

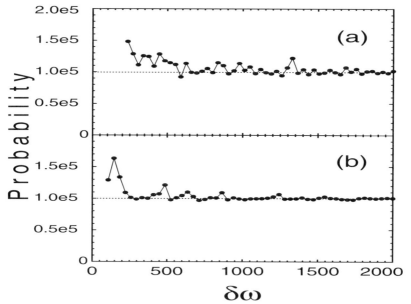

FIGURE 10. The exact solution for the probability of generation of unwanted states, P, for the parameters, which correspond to: (a) curve AB in Fig. 8a; (b) curve AB in Fig. 8b. The dashed lines indicate the solution obtained by the perturbation theory. $N = 10$.

effective classical Hamiltonians can be used for simulation of quantum logic operations even for large number of qubits. We present here the corresponding effective classical Hamiltonian and the Hamiltonian equations of motion in explicit form.

Formally, we can represent the solution of the Schrödinger equation for the Hamiltonian (2)-(4) in the form: $\Psi(t) = \sum_{n=0}^{2^N-1} c_n(t)|n\rangle$, where $|n\rangle$ and E_n are the eigenfunctions and the eigenvalues of the Ising part of the Hamiltonian in (2) in the z-representation,

$$\hat{H}_0|n\rangle = E_n|n\rangle. \tag{62}$$

The complex coefficients, $c_n(t)$, satisfy the equations,

$$i\dot{c}_n(t) = E_n c_n(t) + \sum_{k=0}^{2^N-1} V_{n,k}(t) c_k(t), \tag{63}$$

where,

$$V_{n,k} = \langle n|V(t)|k\rangle, \tag{64}$$

and $V(t)$ is the interaction Hamiltonian in (2).

Equation (63) can be written in Hamiltonian form:

$$i\dot{c}_n = \frac{\delta H_{cl}}{\delta c_n^*}, \qquad i\dot{c}_n^* = -\frac{\delta H_{cl}}{\delta c_n}, \tag{65}$$

where H_{cl} is the Hamiltonian of the equivalent "classical" system,

$$H_{cl} = \sum_{n=0}^{2^N-1} E_n |c_n|^2 + \sum_{n,p=0}^{2^N-1} c_n^* V_{n,p} c_p. \tag{66}$$

Instead of using complex "classical" variables, c_n and c_n^*, one can introduce the following real "classical" variables: a "coordinate", x_n, and a "momentum", p_n, using, for example, the canonical transformation,

$$x_n = \frac{1}{\sqrt{2}}(c_n + c_n^*), \qquad p_n = -\frac{i}{\sqrt{2}}(c_n - c_n^*). \tag{67}$$

Using x_n and p_n, Eqs. (65) take the familiar classical Hamiltonian form,

$$\dot{x}_n = \frac{\delta H_{cl}}{\delta p_n}, \qquad \dot{p}_n = -\frac{\delta H_{cl}}{\delta x_n}, \tag{68}$$

where the equivalent classical Hamiltonian is,

$$H_{cl}(x_n, p_n, t) = \sum_{n=0}^{2^N-1} \frac{E_n}{2}(x_n^2 + p_n^2) + \frac{1}{2} \sum_{n,k=0}^{2^N-1} (x_n - ip_n) V_{n,k}(x_k + ip_k). \tag{69}$$

The Hamiltonian (69) can be written in the form,

$$H_{cl}(x_n, p_n, t) = \sum_{n=0}^{2^N-1} \frac{E_n}{2}(x_n^2 + p_n^2) + \tag{70}$$

$$\frac{1}{2} \sum_{n,k=0}^{2^N-1} [x_n Re(V_{n,k}) x_k + p_n Re(V_{n,k}) p_k + 2 p_n Im(V_{n,k}) x_k],$$

where we used the relations: $V_{n,k} = V_{k,n}^*$, and $Re(V_{n,k}) = Re(V_{k,n})$, $Im(V_{n,k}) = -Im(V_{k,n})$. The corresponding classical Hamiltonian equations follow from (68) and (70),

$$\dot{x}_n = E_n x_n + \sum_{k=0}^{2^N-1} [Re(V_{n,k})p_k + 2Im(V_{n,k})x_k], \quad (71)$$

$$\dot{p}_n = -E_n p_n - \sum_{k=0}^{2^N-1} [Re(V_{n,k})x_k - 2Im(V_{n,k})p_k].$$

For example, the Hamiltonian (70) can be used to calculate the dynamics of the quantum Hamiltonian (2). In this case, the matrix elements, $V_{n,k} \neq 0$ only for those states $|n\rangle$ which differ from the state $|k\rangle$ by a single-spin transitions, and are zero otherwise.

Solutions of Eqs. (70) satisfy the normalization condition,

$$\sum_{n=0}^{2^N-1} [x_n^2(t) + p_n^2(t)] = 1. \quad (72)$$

Eqs. (70) describe the Hamiltonian dynamics of 2^N classical "generalized" one-dimensional oscillators. Each of these oscillators is described by two canonically conjugate variables, x_n and p_n. One can see that 2^N classical harmonic oscillators can simulate the behavior of N interacting quantum qubits, including the dynamics of quantum entanglement, complicated quantum logic gates, and quantum algorithms.

Conclusion

In this paper, we developed an approach, based on small parameters, for the simulation of simple quantum logic operations in a nuclear spin quantum computer with large number of qubits. We considered a quantum computer which is a one-dimensional nuclear spin chain placed in a slightly non-uniform magnetic field, oriented in a direction chosen to suppress the dipole interaction between spins. We took into consideration the Ising interaction between neighboring qubits. Quantum logic operations are implemented by applying resonant electromagnetic pulses to the nuclear spin chain. The electromagnetic pulse which is resonant to a selected qubit is non-resonant for all other qubits. This raises the problem of minimizing the influence of unwanted non-resonant effects in the process of performing a quantum protocol.

We simulate the creation of the long-distance entanglement between remote qubits, $(N-1)$-st and 0-th, in a nuclear spin quantum computer having a large number of qubits (up to 1000). We used two essential assumptions:
1. The nuclear spin chain is prepared initially in its ground state.
2. The frequency difference between the neighboring spins due to the inhomogeneity of the external magnetic field, $\delta\omega$, is much larger than the Ising interaction constant, J, and J is much larger than the Rabi frequency, Ω, i. e. $\Omega \ll J \ll \delta\omega$. Using these assumptions, we developed a numerical method which allowed us to simulate the dynamics of quantum logic operations taking into consideration all quantum states with the probability no less that $P = 10^{-6}$. For the case when the $2\pi k$ condition is satisfied (the π-pulse for the resonant transition is at the same time a $2\pi k$-pulse for non-resonant transitions), we obtained an analytic solution for the evolution of the nuclear spin chain. In the case of small deviations from the

$2\pi k$ condition, the error accumulates and the numerical simulations are required. These results are presented in sections 8 and 9.

The main results of our simulations are the following:

1. The unwanted states exhibit a band structure in their probability distributions. There are two main "bands" in the probability distribution of unwanted states. The unwanted states in these "bands" have significantly different probabilities, $P_{low}/P_{upper} \sim 10^{-3}$. Each of these two bands have their own structures.

2. A typical unwanted state is a state of highly correlated spin excitations. An important fact is that the unwanted states with relatively *high probability* include high energy states of the spin chain (many-spin excitations).

3. The method developed allowed us to study the generation of unwanted states and the probability of the desired states as a function of the distortion of *rf* pulses. This can be used to formulate the requirements for acceptable errors in quantum computation.

The results of this paper can be used to design experimental implementations of quantum logic operations and to estimate (benchmark) the reliability of experimental quantum computer devices. Our approach can be extended to simulations of simple quantum arithmetic operations and fragments of quantum algorithms. These simulations are now in progress.

References

[1] G.P. Berman, G.D. Doolen, V.I. Tsifrinovich, *Superlattices and Microstructures* **27**, 89 (2000).
[2] B.E. Kane, Nature **393**, 133 (1998).
[3] G.P. Berman, G.D. Doolen, V.I. Tsifrinovich, *Phys. Rev. Lett.* **84**, 1615 (2000).
[4] D. Loss, D.P. DiVincenzo, Phys. Rev. A **57**, 120 (1998).
[5] R. Vrijen, E. Yablonovich, K. Wang, H.W. Jiang, A. Balandin, V. Roychowdhury, T. Mor, D. DiVincenzo, *Phys. Rev. A* **62**, 2306 (2000).
[6] Y. Nakamura, C.D. Chen, J.S. Tsai, *Phys. Rev. Lett.* **79**, 2328 (1997).
[7] Y. Nakamura, Yu.A. Pashkin, J.S. Tsai, *Nature* **398**, 786 (1999).
[8] N.A. Gershenfeld, I.L. Chuang, Science **275**, 350 (1997).
[9] D.G. Cory, A.F. Fahmy, T.F. Havel, Proc. Natl. Acad. Sci. USA **94**, 1634 (1997).
[10] F. Yamaguchi, Y. Yamamoto, *Microelectronic Engineering* **47**, 273 (1999).
[11] G.P. Berman, G.D. Doolen, G.D. Holm, V.I. Tsifrinovich, *Phys. Lett. A* **193**, 444 (1994).
[12] I.L. Chuang, N.A. Gershefeld, M. Kubinec, *Phys. Rev. Lett.* **80**, 3408 (1998).
[13] G.P. Berman, G.D. Doolen, R. Mainieri, V.I. Tsifrinovich, *Introduction to Quantum Computers*, World Scientific Publishing Company, Singapore, (1998).
[14] N.N. Bogolubov, *Selected Works, Part I. Dynamical Theory*, Gordon & Breach Science Pub., (1990).
[15] A.J. Lichtenberg and M.A. Liberman, *Regular and Stochastic Motion*, Springer-Verlag, New York, (1983).
[16] G.M. Zaslavsky, *Chaos in Dynamic Systems*, Harwood Academic Pub., (1985).
[17] G.P. Berman, D.K. Campbell, V.I. Tsifrinovich, *Phys. Rev. B* **55**, 5929 (1997).
[18] G.P. Berman, G.D. Doolen, and V.I. Tsifrinovich *Phys. Rev. A* **61**, 2307 (2000).
[19] L. D. Landau, E. M. Lifshits, *Quantum Mechanics: Non-Relativistic Theory*, Pergamon Press, Oxford, New York, (1965).

(G. P. Berman) THEORETICAL DIVISION, T-13, AND CNLS, LOS ALAMOS NATIONAL LABORATORY, LOS ALAMOS, NM 87545

(G. D. Doolen) THEORETICAL DIVISION, T-13, AND CNLS, LOS ALAMOS NATIONAL LABORATORY, LOS ALAMOS, NM 87545

(D. I. Kamenev) THEORETICAL DIVISION, T-13, AND CNLS, LOS ALAMOS NATIONAL LABORATORY, LOS ALAMOS, NM 87545

(G. V. López) DEPARTAMENTO DE FÍSICA, UNIVERSIDAD DE GUADALAJARA, CORREGIDORA 500, S.R. 44420, GUADALAJARA, JALISCO, MÉXICO

(V. I. Tsifrinovich) IDS DEPARTMENT, POLYTECHNIC UNIVERSITY, SIX METROTECH CENTER, BROOKLYN NY 11201

Inconclusive Rate with a Positive Operator Valued Measure

Howard E. Brandt

ABSTRACT. Analysis is performed of explict optical implementations of both a positive operator valued measure (POVM) and of randomly choosing between two ordinary von Neumann projective measures. The POVM is demonstrated to have the lower inconclusive rate. Also, the effect of a general unitary disturbance on the inconclusive rate of the POVM implementation is calculated explicitly.

Contents

1. Introduction
2. Inconclusive rates comparison
3. Disturbed inconclusive rate
4. Consistency
5. Conclusion
6. Acknowledgements

References

1. Introduction

A positive operator valued measure (POVM) [1–7] can be usefully implemented in a quantum key receiver [8–14]. The following set of POVM operators represents the possible measurements performed by the receiver:

$$A_u = (1 + \langle u|v\rangle)^{-1} (1 - |v\rangle \langle v|), \tag{1}$$

$$A_v = (1 + \langle u|v\rangle)^{-1} (1 - |u\rangle \langle u|), \tag{2}$$

$$A_? = 1 - A_u - A_v. \tag{3}$$

Here, the kets $|u\rangle$ and $|v\rangle$ represent the two possible nonorthogonal normalized polarization states of a carrier photon with linear polarizations designated by u and v, respectively. The angle between the corresponding polarization vectors is θ.

2000 *Mathematics Subject Classification.* Primary 81P68, 94-02, 94A60.

Key words and phrases. Quantum cryptography, quantum key distribution, quantum communication, quantum information.

The photon is a spin-one representation of the Lorentz group, and it follows that the Dirac bracket between the two states is [11]

(4) $$\langle u|v\rangle = \sin 2\alpha,$$

where

(5) $$\alpha = \frac{1}{2}\left(\frac{\pi}{2} - \theta\right).$$

(The use of the angle α instead of θ is convenient in the following.) The states $|u\rangle$ and $|v\rangle$ may encode bit values 0 and 1, respectively. The POVM operators, Eqs. (1)–(3), are nonnegative and their sum is unity. The operators A_u and A_v measure the probability of outcomes u and v, respectively. The operator $A_?$ measures the probability of an inconclusive measurement.

The advantage of a POVM over an ordinary von Neumann projective measurement is that, for the POVM, the probability of getting an inconclusive result can be lower [8,14,15]. To see this, first consider, for comparison of a projective valued (PV) receiver with the POVM receiver, the simple all-optical PV receiver depicted in Figure 1. Effectively, the device randomly chooses between two ordinary von Neumann projective measures. (The all-optical POVM receiver is already exposited elsewhere [9–13].) The PV receiver consists of an incoming carrier photon in polarization state

(6) $$|\psi\rangle = \bar{\alpha}|u\rangle + \bar{\beta}|v\rangle$$

for complex numbers $\bar{\alpha}$ and $\bar{\beta}$, a 50-50 beam splitter BS, two Wollaston prisms W_u and W_v, and four photodetectors D_u, $D_{u_\perp}$, $D_{v_\perp}$, and D_v. The Wollaston prism W_u is aligned so that a photon in state $|u\rangle$ would take the path labeled by the state $|\psi_3\rangle$ and polarization vector $\hat{e}_u$, and not the path labeled by the state $|\psi_4\rangle$ and polarization vector $\hat{e}_{u_\perp}$. Here, $\hat{e}_u$ denotes a unit polarization vector corresponding to the polarization state $|u\rangle$ and is perpendicular to the polarization vector $\hat{e}_{u_\perp}$ corresponding to the polarization state $|u_\perp\rangle$ orthogonal to $|u\rangle$. Analogously, the Wollaston prism W_v is aligned so that a photon in state $|v\rangle$ would take the path labeled by the state $|\psi_6\rangle$ and polarization vector $\hat{e}_v$, and not the path labeled by the state $|\psi_5\rangle$ and polarization vector $\hat{e}_{v_\perp}$ (perpendicular to $\hat{e}_v$). The beamsplitter BS serves to choose at random between two projective valued measurements, the first having projections $\{|u\rangle\langle u|, |u_\perp\rangle\langle u_\perp|\}$ and the second having projections $\{|v\rangle\langle v|, |v_\perp\rangle\langle v_\perp|\}$. It is immediately evident from Figure 1 that

(7) $$|\psi_1\rangle = 2^{-1/2}i\left(\bar{\alpha}|u\rangle + \bar{\beta}|v\rangle\right),$$

(8) $$|\psi_2\rangle = 2^{-1/2}\left(\bar{\alpha}|u\rangle + \bar{\beta}|v\rangle\right),$$

(9) $$|\psi_3\rangle = 2^{-1/2}i\left(\bar{\alpha} + \bar{\beta}\sin 2\alpha\right)\left|\hat{e}_u\right\rangle,$$

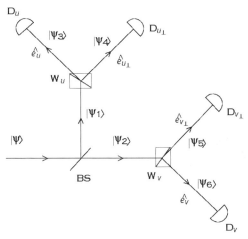

Figure 1. PV receiver.

$$|\psi_4\rangle = 2^{-1/2} i \bar{\beta} \cos 2\alpha \left|\hat{e}_{u_\perp}\right\rangle, \tag{10}$$

$$|\psi_5\rangle = 2^{-1/2} \bar{\alpha} \cos 2\alpha \left|\hat{e}_{v_\perp}\right\rangle, \tag{11}$$

$$|\psi_6\rangle = 2^{-1/2} \left(\bar{\alpha} \sin 2\alpha + \bar{\beta}\right) \left|\hat{e}_v\right\rangle, \tag{12}$$

where $\left|\hat{e}_u\right\rangle$, $\left|\hat{e}_{u_\perp}\right\rangle$, $\left|\hat{e}_v\right\rangle$, and $\left|\hat{e}_{v_\perp}\right\rangle$ represent unit kets corresponding to polarization vectors $\hat{e}_u$, $\hat{e}_{u_\perp}$, $\hat{e}_v$, and $\hat{e}_{v_\perp}$, respectively. It then follows that the probability $P_{\psi u}$ that the photon in state $|\psi\rangle$ is detected by ideal detector D_u is given by

$$P_{\psi u} = |\psi_3|^2 = \frac{1}{2} \left|\bar{\alpha} + \bar{\beta} \sin 2\alpha\right|^2. \tag{13}$$

Analogously, for detectors $D_{u_\perp}$, $D_{v_\perp}$, and D_v, one has

$$P_{\psi u_\perp} = |\psi_4|^2 = \frac{1}{2} \left|\bar{\beta}\right|^2 \cos^2 2\alpha, \tag{14}$$

$$P_{\psi v_\perp} = |\psi_5|^2 = \frac{1}{2} |\bar{\alpha}|^2 \cos^2 2\alpha, \tag{15}$$

$$P_{\psi v} = |\psi_6|^2 = \frac{1}{2} \left|\bar{\alpha} \sin 2\alpha + \bar{\beta}\right|^2. \tag{16}$$

From Eqs. (13)–(16), it follows that

$$P_{\psi u} + P_{\psi u_\perp} + P_{\psi v} + P_{\psi v_\perp} = |\bar{\alpha}|^2 + \bar{\alpha}^* \bar{\beta} \sin 2\alpha + \bar{\alpha} \bar{\beta}^* \sin 2\alpha + \left|\bar{\beta}\right|^2 = 1, \tag{17}$$

as must be the case, provided that the state $|\psi\rangle$, Eq. (6), is normalized to unity, and probability is conserved.

2. Inconclusive rates comparison

If the incoming photon state is $|\psi\rangle = |u\rangle$, one has $\{\bar{\alpha}, \bar{\beta}\} = \{1, 0\}$ and Eqs. (13)–(16) yield $P_{uu} = \frac{1}{2}$, $P_{uu_\perp} = 0$, $P_{uv_\perp} = \frac{1}{2}\cos^2 2\alpha$, $P_{uv} = \frac{1}{2}\sin^2 2\alpha$. Analogously, in the case where the incoming photon state is $|\psi\rangle = |v\rangle$, one has $\{\bar{\alpha}, \bar{\beta}\} = \{0, 1\}$ and $P_{vu} = \frac{1}{2}\sin^2 2\alpha$, $P_{vu_\perp} = \frac{1}{2}\cos^2 2\alpha$, $P_{vv_\perp} = 0$, $P_{vv} = \frac{1}{2}$. If states $|u\rangle$ and $|v\rangle$ are equiprobably incident on the receiver, then since detector D_u or D_v can be triggered by both states $|u\rangle$ and $|v\rangle$, it follows that the probability $P_?^{\text{PV}}$ of an inconclusive measurement is given by

$$P_?^{\text{PV}} = P_{uu} + P_{uv} = \frac{1}{2}\left(1 + \sin^2 2\alpha\right), \tag{18}$$

or equivalently,

$$P_?^{\text{PV}} = P_{vv} + P_{vu} = \frac{1}{2}\left(1 + \sin^2 2\alpha\right). \tag{19}$$

One can conclude that for the two-state quantum key distribution protocol, in which a photon is incident equiprobably in state $|u\rangle$ or $|v\rangle$, the inconclusive rate $P_?^{\text{PV}}$ for the projective receiver is

$$P_?^{\text{PV}} = \frac{1}{2}\left(1 + \sin^2 2\alpha\right). \tag{20}$$

One can also obtain Eq. (20) by reasoning that the inconclusive rate for the PV measure is given by

$$P_?^{\text{PV}} = 1 - P_{con}^{\text{PV}}, \tag{21}$$

where P_{con}^{PV} is the probability of obtaining a conclusive result. From Fig. 1, it is evident that

$$P_{con}^{\text{PV}} = P_{uv_\perp} = \frac{1}{2}\langle u|\left(|v_\perp\rangle\langle v_\perp|\right)|u\rangle = \frac{1}{2}|\langle u|v_\perp\rangle|^2$$
$$= \frac{1}{2}\sin^2\theta = \frac{1}{2}\left(1 - \sin^2 2\alpha\right). \tag{22}$$

Equation (22) follows, since the ideal detector $D_{v_\perp}$ cannot have been excited by the state $|v\rangle$, and therefore can only have been excited by the state $|u\rangle$, and the measurement operator for the state $|v_\perp\rangle$ is $|v_\perp\rangle\langle v_\perp|$ [11]. (The overall factor of $\frac{1}{2}$ appears in Eq. (22), because the probability is $\frac{1}{2}$ that the photon takes the path from the beamsplitter leading to the Wollaston prism W_v. Also note that Eq. (22) is consistent with Eq. (15) for $\bar{\alpha} = 1$.) If one substitutes Eq. (22) in Eq. (21), then Eq. (20) again follows. Of course, Eq. (22) also follows analogously from

$$P_{con}^{\text{PV}} = P_{vu_\perp} = \frac{1}{2}\langle v|\left(|u_\perp\rangle\langle u_\perp|\right)|v\rangle = \frac{1}{2}|\langle v|u_\perp\rangle|^2$$
$$= \frac{1}{2}\sin^2\theta = \frac{1}{2}\left(1 - \sin^2 2\alpha\right). \tag{23}$$

It has been demonstrated in previous work that the inconclusive rate $P_{\psi?}^{\text{POVM}}$ of the POVM receiver for the arbitrary incoming state, Eq. (6), is given by [9,11–13]

$$P_{\psi?}^{\text{POVM}} = \langle\psi|A_?|\psi\rangle = \left|\bar{\alpha} + \bar{\beta}\right|^2 \sin 2\alpha. \tag{24}$$

(The second equality in Eq. (24) is also consistent with the first, as can be seen by substituting Eqs. (3) and (6) in the first.)

For incoming state $|\psi\rangle = |u\rangle$, one then has $\{\bar{\alpha},\bar{\beta}\} = \{1,0\}$, and Eq. (24) becomes

$$P_{u?}^{\text{POVM}} = \sin 2\alpha. \tag{25}$$

For incoming state $|\psi\rangle = |v\rangle$, one has $\{\bar{\alpha},\bar{\beta}\} = \{0,1\}$, and

$$P_{v?}^{\text{POVM}} = P_{u?}^{\text{POVM}} = \sin 2\alpha. \tag{26}$$

It follows that the inconclusive rate $P_?^{\text{POVM}}$ of the ideal POVM receiver for the equiprobable two-state protocol is given by

$$P_?^{\text{POVM}} = \sin 2\alpha. \tag{27}$$

Using Eqs. (20) and (27), one then obtains

$$\frac{P_?^{\text{POVM}}}{P_?^{\text{PV}}} = \frac{2\sin 2\alpha}{1 + \sin^2 2\alpha} < 1, \tag{28}$$

as depicted in Figure 2. Thus, in fact, the inconclusive rate for the POVM receiver is less than that of the PV receiver, and the rate ratio is determined by the angle between the two polarization states (see Eq. (5)).

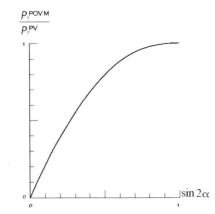

Figure 2. Inconclusive rate comparison for POVM and PV receivers.

3. Disturbed inconclusive rate

The Fuchs Peres model of eavesdropping on the two-state key-distribution protocol represents the most general possible unitary disturbance of each encoded photon incident on the receiver [16]. In this model, an incoming carrier state $|u\rangle$ and the state $|w\rangle$ of a disturbing probe undergo joint unitary evolution represented by a unitary operator U, resulting in the entangled state [10,16,17]:

$$\begin{aligned} U|u \otimes w\rangle = \frac{1}{2}\big[&(1 + \sec 2\alpha)|\Phi_{00}\rangle + \tan 2\alpha |\Phi_{10}\rangle - \tan 2\alpha |\Phi_{01}\rangle \\ &+ (1 - \sec 2\alpha)|\Phi_{11}\rangle\big] \otimes |u\rangle - \frac{1}{2}\big[\tan 2\alpha |\Phi_{00}\rangle - (1 - \sec 2\alpha)|\Phi_{10}\rangle \\ &- (1 + \sec 2\alpha)|\Phi_{01}\rangle - \tan 2\alpha |\Phi_{11}\rangle\big] \otimes |v\rangle. \end{aligned} \tag{29}$$

Here $|\Phi_{mn}\rangle$ are states in the Hilbert space of the disturbing probe, and are neither normalized nor orthogonal. Equation (29) follows from Eqs. (1) and (2) of Slutsky et al [17]. Similarly, for an incoming state $|v\rangle$, one has

$$U|v \otimes w\rangle = \frac{1}{2}\big[\tan 2\alpha\,|\Phi_{00}\rangle + (1+\sec 2\alpha)\,|\Phi_{10}\rangle + (1-\sec 2\alpha)\,|\Phi_{01}\rangle$$
$$- \tan 2\alpha\,|\Phi_{11}\rangle\big] \otimes |u\rangle + \frac{1}{2}\big[(1-\sec 2\alpha)\,|\Phi_{00}\rangle - \tan 2\alpha\,|\Phi_{10}\rangle$$
(30) $$+ \tan 2\alpha\,|\Phi_{01}\rangle + (1+\sec 2\alpha)\,|\Phi_{11}\rangle\big] \otimes |v\rangle.$$

The probe states $|\Phi_{mn}\rangle$ have certain symmetry properties that arise from the random equiprobable selection of carrier states $|u\rangle$ and $|v\rangle$ by the key transmitter, and the resulting symmetry of the probe under interchange of $|u\rangle$ and $|v\rangle$. Specifically, one has [16,17]

(31) $$|\Phi_{00}| = |\Phi_{11}|,$$

(32) $$|\Phi_{01}| = |\Phi_{10}|,$$

(33) $$\langle\Phi_{00}|\Phi_{01}\rangle = \langle\Phi_{11}|\Phi_{10}\rangle,$$

(34) $$\langle\Phi_{00}|\Phi_{10}\rangle = \langle\Phi_{11}|\Phi_{01}\rangle,$$

(35) $$\langle\Phi_{01}|\Phi_{10}\rangle = \langle\Phi_{10}|\Phi_{01}\rangle,$$

(36) $$\langle\Phi_{01}|\Phi_{00}\rangle = \langle\Phi_{10}|\Phi_{11}\rangle,$$

(37) $$\langle\Phi_{01}|\Phi_{11}\rangle = \langle\Phi_{10}|\Phi_{00}\rangle,$$

(38) $$\langle\Phi_{11}|\Phi_{00}\rangle = \langle\Phi_{00}|\Phi_{11}\rangle.$$

According to Eq. (24), the inconclusive rate $R_?$ induced by the disturbing probe in the POVM receiver is given by

(39) $$R_? = P_{u?} = \langle u \otimes w|\,U^\dagger A_? U\,|u \otimes w\rangle,$$

where $P_{u?}$ is the probability that if a photon in polarization state $|u\rangle$ is transmitted, then the measurement by the POVM receiver is inconclusive. Alternatively, one also has

(40) $$R_? = P_{v?} = \langle v \otimes w|\,U^\dagger A_? U\,|v \otimes w\rangle,$$

because of the symmetry of the two-state protocol. Equivalently, using Eq. (3) in Eq. (39), one also has for the induced inconclusive rate:

(41) $$R_? = 1 - P_{uu} - P_{uv},$$

where P_{uu} and P_{uv} are the probabilities that if the carrier is a $|u\rangle$ state, then the detectors D_u and D_v, respectively, respond. Here, one has

(42) $$P_{uu} = \langle u \otimes w|\,U^\dagger A_u U\,|u \otimes w\rangle,$$

(43) $$P_{uv} = \langle u \otimes w|\,U^\dagger A_v U\,|u \otimes w\rangle.$$

Substituting Eqs. (2), (29) and (31)–(38) in Eq. (43), one obtains

$$P_{uv} = (1 + \sin 2\alpha)^{-1} \Bigg[\left(1 - \sin^4 \alpha - \cos^4 \alpha\right) |\Phi_{00}|^2 + \left(1 - \frac{1}{2} \sin^2 2\alpha\right) |\Phi_{01}|^2$$
$$+ \frac{1}{2} \sin 2\alpha \langle \Phi_{01} | \Phi_{11} \rangle + \frac{1}{2} \sin 2\alpha \langle \Phi_{00} | \Phi_{10} \rangle - \frac{1}{2} \sin 2\alpha \langle \Phi_{00} | \Phi_{01} \rangle$$
(44)
$$- \frac{1}{2} \sin^2 2\alpha \langle \Phi_{00} | \Phi_{11} \rangle - \frac{1}{2} \sin 2\alpha \langle \Phi_{01} | \Phi_{00} \rangle - \frac{1}{2} \sin^2 2\alpha \langle \Phi_{01} | \Phi_{10} \rangle \Bigg].$$

The probe states $|\Phi_{mn}\rangle$, expanded in terms of orthonormal basis vectors $|w_\beta\rangle$, are given by Eqs. (3a), (3b), and (4) of Slutsky et al [17], namely,

(45) $$|\Phi_{00}\rangle = X_0 |w_0\rangle + X_1 |w_1\rangle + X_2 |w_2\rangle + X_3 |w_3\rangle,$$

(46) $$|\Phi_{11}\rangle = X_3 |w_0\rangle + X_2 |w_1\rangle + X_1 |w_2\rangle + X_0 |w_3\rangle,$$

(47) $$|\Phi_{01}\rangle = X_5 |w_1\rangle + X_6 |w_2\rangle,$$

(48) $$|\Phi_{10}\rangle = X_6 |w_1\rangle + X_5 |w_2\rangle.$$

Here the real coefficients $\{X_0, X_1, X_2, X_3, X_5, X_6\}$, expressed in terms of the probe parameters $\{\lambda, \mu, \theta, \phi\}$, are [16,17]

(49) $$X_0 = \sin \lambda \cos \mu,$$

(50) $$X_1 = \cos \lambda \cos \theta \cos \phi,$$

(51) $$X_2 = \cos \lambda \cos \theta \sin \phi,$$

(52) $$X_3 = \sin \lambda \sin \mu,$$

(53) $$X_5 = \cos \lambda \sin \theta \cos \phi,$$

(54) $$X_6 = -\cos \lambda \sin \theta \sin \phi,$$

consistent with the assumed unitarity of the disturbing probe.

Next, substituting Eqs. (45)–(48) in Eq. (44), one gets

$$P_{uv} = (1 + \sin 2\alpha)^{-1} \Bigg[\left(1 - \sin^4 \alpha - \cos^4 \alpha\right) \left(X_0^2 + X_1^2 + X_2^2 + X_3^2\right)$$
$$+ \left(1 - \frac{1}{2} \sin^2 2\alpha\right) \left(X_5^2 + X_6^2\right) + \sin 2\alpha \left(X_1 X_6 + X_2 X_5\right)$$
$$- \sin 2\alpha \left(X_1 X_5 + X_2 X_6\right)$$
(55)
$$- \sin^2 2\alpha \left(X_0 X_3 + X_1 X_2 + X_5 X_6\right) \Bigg].$$

Then if one substitutes Eqs. (49)–(54) in Eq. (55), the latter becomes

$$P_{uv} = \frac{1}{4}(1 + \sin 2\alpha)^{-1} \Bigg[1 - \cos 4\alpha + 2(1 + \cos 4\alpha) \cos^2 \lambda \sin^2 \theta$$
$$- 2 \sin 2\alpha \cos^2 \lambda \sin 2\theta \cos 2\phi - 2 \sin^2 2\alpha \sin^2 \lambda \sin 2\mu$$
(56)
$$- 2 \sin^2 2\alpha \cos^2 \lambda \cos 2\theta \sin 2\phi \Bigg].$$

Analogously, it can be shown that Eq. (42) becomes

$$P_{uu} = \frac{1}{2}(1 - \sin 2\alpha) \Big[2\sin^2\lambda + 2\cos^2\lambda \cos^2\theta + \tan^2 2\alpha$$
$$- \tan 2\alpha \sec 2\alpha \cos^2\lambda \sin 2\theta \cos 2\phi$$
(57)
$$- \tan^2 2\alpha \left(\sin^2\lambda \sin 2\mu + \cos^2\lambda \cos 2\theta \sin 2\phi \right) \Big].$$

Next, substituting Eqs. (56) and (57) in Eq. (41), one obtains, after extensive algebraic reduction, the following expression for the inconclusive rate induced by the disturbing probe:

(58)
$$R_? = \frac{\sin 2\alpha \, (1 + c + a \sin 2\alpha)}{1 + \sin 2\alpha},$$

where (in the notation of Slutsky et al [17]),

(59)
$$a = \sin^2\lambda \sin 2\mu + \cos^2\lambda \cos 2\theta \sin 2\phi,$$

(60)
$$c = \cos^2\lambda \sin 2\theta \cos 2\phi,$$

expressed in terms of the probe parameters λ, μ, θ, and ϕ.

4. Consistency

It is well to check the consistency of Eq. (58) with the analogue of the second equality of Eq. (24) in which $\bar{\alpha}$ and $\bar{\beta}$ correspond to the correlated probe states of Eq. (29). Specifically, if

(61)
$$|\psi\rangle = |C_u\rangle \otimes |u\rangle + |C_v\rangle \otimes |v\rangle$$

for generic correlated states $|C_u\rangle$ and $|C_v\rangle$, then it can be shown, by using the first equality of Eq. (24) together with Eq. (3), that

(62)
$$P^{\text{POVM}}_{\psi?} = (\langle C_u| + \langle C_v|)(|C_u\rangle + |C_v\rangle) \sin 2\alpha.$$

Comparing Eqs. (61) and (29), and using Eq. (62), one then also has

(63)
$$R_? = P_{u?} = \langle \Phi_u | \Phi_u \rangle \sin 2\alpha,$$

where

$$|\Phi_u\rangle = \frac{1}{2}\Big[(1 + \sec 2\alpha)|\Phi_{00}\rangle + \tan 2\alpha |\Phi_{10}\rangle - \tan 2\alpha |\Phi_{01}\rangle$$
$$+ (1 - \sec 2\alpha)|\Phi_{11}\rangle - \tan 2\alpha |\Phi_{00}\rangle + (1 - \sec 2\alpha)|\Phi_{10}\rangle$$
(64)
$$+ (1 + \sec 2\alpha)|\Phi_{01}\rangle + \tan 2\alpha |\Phi_{11}\rangle \Big].$$

Using Eqs. (45)–(48) in Eq. (63), the latter becomes

$$R_? = \sin 2\alpha \, (\sec 2\alpha - \tan 2\alpha) \Big[\sec 2\alpha \left(X_0^2 + X_1^2 + X_2^2 + X_3^2 \right)$$
$$+ 2\tan 2\alpha \, (X_1 X_6 + X_2 X_5) + 2\sec 2\alpha \, (X_1 X_5 + X_2 X_6)$$
$$+ 2\tan 2\alpha \, (X_0 X_3 + X_1 X_2) + \sec 2\alpha \left(X_6^2 + X_5^2 \right)$$
(65)
$$+ 2\tan 2\alpha X_5 X_6 \Big].$$

Next, substituting Eqs. (49)–(54) in Eq. (65), and using trigonometric identities, one obtains

$$R_? = \sin 2\alpha \left(1 + \sin 2\alpha\right)^{-1} \left[1 + \cos^2 \lambda \sin 2\theta \cos 2\phi \right.$$
$$\left. + \left(\sin^2 \lambda \sin 2\mu + \cos^2 \lambda \cos 2\theta \sin 2\phi\right) \sin 2\alpha\right], \quad (66)$$

which agrees with Eqs. (58)–(60).

Equivalently, one also has, using Eqs. (62) and (30),

$$R_? = P_{v?} = \langle \Phi_v | \Phi_v \rangle \sin 2\alpha, \quad (67)$$

where

$$|\Phi_v\rangle = \frac{1}{2}\Big[\tan 2\alpha\, |\Phi_{00}\rangle + (1 + \sec 2\alpha)\,|\Phi_{10}\rangle + (1 - \sec 2\alpha)\,|\Phi_{01}\rangle$$
$$- \tan 2\alpha\, |\Phi_{11}\rangle + (1 - \sec 2\alpha)\,|\Phi_{00}\rangle - \tan 2\alpha\, |\Phi_{10}\rangle$$
$$+ \tan 2\alpha\, |\Phi_{01}\rangle + (1 + \sec 2\alpha)\,|\Phi_{11}\rangle\Big], \quad (68)$$

and if one substitutes Eqs. (45)–(54) in Eq. (67), it can be shown that Eq. (67) also reduces to Eqs. (58)–(60).

Equation (58) can be used in optimizing the disturbing probe parameters for maximum Renyi information gain by the probe with both a fixed induced inconclusive rate and a fixed induced error rate on corrected bits [18,19]. This is a challenging nonlinear optimization problem.

5. Conclusion

For an optical implementation of a projective measure and the POVM implementation of other works [9-13], the unperturbed inconclusive rates are calculated, and the POVM is shown explicitly to have the lower inconclusive rate. The ratio of the two rates is given by Eq. (28). Also, the disturbed inconclusive rate of the POVM receiver due to a general unitary disturbance of the carrier by a probe is calculated in three different ways, and is shown to be given by Eqs. (58)–(60).

6. Acknowledgements

This work was supported by the U.S. Army Research Laboratory. The hospitality and stimulation of the Isaac Newton Institute for Mathematical Sciences at the University of Cambridge is gratefully acknowledged. The author wishes to especially thank Prof. Peter Knight, FRS, for inviting him to participate in the programme *Complexity, Computation, and the Physics of Information* at the Newton Institute, where much of this work was completed. Useful communications with J. M. Myers, J. D. Franson, B. A. Slutsky, A. Peres, S. J. Lomonaco, J. D. Murley, and M. Kruger are gratefully acknowledged.

References

[1] C. W. Helstrom, *Quantum Detection and Estimation Theory*, Academic Press, New York (1976).
[2] J. M. Jauch and C. Piron, "Generalized Localizability," Helv. Phys. Acta **40**, 559–570 (1967).
[3] E. B. Davies and J. T. Lewis, "An Operational Approach to Quantum Probability," Commun. Math Phys. **17**, 239–260 (1970).
[4] E. B. Davies, *Quantum Theory of Open Systems*, Academic, New York (1976).
[5] P. A. Benioff, "Operator Valued Measures in Quantum Mechanics: Finite and Infinite Processes," J. Math. Phys. **13**, 231–242 (1972).

[6] P. Busch, P. J. Lahti, and P. Mittelstaedt, *The Quantum Theory of Measurement*, 2nd Ed., Springer, Berlin (1996).
[7] P. Busch, M. Grabowski, and P. J. Lahti, *Operational Quantum Physics*, Springer, Berlin (1995).
[8] A. K. Ekert, B. Huttner, G. M. Palma, and A. Peres, "Eavesdropping on Quantum-Cryptographical Systems," Phys. Rev. A **50**, 1047–1056 (1994).
[9] H. E. Brandt, J. M. Myers, and S. J. Lomonaco, Jr., "Aspects of Entangled Translucent Eavesdropping in Quantum Cryptography," Phys. Rev. A **56**, 4456–4465 (1997); **58**, 2617 (1998).
[10] H. E. Brandt, "Eavesdropping Optimization for Quantum Cryptography Using a Positive Operator Valued Measure," Phys. Rev. A **59**, 2665–2669 (1999).
[11] H. E. Brandt, "Positive Operator Valued Measure in Quantum Information Processing," Am. J. Phys. **67**, 434–439 (1999).
[12] J. M. Myers and H. E. Brandt, "Converting a Positive Operator-Valued Measure to a Design for a Measuring Instrument on the Laboratory Bench," Meas. Sci. Technol. **8**, 1222–1227 (1997).
[13] H. E. Brandt, "Qubit Devices and the Issue of Qauntum Decoherence," Progr. Quantum Electronics **22**, 257–370 (1998).
[14] A. Peres, *Quantum Theory: Concepts and Methods,* Kluwer, Dordrecht (1993).
[15] A. Peres, "How to Differentiate Between Non-Orthogonal States," Phys. Lett. A **128**, 19 (1998).
[16] C. A. Fuchs and A. Peres, "Quantum-State Disturbance Versus Information Gain: Uncertainty Relations for Quantum Information," Phys. Rev. A **53**, 2038–2045 (1996).
[17] B. A. Slutsky, R. Rao, P. C. Sun, and Y. Fainman, "Security of Quantum Cryptography Against Individual Attacks," Phys. Rev. A **57**, 2383–2398 (1998).
[18] H. E. Brandt, "Inconclusive Rate as a Disturbance Measure in Quantum Cryptography," Phys. Rev. A **62**, 042310-1-14 (2000).
[19] H. E. Brandt, "Inconclusive Rate in Quantum Key Distribution," Phys. Rev. A **64**, 042316-1-5 (2001).

(Howard E. Brandt) U.S. ARMY RESEARCH LABORATORY, ADELPHI, MD 20783, AND, UNIVERSITY OF CAMBRIDGE, ISAAC NEWTON INSTITUTE FOR THE MATHEMATICAL SCIENCES, CAMBRIDGE, U.K.

E-mail address, Howard E. Brandt: `hbrandt@arl.army.mil`

Quantum Amplitude Amplification and Estimation

Gilles Brassard, Peter Høyer, Michele Mosca, and Alain Tapp

ABSTRACT. Consider a Boolean function $\chi : X \to \{0, 1\}$ that partitions set X between its *good* and *bad* elements, where x is good if $\chi(x) = 1$ and bad otherwise. Consider also a quantum algorithm $\mathcal{A}$ such that $\mathcal{A}|0\rangle = \sum_{x \in X} \alpha_x |x\rangle$ is a quantum superposition of the elements of X, and let a denote the probability that a good element is produced if $\mathcal{A}|0\rangle$ is measured. If we repeat the process of running $\mathcal{A}$, measuring the output, and using χ to check the validity of the result, we shall expect to repeat $1/a$ times on the average before a solution is found. *Amplitude amplification* is a process that allows to find a good x after an expected number of applications of $\mathcal{A}$ and its inverse which is proportional to $1/\sqrt{a}$, assuming algorithm $\mathcal{A}$ makes no measurements. This is a generalization of Grover's searching algorithm in which $\mathcal{A}$ was restricted to producing an equal superposition of all members of X and we had a promise that a single x existed such that $\chi(x) = 1$. Our algorithm works whether or not the value of a is known ahead of time. In case the value of a is known, we can find a good x after a number of applications of $\mathcal{A}$ and its inverse which is proportional to $1/\sqrt{a}$ even in the worst case. We show that this quadratic speedup can also be obtained for a large family of search problems for which good classical heuristics exist. Finally, as our main result, we combine ideas from Grover's and Shor's quantum algorithms to perform *amplitude estimation*, a process that allows to estimate the value of a. We apply amplitude estimation to the problem of *approximate counting*, in which we wish to estimate the number of $x \in X$ such that $\chi(x) = 1$. We obtain optimal quantum algorithms in a variety of settings.

1. Introduction

Quantum computing is a field at the junction of theoretical modern physics and theoretical computer science. Practical experiments involving a few quantum bits have been successfully performed, and much progress has been achieved in

---

2000 *Mathematics Subject Classification*. Primary 81P68.
Key words and phrases. Quantum computation. Searching. Counting. Lower bound.
Supported in part by Canada's NSERC, Québec's FCAR and the Canada Research Chair Programme.
Part of this work was done while at Département IRO, Université de Montréal. Basic Research in Computer Science is supported by the Danish National Research Foundation.
Most of this work was done while at Centre for Quantum Computation, Clarendon Laboratory, University of Oxford. Supported in part by Canada's NSERC and UK's CESG.
Supported in part by Canada's NSERC and Québec's FCAR.

©2002 American Mathematical Society

quantum information theory, quantum error correction and fault tolerant quantum computation. Although we are still far from having desktop quantum computers in our offices, the quantum computational paradigm could soon be more than mere theoretical exercise.

The discovery by Peter Shor [15] of a polynomial-time quantum algorithm for factoring and computing discrete logarithms was a major milestone in the history of quantum computing. Another significant result is Lov Grover's quantum search algorithm [8, 9]. Grover's algorithm does not solve **NP**–complete problems in polynomial time, but the wide range of its applications more than compensates for this.

In this paper, we generalize Grover's algorithm in a variety of directions. Consider a problem that is characterized by a Boolean function $\chi(x, y)$ in the sense that y is a good solution to instance x if and only if $\chi(x, y) = 1$. (There could be more than one good solution to a given instance.) If we have a probabilistic algorithm $\mathcal{P}$ that outputs a guess $\mathcal{P}(x)$ on input x, we can call $\mathcal{P}$ and χ repeatedly until a solution to instance x is found. If $\chi(x, \mathcal{P}(x)) = 1$ with probability $p_x > 0$, we expect to repeat this process $1/p_x$ times on the average. Consider now the case when we have a quantum algorithm $\mathcal{A}$ instead of the probabilistic algorithm. Assume $\mathcal{A}$ makes no measurements: instead of a classical answer, it produces quantum superposition $|\Psi_x\rangle$ when run on input x. Let a_x denote the probability that $|\Psi_x\rangle$, *if measured*, would be a good solution. If we repeat the process of running $\mathcal{A}$ on x, measuring the output, and using χ to check the validity of the result, we shall expect to repeat $1/a_x$ times on the average before a solution is found. This is no better than the classical probabilistic paradigm.

In Section 2, we describe a more efficient approach to this problem, which we call amplitude amplification. Intuitively, the probabilistic paradigm increases the probability of success roughly by a constant on each iteration; by contrast, amplitude amplification increases the *amplitude* of success roughly by a constant on each iteration. Because amplitudes correspond to square roots of probabilities, it suffices to repeat the amplitude amplification process approximately $1/\sqrt{a_x}$ times to achieve success with overwhelming probability. For simplicity, we assume in the rest of this paper that there is a single instance for which we seek a good solution, which allows us to dispense with input x, but the generalization to the paradigm outlined above is straightforward. Grover's original database searching quantum algorithm is a special case of this process, in which χ is given by a function $f : \{0, 1, \ldots, N-1\} \to \{0, 1\}$ for which we are promised that there exists a unique x_0 such that $f(x_0) = 1$. If we use the Fourier transform as quantum algorithm $\mathcal{A}$—or more simply the Walsh–Hadamard transform in case N is a power of 2—an equal superposition of all possible x's is produced, whose success probability would be $1/N$ if measured. Classical repetition would succeed after an expected number N of evaluations of f. Amplitude amplification corresponds to Grover's algorithm: it succeeds after approximately $\sqrt{N}$ evaluations of the function.

We generalize this result further to the case when the probability of success a of algorithm $\mathcal{A}$ is not known ahead of time: it remains sufficient to evaluate $\mathcal{A}$ and χ an expected number of times that is proportional to $1/\sqrt{a}$. Moreover, in the case a is known ahead of time, we give two different techniques that are guaranteed to find a good solution after a number of iterations that is proportional to $1/\sqrt{a}$ in the worst case.

It can be proven that Grover's algorithm goes quadratically faster than any possible classical algorithm when function f is given as a black box. However, it is usually the case in practice that information is known about f that allows us to solve the problem much more efficiently than by exhaustive search. The use of classical *heuristics*, in particular, will often yield a solution significantly more efficiently than straight quantum amplitude amplification would. In Section 3, we consider a broad class of classical heuristics and show how to apply amplitude amplification to obtain quadratic speedup compared to any such heuristic.

Finally, Section 4 addresses the question of estimating the success probability a of quantum algorithm $\mathcal{A}$. We call this process *amplitude estimation*. As a special case of our main result (Theorem 12), an estimate for a is obtained after any number M of iterations which is within $2\pi\sqrt{a(1-a)}/M + \pi^2/M^2$ of the correct value with probability at least $8/\pi^2$, where one iteration consists of running algorithm $\mathcal{A}$ once forwards and once backwards, and of computing function χ once. As an application of this technique, we show how to approximately count the number of x such that $f(x) = 1$ given a function $f : \{0, 1, \ldots, N-1\} \to \{0, 1\}$. If the correct answer is $t > 0$, it suffices to compute the function $\sqrt{N}$ times to obtain an estimate roughly within $\sqrt{t}$ of the correct answer. A number of evaluations of f proportional to $\frac{1}{\varepsilon}\sqrt{N/t}$ yields a result that is likely to be within εt of the correct answer. (We can do slightly better in case ε is not fixed.) If it is known ahead of time that the correct answer is either $t = 0$ or $t = t_0$ for some fixed t_0, we can determine which is the case with certainty using a number of evaluations of f proportional to $\sqrt{N/t_0}$. If we have no prior knowledge about t, the exact count can be obtained with high probability after a number of evaluations of f that is proportional to $\sqrt{t(N-t)}$ when $0 < t < N$ and $\sqrt{N}$ otherwise. Most of these results are optimal.

We assume in this paper that the reader is familiar with basic notions of quantum computing.

2. Quantum amplitude amplification

Suppose we have a classical randomized algorithm that succeeds with some probability p. If we repeat the algorithm, say, j times, then our probability of success increases to roughly jp (assuming $jp \ll 1$). Intuitively, we can think of this strategy as each additional run of the given algorithm boosting the probability of success by an additive amount of roughly p.

A quantum analogue of boosting the probability of success would be to boost the *amplitude* of being in a certain subspace of a Hilbert space. The general concept of amplifying the amplitude of a subspace was discovered by Brassard and Høyer [4] as a generalization of the boosting technique applied by Grover in his original quantum searching paper [8]. Following [4] and [3], we refer to their idea as *amplitude amplification* and detail the ingredients below.

Let $\mathcal{H}$ denote the Hilbert space representing the state space of a quantum system. Every Boolean function $\chi : \mathbb{Z} \to \{0, 1\}$ induces a partition of $\mathcal{H}$ into a direct sum of two subspaces, a good subspace and a bad subspace. The *good subspace* is the subspace spanned by the set of basis states $|x\rangle \in \mathcal{H}$ for which $\chi(x) = 1$, and the *bad subspace* is its orthogonal complement in $\mathcal{H}$. We say that the elements of the good subspace are *good*, and that the elements of the bad subspace are *bad*.

Every pure state $|\Upsilon\rangle$ in $\mathcal{H}$ has a unique decomposition as $|\Upsilon\rangle = |\Upsilon_1\rangle + |\Upsilon_0\rangle$, where $|\Upsilon_1\rangle$ denotes the projection onto the good subspace, and $|\Upsilon_0\rangle$ denotes the projection onto the bad subspace. Let $a_\Upsilon = \langle \Upsilon_1 | \Upsilon_1 \rangle$ denote the probability that measuring $|\Upsilon\rangle$ produces a good state, and similarly, let $b_\Upsilon = \langle \Upsilon_0 | \Upsilon_0 \rangle$. Since $|\Upsilon_1\rangle$ and $|\Upsilon_0\rangle$ are orthogonal, we have $a_\Upsilon + b_\Upsilon = 1$.

Let $\mathcal{A}$ be any quantum algorithm that acts on $\mathcal{H}$ and uses no measurements. Let $|\Psi\rangle = \mathcal{A}|0\rangle$ denote the state obtained by applying $\mathcal{A}$ to the initial zero state. The amplification process is realized by repeatedly applying the following unitary operator [4] on the state $|\Psi\rangle$,

$$(1) \qquad \mathbf{Q} = \mathbf{Q}(\mathcal{A}, \chi) = -\mathcal{A} \mathbf{S}_0 \mathcal{A}^{-1} \mathbf{S}_\chi.$$

Here, the operator $\mathbf{S}_\chi$ conditionally changes the sign of the amplitudes of the good states,

$$|x\rangle \longmapsto \begin{cases} -|x\rangle & \text{if } \chi(x) = 1 \\ |x\rangle & \text{if } \chi(x) = 0, \end{cases}$$

while the operator $\mathbf{S}_0$ changes the sign of the amplitude if and only if the state is the zero state $|0\rangle$. The operator $\mathbf{Q}$ is well-defined since we assume that $\mathcal{A}$ uses no measurements and, therefore, $\mathcal{A}$ has an inverse.

The usefulness of operator $\mathbf{Q}$ stems from its simple action on the subspace $\mathcal{H}_\Psi$ spanned by the vectors $|\Psi_1\rangle$ and $|\Psi_0\rangle$.

LEMMA 1. *We have that*

$$\mathbf{Q}|\Psi_1\rangle = (1-2a)|\Psi_1\rangle - 2a|\Psi_0\rangle$$
$$\mathbf{Q}|\Psi_0\rangle = 2(1-a)|\Psi_1\rangle + (1-2a)|\Psi_0\rangle,$$

where $a = \langle \Psi_1 | \Psi_1 \rangle$.

It follows that the subspace $\mathcal{H}_\Psi$ is stable under the action of $\mathbf{Q}$, a property that was first observed by Brassard and Høyer [4] and rediscovered by Grover [10].

Suppose $0 < a < 1$. Then $\mathcal{H}_\Psi$ is a subspace of dimension 2, and otherwise $\mathcal{H}_\Psi$ has dimension 1. The action of $\mathbf{Q}$ on $\mathcal{H}_\Psi$ is also realized by the operator

$$(2) \qquad \mathbf{U}_\Psi \mathbf{U}_{\Psi_0},$$

which is composed of 2 reflections. The first operator, $\mathbf{U}_{\Psi_0} = \mathbf{I} - \frac{2}{1-a}|\Psi_0\rangle\langle\Psi_0|$, implements a reflection through the ray spanned by the vector $|\Psi_0\rangle$, while the second operator $\mathbf{U}_\Psi = \mathbf{I} - 2|\Psi\rangle\langle\Psi|$ implements a reflection through the ray spanned by the vector $|\Psi\rangle$.

Consider the orthogonal complement $\mathcal{H}_\Psi^\perp$ of $\mathcal{H}_\Psi$ in $\mathcal{H}$. Since the operator $\mathcal{A}\mathbf{S}_0\mathcal{A}^{-1}$ acts as the identity on $\mathcal{H}_\Psi^\perp$, operator $\mathbf{Q}$ acts as $-\mathbf{S}_\chi$ on $\mathcal{H}_\Psi^\perp$. Thus, $\mathbf{Q}^2$ acts as the identity on $\mathcal{H}_\Psi^\perp$, and every eigenvector of $\mathbf{Q}$ in $\mathcal{H}_\Psi^\perp$ has eigenvalue $+1$ or -1. It follows that to understand the action of $\mathbf{Q}$ on an arbitrary initial vector $|\Upsilon\rangle$ in $\mathcal{H}$, it suffices to consider the action of $\mathbf{Q}$ on the projection of $|\Upsilon\rangle$ onto $\mathcal{H}_\Psi$.

Since operator $\mathbf{Q}$ is unitary, the subspace $\mathcal{H}_\Psi$ has an orthonormal basis consisting of two eigenvectors of $\mathbf{Q}$,

$$(3) \qquad |\Psi_\pm\rangle = \frac{1}{\sqrt{2}} \left(\frac{1}{\sqrt{a}}|\Psi_1\rangle \pm \frac{\iota}{\sqrt{1-a}}|\Psi_0\rangle \right),$$

provided $0 < a < 1$, where $\imath = \sqrt{-1}$ denotes the principal square root of -1. The corresponding eigenvalues are

(4) $$\lambda_\pm = e^{\pm \imath 2\theta_a},$$

where the angle θ_a is defined so that

(5) $$\sin^2(\theta_a) = a = \langle \Psi_1 | \Psi_1 \rangle$$

and $0 \leq \theta_a \leq \pi/2$.

We use operator $\mathbf{Q}$ to boost the success probability a of the quantum algorithm $\mathcal{A}$. First, express $|\Psi\rangle = \mathcal{A}|0\rangle$ in the eigenvector basis,

(6) $$\mathcal{A}|0\rangle = |\Psi\rangle = \frac{-\imath}{\sqrt{2}} \left(e^{\imath \theta_a} |\Psi_+\rangle - e^{-\imath \theta_a} |\Psi_-\rangle \right).$$

It is now immediate that after j applications of operator $\mathbf{Q}$, the state is

(7) $$\mathbf{Q}^j |\Psi\rangle = \frac{-\imath}{\sqrt{2}} \left(e^{(2j+1)\imath \theta_a} |\Psi_+\rangle - e^{-(2j+1)\imath \theta_a} |\Psi_-\rangle \right)$$

(8) $$= \frac{1}{\sqrt{a}} \sin((2j+1)\theta_a) |\Psi_1\rangle + \frac{1}{\sqrt{1-a}} \cos((2j+1)\theta_a) |\Psi_0\rangle.$$

It follows that if $0 < a < 1$ and if we compute $\mathbf{Q}^m |\Psi\rangle$ for some integer $m \geq 0$, then a final measurement will produce a good state with probability equal to $\sin^2((2m+1)\theta_a)$.

If the initial success probability a is either 0 or 1, then the subspace $\mathcal{H}_\Psi$ spanned by $|\Psi_1\rangle$ and $|\Psi_0\rangle$ has dimension 1 only, but the conclusion remains the same: If we measure the system after m rounds of amplitude amplification, then the outcome is good with probability $\sin^2((2m+1)\theta_a)$, where the angle θ_a is defined so that Equation 5 is satisfied and so that $0 \leq \theta_a \leq \pi/2$.

Therefore, assuming $a > 0$, to obtain a high probability of success, we want to choose integer m such that $\sin^2((2m+1)\theta_a)$ is close to 1. Unfortunately, our ability to choose m wisely depends on our knowledge about θ_a, which itself depends on a. The two extreme cases are when we know the exact value of a, and when we have no prior knowledge about a whatsoever.

Suppose the value of a is known. If $a > 0$, then by letting $m = \lfloor \pi/4\theta_a \rfloor$, we have that $\sin^2((2m+1)\theta_a) \geq 1-a$, as shown in [3]. The next theorem is immediate.

THEOREM 2 (Quadratic speedup). *Let $\mathcal{A}$ be any quantum algorithm that uses no measurements, and let $\chi : \mathbb{Z} \to \{0,1\}$ be any Boolean function. Let a the initial success probability of $\mathcal{A}$. Suppose $a > 0$, and set $m = \lfloor \pi/4\theta_a \rfloor$, where θ_a is defined so that $\sin^2(\theta_a) = a$ and $0 < \theta_a \leq \pi/2$. Then, if we compute $\mathbf{Q}^m \mathcal{A} |0\rangle$ and measure the system, the outcome is good with probability at least $\max(1-a, a)$.*

Note that any implementation of algorithm $\mathbf{Q}^m \mathcal{A}|0\rangle$ requires that the value of a is known so that the value of m can be computed. We refer to Theorem 2 as a quadratic speedup, or the square-root running-time result. The reason for this is that if an algorithm $\mathcal{A}$ has success probability $a > 0$, then after an expected number of $1/a$ applications of $\mathcal{A}$, we will find a good solution. Applying the above theorem reduces this to an expected number of at most $(2m+1)/\max(1-a, a) \in \Theta(\frac{1}{\sqrt{a}})$ applications of $\mathcal{A}$ and $\mathcal{A}^{-1}$.

As an application of Theorem 2, consider the search problem [9] in which we are given a Boolean function $f : \{0, 1, \ldots, N-1\} \to \{0, 1\}$ satisfying the promise that there exists a unique $x_0 \in \{0, 1, \ldots, N-1\}$ on which f takes value 1, and we are asked to find x_0. If f is given as a black box, then on a classical computer, we need to evaluate f on an expected number of roughly half the elements of the domain in order to determine x_0.

By contrast, Grover [9] discovered a quantum algorithm that only requires an expected number of evaluations of f in the order of $\sqrt{N}$. In terms of amplitude amplification, Grover's algorithm reads as follows: Let $\chi = f$, and let $\mathcal{A} = \mathbf{W}$ be the Walsh-Hadamard transform on n qubits that maps the initial zero state $|0\rangle$ to $\frac{1}{\sqrt{N}}\sum_{x=0}^{N-1}|x\rangle$, an equally-weighted superposition of all $N = 2^n$ elements in the domain of f. Then the operator $\mathbf{Q} = -\mathcal{A}\mathbf{S}_0\mathcal{A}^{-1}\mathbf{S}_\chi$ is equal to the iterate $-\mathbf{W}\mathbf{S}_0\mathbf{W}\mathbf{S}_f$ applied by Grover in his searching paper [9]. The initial success probability a of $\mathcal{A}$ is exactly $1/N$, and if we measure after $m = \lfloor \pi/4\theta_a \rfloor$ iterations of $\mathbf{Q}$, the probability of measuring x_0 is lower bounded by $1 - 1/N$ [3].

Now, suppose that the value of a is not known. In Section 4, we discuss techniques for finding an estimate of a, whereafter one then can apply a weakened version of Theorem 2 in which the exact value of a is replaced by an estimate of it. Another idea is to try to find a good solution without prior computation of an estimate of a. Within that approach, by adapting the ideas in Section 6 in [3] we can still obtain a quadratic speedup.

THEOREM 3 (Quadratic speedup without knowing a). *There exists a quantum algorithm* **QSearch** *with the following property. Let $\mathcal{A}$ be any quantum algorithm that uses no measurements, and let $\chi : \mathbb{Z} \to \{0,1\}$ be any Boolean function. Let a denote the initial success probability of $\mathcal{A}$. Algorithm* **QSearch** *finds a good solution using an expected number of applications of $\mathcal{A}$ and $\mathcal{A}^{-1}$ which are in $\Theta(\frac{1}{\sqrt{a}})$ if $a > 0$, and otherwise runs forever.*

The algorithm in the above theorem utilizes the given quantum algorithm $\mathcal{A}$ as a subroutine and the operator $\mathbf{Q}$. The complete algorithm is as follows:

Algorithm(QSearch($\mathcal{A}, \chi$))
1. Set $l = 0$ and let c be any constant such that $1 < c < 2$.
2. Increase l by 1 and set $M = \lceil c^l \rceil$.
3. Apply $\mathcal{A}$ on the initial state $|0\rangle$, and measure the system. If the outcome $|z\rangle$ is good, that is, if $\chi(z) = 1$, then output z and stop.
4. Initialize a register of appropriate size to the state $\mathcal{A}|0\rangle$.
5. Pick an integer j between 1 and M uniformly at random.
6. Apply $\mathbf{Q}^j$ to the register, where $\mathbf{Q} = \mathbf{Q}(\mathcal{A}, \chi)$.
7. Measure the register. If the outcome $|z\rangle$ is good, then output z and stop. Otherwise, go to step 2.

The intuition behind this algorithm is as follows. In a 2-dimensional real vector space, if we pick a unit vector $(x, y) = (\cos(\cdot), \sin(\cdot))$ uniformly at random then the expected value of y^2 is $1/2$. Consider Equation 8. If we pick j at random between 1 and M for some integer M such that $M\theta_a$ is larger than, say, 100π, then we have a good approximation to a random unit vector, and we will succeed with probability close to $1/2$.

To turn this intuition into an algorithm, the only obstacle left is that we do not know the value of θ_a, and hence do not know an appropriate value for M. However, we can overcome this by using exponentially increasing values of M, an idea similar to the one used in "exponential searching" (which is a term that does not refer to the running time of the method, but rather to an exponentially increasing growth of the size of the search space).

The correctness of algorithm **QSearch** is immediate and thus to prove the theorem, it suffices to show that the expected number of applications of $\mathcal{A}$ and $\mathcal{A}^{-1}$ is in the order of $1/\sqrt{a}$. This can be proven by essentially the same techniques applied in the proof of Theorem 3 in [**3**] and we therefore only give a very brief sketch of the proof.

On the one hand, if the initial success probability a is at least $3/4$, then step 3 ensures that we soon will measure a good solution. On the other hand, if $0 < a < 3/4$ then, for any given value of M, the probability of measuring a good solution in step 7 is lower bounded by

$$(9) \qquad \frac{1}{2}\left(1 - \frac{1}{2M\sqrt{a}}\right).$$

Let $c_0 > 0$ be such that $c = 2(1 - c_0)$ and let $M_0 = 1/(2c_0\sqrt{a})$. The expected number of applications of $\mathcal{A}$ is upper bounded by $T_1 + T_2$, where T_1 denotes the maximum number of applications of $\mathcal{A}$ the algorithm uses while $M < M_0$, and where T_2 denotes the expected number of applications of $\mathcal{A}$ the algorithm uses while $M \geq M_0$. Clearly $T_1 \in O(M_0) = O(\frac{1}{\sqrt{a}})$ and we now show that $T_2 \in O(\frac{1}{\sqrt{a}})$ as well.

For all $M \geq M_0$, the measurement in step 7 yields a good solution with probability at least $\frac{1}{2}(1 - c_0)$, and hence it fails to yield a good solution with probability at most $p_0 = \frac{1}{2}(1 + c_0)$. Thus for all $i \geq 0$, with probability at most p_0^i, we have that $M \geq M_0 c^i$ at some point after step 2 while running the algorithm. Hence T_2 is at most on the order of $\sum_{i \geq 0} M_0(cp_0)^i$ which is in $O(M_0)$ since $cp_0 < 1$. The total expected number of applications of $\mathcal{A}$ is thus in $O(M_0)$, which is $O(\frac{1}{\sqrt{a}})$.

For the lower bound, if M were in $o(\frac{1}{\sqrt{a}})$, then the probability that we measure a good solution in step 7 would be vanishingly small. This completes our sketch of the proof of Theorem 3.

2.1. Quantum de-randomization when the success probability is known. We now consider the situation where the success probability a of the quantum algorithm $\mathcal{A}$ is known. If $a = 0$ or $a = 1$, then amplitude amplification will not change the success probability, so in the rest of this section, we assume that $0 < a < 1$. Theorem 2 allows us to boost the probability of success to at least $\max(1 - a, a)$. A natural question to ask is whether it is possible to improve this to certainty, still given the value of a. It turns out that the answer is positive. This is unlike classical computers, where no such general de-randomization technique is known. We now describe 2 optimal methods for obtaining this, but other approaches are possible.

The first method is by applying amplitude amplification, not on the original algorithm $\mathcal{A}$, but on a slightly modified version of it. By Equation 8, if we measure the state $\mathbf{Q}^m \mathcal{A}|0\rangle$, then the outcome is good with probability $\sin^2((2m+1)\theta_a)$. In particular, if $\tilde{m} = \pi/4\theta_a - 1/2$ happens to be an integer, then we would succeed with certainty after $\tilde{m}$ applications of $\mathbf{Q}$. In general, $\overline{m} = \lceil \tilde{m} \rceil$ iterations is a fraction of 1 iteration too many, but we can compensate for that by choosing $\overline{\theta_a} = \pi/(4\overline{m} + 2)$, an angle slightly smaller than θ_a. Any quantum algorithm that succeeds with probability $\overline{a}$ such that $\sin^2(\overline{\theta_a}) = \overline{a}$, will succeed with certainty after $\overline{m}$ iterations of amplitude amplification. Given $\mathcal{A}$ and its initial success probability a, it is easy to construct a new quantum algorithm that succeeds with probability $\bar{a} \leq a$: Let $\mathcal{B}$ denote the quantum algorithm that takes a single qubit in the initial state $|0\rangle$ and rotates it to the superposition $\sqrt{1-\bar{a}/a}\,|0\rangle + \sqrt{\bar{a}/a}\,|1\rangle$. Apply both $\mathcal{A}$ and $\mathcal{B}$, and define a good solution as one in which $\mathcal{A}$ produces a good solution, and the outcome of $\mathcal{B}$ is the state $|1\rangle$. Theorem 4 follows.

THEOREM 4 (Quadratic speedup with known a). *Let $\mathcal{A}$ be any quantum algorithm that uses no measurements, and let $\chi : \mathbb{Z} \to \{0,1\}$ be any Boolean function. There exists a quantum algorithm that given the initial success probability $a > 0$ of $\mathcal{A}$, finds a good solution with certainty using a number of applications of $\mathcal{A}$ and $\mathcal{A}^{-1}$ which is in $\Theta(\frac{1}{\sqrt{a}})$ in the worst case.*

The second method to obtain success probability 1 requires a generalization of operator $\mathbf{Q}$. Given angles $0 \leq \phi, \varphi < 2\pi$, redefine $\mathbf{Q}$ as follows,

$$(10) \qquad \mathbf{Q} = \mathbf{Q}(\mathcal{A}, \chi, \phi, \varphi) = -\mathcal{A}\mathbf{S}_0(\phi)\mathcal{A}^{-1}\mathbf{S}_\chi(\varphi).$$

Here, the operator $\mathbf{S}_\chi(\varphi)$ is the natural generalization of the $\mathbf{S}_\chi$ operator,

$$|x\rangle \longmapsto \begin{cases} e^{i\varphi}|x\rangle & \text{if } \chi(x) = 1 \\ |x\rangle & \text{if } \chi(x) = 0. \end{cases}$$

Similarly, the operator $\mathbf{S}_0(\phi)$ multiplies the amplitude by a factor of $e^{i\phi}$ if and only if the state is the zero state $|0\rangle$. The action of operator $\mathbf{Q}(\mathcal{A}, \chi, \phi, \varphi)$ is also realized by applying an operator that is composed of two pseudo-reflections: the operator $\mathcal{A}\mathbf{S}_0(\phi)\mathcal{A}^{-1}$ and the operator $-\mathbf{S}_\chi(\varphi)$.

The next lemma shows that the subspace $\mathcal{H}_\Psi$ spanned by $|\Psi_1\rangle$ and $|\Psi_0\rangle$ is stable under the action of $\mathbf{Q}$, just as in the special case $\mathbf{Q}(\mathcal{A}, \chi, \pi, \pi)$ studied above.

LEMMA 5. *Let $\mathbf{Q} = \mathbf{Q}(\mathcal{A}, \chi, \phi, \varphi)$. Then*

$$\mathbf{Q}|\Psi_1\rangle = e^{i\varphi}((1-e^{i\phi})a - 1)|\Psi_1\rangle + \quad e^{i\varphi}(1-e^{i\phi})a|\Psi_0\rangle$$
$$\mathbf{Q}|\Psi_0\rangle = \quad (1-e^{i\phi})(1-a)|\Psi_1\rangle - ((1-e^{i\phi})a + e^{i\phi})|\Psi_0\rangle,$$

where $a = \langle\Psi_1|\Psi_1\rangle$.

Let $\tilde{m} = \pi/4\theta_a - 1/2$, and suppose that $\tilde{m}$ is not an integer. In the second method to obtain a good solution with certainty, we also apply $\lceil\tilde{m}\rceil$ iterations of amplitude amplification, but now we slow down the speed of the very last iteration only, as opposed to of all iterations as in the first method. For the case $\tilde{m} < 1$, this second method has also been suggested by Chi and Kim [6]. We start by applying the operator $\mathbf{Q}(\mathcal{A}, \chi, \phi, \varphi)$ with $\phi = \varphi = \pi$ a number of $\lfloor\tilde{m}\rfloor$ times to the initial state $|\Psi\rangle = \mathcal{A}|0\rangle$. By Equation 8, this produces the superposition

$$\frac{1}{\sqrt{a}}\sin\big((2\lfloor\tilde{m}\rfloor+1)\theta_a\big)|\Psi_1\rangle + \frac{1}{\sqrt{1-a}}\cos\big((2\lfloor\tilde{m}\rfloor+1)\theta_a\big)|\Psi_0\rangle.$$

Then, we apply operator $\mathbf{Q}$ one more time, but now using angles ϕ and φ, both between 0 and 2π, satisfying

$$(11) \quad e^{i\varphi}(1-e^{i\phi})\sqrt{a}\sin\big((2\lfloor\tilde{m}\rfloor+1)\theta_a\big)$$
$$= ((1-e^{i\phi})a + e^{i\phi})\frac{1}{\sqrt{1-a}}\cos\big((2\lfloor\tilde{m}\rfloor+1)\theta_a\big).$$

By Lemma 5, this ensures that the resulting superposition has inner product zero with $|\Psi_0\rangle$, and thus a subsequent measurement will yield a good solution with certainty.

The problem of choosing $\phi, \varphi \in \mathbb{R}$ such that Equation 11 holds is equivalent to requiring that

$$(12) \quad \cot\left((2\lfloor \tilde{m} \rfloor + 1)\theta_a\right) = e^{\imath\varphi} \sin(2\theta_a)\bigl(-\cos(2\theta_a) + \imath \cot(\phi/2)\bigr)^{-1}.$$

By appropriate choices of ϕ and φ, the right hand side of Equation 12 can be made equal to any nonzero complex number of norm at most $\tan(2\theta_a)$. Thus, since the left hand side of this equation is equal to some real number smaller than $\tan(2\theta_a)$, there exist $\phi, \varphi \in \mathbb{R}$ such that Equation 12 is satisfied, and hence also such that the expression in Equation 11 vanishes. In conclusion, applying $\mathbf{Q}(\mathcal{A}, \chi, \phi, \varphi)$ with such $\phi, \varphi \in \mathbb{R}$ at the very last iteration allows us to measure a good solution with certainty.

3. Heuristics

As explained in the previous section, using the amplitude amplification technique to search for a solution to a search problem, one obtains a quadratic speedup compared to a brute force search. For many problems, however, good heuristics are known for which the expected running time, when applied to a "real-life" problem, is in $o(\sqrt{N})$, where N is the size of the search space. This fact would make amplitude amplification much less useful unless a quantum computer is somehow able to take advantage of these classical heuristics. In this section we concentrate on a large family of classical heuristics that can be applied to search problems. We show how these heuristics can be incorporated into the general amplitude amplification process.

By a heuristic, we mean a probabilistic algorithm, running in polynomial time, that outputs what one is searching for with some non-negligible probability.

Suppose we have a family $\mathcal{F}$ of functions such that each $f \in \mathcal{F}$ is of the form $f : X \to \{0, 1\}$. For a given function f we seek an input $x \in X$ such that $f(x) = 1$. A *heuristic* is a function $G : \mathcal{F} \times R \to X$, for an appropriate finite set R. The heuristic G uses a random seed $r \in R$ to generate a guess for an x such that $f(x) = 1$. For every function $f \in \mathcal{F}$, let $t_f = |\{x \in X \mid f(x) = 1\}|$, the number of good inputs x, and let $h_f = |\{r \in R \mid f(G(f, r)) = 1\}|$, the number of good seeds. We say that the heuristic is *efficient* for a given f if $h_f/|R| > t_f/|X|$, that is, if using G and a random seed to generate inputs to f succeeds with a higher probability than directly guessing inputs to f uniformly at random. The heuristic is *good* in general if

$$\mathrm{E}_{\mathcal{F}}\left(\frac{h_f}{|R|}\right) > \mathrm{E}_{\mathcal{F}}\left(\frac{t_f}{|X|}\right).$$

Here $\mathrm{E}_{\mathcal{F}}$ denotes the expectation over all f according to some fixed distribution. Note that for some f, h_f might be small but repeated uses of the heuristic, with seeds uniformly chosen in R, will increase the probability of finding a solution.

THEOREM 6. *Let $\mathcal{F} \subseteq \{f \mid f : X \to \{0,1\}\}$ be a family of Boolean functions and $\mathcal{D}$ be a probability distribution over $\mathcal{F}$. If on a classical computer, using heuristic $G : \mathcal{F} \times R \to X$, one finds $x_0 \in X$ such that $f(x_0) = 1$ for random f taken from distribution D in expected time T then using a quantum computer, a solution can be found in expected time in $O(\sqrt{T})$.*

PROOF. A simple solution to this problem is to embed the classical heuristic G into the function used in the algorithm **QSearch**. Let $\chi(r) = f(G(f,r))$ and $x = G(f, \mathbf{QSearch}(\mathbf{W}, \chi))$, so that $f(x) = 1$. By Theorem 3, for each function $f \in \mathcal{F}$, we have an expected running time in $\Theta(\sqrt{|R|/h_f})$. Let P_f denote the probability that f occurs. Then $\sum_{f \in \mathcal{F}} P_f = 1$, and we have that the expected running time is in the order of $\sum_{f \in \mathcal{F}} \sqrt{|R|/h_f}\, P_f$, which can be rewritten as

$$\sum_{f \in \mathcal{F}} \sqrt{\frac{|R|}{h_f} P_f} \sqrt{P_f} \leq \left(\sum_{f \in \mathcal{F}} \frac{|R|}{h_f} P_f \right)^{1/2} \left(\sum_{f \in \mathcal{F}} P_f \right)^{1/2} = \left(\sum_{f \in \mathcal{F}} \frac{|R|}{h_f} P_f \right)^{1/2}$$

by Cauchy–Schwarz's inequality. □

An alternative way to prove Theorem 6 is to incorporate the heuristic into the operator $\mathcal{A}$ and do a minor modification to f. Let $\mathcal{A}$ be the quantum implementation of G. It is required that the operator $\mathcal{A}$ be unitary, but clearly in general the classical heuristic does not need to be reversible. As usual in quantum algorithms one will need first to modify the heuristic $G : \mathcal{F} \times R \to X$ to make it reversible, which can be done efficiently using standard techniques [2]. We obtain a reversible function $G'_f : R \times \mathbf{0} \to R \times X$. Let $\mathcal{A}$ be the natural unitary operation implementing G'_f and let us modify χ (the good set membership function) to consider only the second part of the register, that is $\chi((r,x)) = 1$ if and only if $f(x) = 1$. We then have that $a = h_f/|R|$ and by Theorem 3, for each function $f \in \mathcal{F}$, we have an expected running time in $\Theta(\sqrt{|R|/h_f})$. The rest of the reasoning is similar. This alternative technique shows, using a simple example, the usefulness of the general scheme of amplitude amplification described in the preceding section, although it is clear that from a computational point of view this is strictly equivalent to the technique given in the earlier proof of the theorem.

4. Quantum amplitude estimation

Section 2 dealt in a very general way with combinatorial search problems, namely, given a Boolean function $f : X \to \{0,1\}$ find an $x \in X$ such that $f(x) = 1$. In this section, we deal with the related problem of estimating $t = |\{x \in X \mid f(x) = 1\}|$, the number of inputs on which f takes the value 1.

We can describe this counting problem in terms of amplitude estimation. Using the notation of Section 2, given a unitary transformation $\mathcal{A}$ and a Boolean function χ, let $|\Psi\rangle = \mathcal{A}|0\rangle$. Write $|\Psi\rangle = |\Psi_1\rangle + |\Psi_0\rangle$ as a superposition of the good and bad components of $|\Psi\rangle$. Then *amplitude estimation* is the problem of estimating $a = \langle \Psi_1 | \Psi_1 \rangle$, the probability that a measurement of $|\Psi\rangle$ yields a good state.

The problem of estimating $t = |\{x \in X \mid f(x) = 1\}|$ can be formulated in these terms as follows. For simplicity, we take $X = \{0, 1, \ldots, N-1\}$. If N is a power of 2, then we set $\chi = f$ and $\mathcal{A} = \mathbf{W}$. If N is not a power of 2, we set $\chi = f$ and $\mathcal{A} = \mathbf{F}_N$, the quantum Fourier transform which, for every integer $M \geq 1$, is defined by

$$(13) \qquad \mathbf{F}_M : |x\rangle \longmapsto \frac{1}{\sqrt{M}} \sum_{y=0}^{M-1} e^{2\pi i x y / M} |y\rangle \qquad (0 \leq x < M).$$

Then in both cases we have $a = t/N$, and thus an estimate for a directly translates into an estimate for t.

To estimate a, we make good use of the properties of operator $\mathbf{Q} = -\mathcal{A}\mathbf{S}_0\mathcal{A}^{-1}\mathbf{S}_f$. By Equation 8 in Section 2, we have that the amplitudes of $|\Psi_1\rangle$ and $|\Psi_0\rangle$ as functions of the number of applications of $\mathbf{Q}$, are sinusoidal functions, both of period $\frac{\pi}{\theta_a}$. Recall that $0 \leq \theta_a \leq \pi/2$ and $a = \sin^2(\theta_a)$, and thus an estimate for θ_a also gives an estimate for a.

To estimate this period, it is a natural approach [5] to apply Fourier analysis like Shor [15] does for a classical function in his factoring algorithm. This approach can also be viewed as an eigenvalue estimation [12, 7] and is best analysed in the basis of eigenvectors of the operator at hand [13]. By Equation 4, the eigenvalues of $\mathbf{Q}$ on the subspace spanned by $|\Psi_1\rangle$ and $|\Psi_0\rangle$ are $\lambda_+ = e^{\imath 2\theta_a}$ and $\lambda_- = e^{-\imath 2\theta_a}$. Thus we can estimate a simply by estimating one of these two eigenvalues. Errors in our estimate $\tilde{\theta}_a$ for θ_a translate into errors in our estimate $\tilde{a} = \sin^2(\tilde{\theta}_a)$ for a, as described in the next lemma.

LEMMA 7. *Let $a = \sin^2(\theta_a)$ and $\tilde{a} = \sin^2(\tilde{\theta}_a)$ with $0 \leq \theta_a, \tilde{\theta}_a \leq 2\pi$ then*
$$|\tilde{\theta}_a - \theta_a| \leq \varepsilon \Rightarrow |\tilde{a} - a| \leq 2\varepsilon\sqrt{a(1-a)} + \varepsilon^2.$$

PROOF. For $\varepsilon \geq 0$, using standard trigonometric identities, we obtain
$$\sin^2(\theta_a + \varepsilon) - \sin^2(\theta_a) = \sqrt{a(1-a)}\sin(2\varepsilon) + (1-2a)\sin^2(\varepsilon) \text{ and}$$
$$\sin^2(\theta_a) - \sin^2(\theta_a - \varepsilon) = \sqrt{a(1-a)}\sin(2\varepsilon) + (2a-1)\sin^2(\varepsilon).$$

The inequality follows directly. □

We want to estimate one of the eigenvalues of $\mathbf{Q}$. For this purpose, we utilize the following operator Λ. For any positive integer M and any unitary operator $\mathbf{U}$, the operator $\Lambda_M(\mathbf{U})$ is defined by

(14) $$|j\rangle|y\rangle \longmapsto |j\rangle(\mathbf{U}^j|y\rangle) \qquad (0 \leq j < M).$$

Note that if $|\Phi\rangle$ is an eigenvector of $\mathbf{U}$ with eigenvalue $e^{2\pi \imath \omega}$, then $\Lambda_M(\mathbf{U})$ maps $|j\rangle|\Phi\rangle$ to $e^{2\pi \imath \omega j}|j\rangle|\Phi\rangle$.

DEFINITION 8. *For any integer $M > 0$ and real number $0 \leq \omega < 1$, let*
$$|\mathcal{S}_M(\omega)\rangle = \frac{1}{\sqrt{M}}\sum_{y=0}^{M-1} e^{2\pi \imath \omega y}|y\rangle.$$

We then have, for all $0 \leq x \leq M - 1$
$$\mathbf{F}_M|x\rangle = |\mathcal{S}_M(x/M)\rangle.$$

The state $|\mathcal{S}_M(\omega)\rangle$ encodes the angle $2\pi\omega$ ($0 \leq \omega < 1$) in the phases of an equally weighted superposition of all basis states. Different angles have different encodings, and the overlap between $|\mathcal{S}_M(\omega_0)\rangle$ and $|\mathcal{S}_M(\omega_1)\rangle$ is a measure for the distance between the two angles ω_0 and ω_1.

DEFINITION 9. *For any two real numbers $\omega_0, \omega_1 \in \mathbb{R}$, let $d(\omega_0, \omega_1) = \min_{z \in \mathbb{Z}}\{|z + \omega_1 - \omega_0|\}$.*

Thus $2\pi d(\omega_0, \omega_1)$ is the length of the shortest arc on the unit circle going from $e^{2\pi \imath \omega_0}$ to $e^{2\pi \imath \omega_1}$.

LEMMA 10. *For $0 \leq \omega_0 < 1$ and $0 \leq \omega_1 < 1$ let $\Delta = d(\omega_0, \omega_1)$. If $\Delta = 0$ we have $|\langle \mathcal{S}_M(\omega_0) | \mathcal{S}_M(\omega_1) \rangle|^2 = 1$. Otherwise*

$$|\langle \mathcal{S}_M(\omega_0) | \mathcal{S}_M(\omega_1) \rangle|^2 = \frac{\sin^2(M\Delta\pi)}{M^2 \sin^2(\Delta\pi)}.$$

PROOF.

$$\begin{aligned}
|\langle \mathcal{S}_M(\omega_0) | \mathcal{S}_M(\omega_1) \rangle|^2 &= \left| \left(\frac{1}{\sqrt{M}} \sum_{y=0}^{M-1} e^{-2\pi i \omega_0 y} \langle y| \right) \left(\frac{1}{\sqrt{M}} \sum_{y=0}^{M-1} e^{2\pi i \omega_1 y} |y\rangle \right) \right|^2 \\
&= \frac{1}{M^2} \left| \sum_{y=0}^{M-1} e^{2\pi i \Delta y} \right|^2 = \frac{\sin^2(M\Delta\pi)}{M^2 \sin^2(\Delta\pi)}.
\end{aligned}$$

□

Consider the problem of estimating ω where $0 \leq \omega < 1$, given the state $|\mathcal{S}_M(\omega)\rangle$. If $\omega = x/M$ for some integer $0 \leq x < M$, then $\mathbf{F}_M^{-1} |\mathcal{S}_M(x/M)\rangle = |x\rangle$ by definition, and thus we have a perfect phase estimator. If $M\omega$ is not an integer, then observing $\mathbf{F}_M^{-1} |\mathcal{S}_M(\omega)\rangle$ still provides a good estimation of ω, as shown in the following theorem.

THEOREM 11. *Let X be the discrete random variable corresponding to the classical result of measuring $\mathbf{F}_M^{-1} |\mathcal{S}_M(\omega)\rangle$ in the computational basis. If $M\omega$ is an integer then $\mathrm{Prob}(X = M\omega) = 1$. Otherwise, letting $\Delta = d(\omega, x/M)$,*

$$\mathrm{Prob}(X = x) = \frac{\sin^2(M\Delta\pi)}{M^2 \sin^2(\Delta\pi)} \leq \frac{1}{(2M\Delta)^2}.$$

For any $k > 1$ we also have

$$\mathrm{Prob}\left(d(X/M, \omega) \leq k/M\right) \geq 1 - \frac{1}{2(k-1)}$$

and, in the case $k = 1$ and $M > 2$,

$$\mathrm{Prob}\left(d(X/M, \omega) \leq 1/M\right) \geq \frac{8}{\pi^2}.$$

PROOF. Clearly

$$\begin{aligned}
\mathrm{Prob}(X = x) &= \left| \langle x | \mathbf{F}^{-1} | \mathcal{S}_M(\omega) \rangle \right|^2 \\
&= \left| (\mathbf{F}|x\rangle)^\dagger | \mathcal{S}_M(\omega) \rangle \right|^2 \\
&= \left| \langle \mathcal{S}_M(x/M) | \mathcal{S}_M(\omega) \rangle \right|^2
\end{aligned}$$

thus using Lemma 10 we directly obtain the first part of the theorem. We use this fact to prove the next part of the theorem.

$$\begin{aligned}
\mathrm{Prob}\left(d(X/M, \omega) \leq k/M\right) &= 1 - \mathrm{Prob}(d(X/M, \omega) > k/M) \\
&\geq 1 - 2 \sum_{j=k}^{\infty} \frac{1}{4M^2 (\frac{j}{M})^2} \\
&\geq 1 - \frac{1}{2(k-1)}.
\end{aligned}$$

For the last part, we use the fact that for $M > 2$, the given expression attains its minimum at $\Delta = 1/(2M)$ in the range $0 \leq \Delta \leq 1/M$.

$$\begin{aligned}
\text{Prob}\left(d(X/M,\omega) \leq 1/M\right) &= \text{Prob}(X = \lfloor M\omega \rfloor) + \text{Prob}(X = \lceil M\omega \rceil) \\
&= \frac{\sin^2(M\Delta\pi)}{M^2 \sin^2(\Delta\pi)} + \frac{\sin^2(M(\frac{1}{M}-\Delta)\pi)}{M^2 \sin^2((\frac{1}{M}-\Delta)\pi)} \\
&\geq \frac{8}{\pi^2}.
\end{aligned}$$

$\square$

The following algorithm computes an estimate for a, via an estimate for θ_a.

Algorithm(Est_Amp$(\mathcal{A}, \chi, M)$)
1. Initialize two registers of appropriate sizes to the state $|0\rangle \mathcal{A}|0\rangle$.
2. Apply $\mathbf{F}_M$ to the first register.
3. Apply $\Lambda_M(\mathbf{Q})$ where $\mathbf{Q} = -\mathcal{A}\mathbf{S}_0 \mathcal{A}^{-1} \mathbf{S}_\chi$.
4. Apply $\mathbf{F}_M^{-1}$ to the first register.
5. Measure the first register and denote the outcome $|y\rangle$.
6. Output $\tilde{a} = \sin^2(\pi \frac{y}{M})$.

Steps 1 to 5 are illustrated on Figure 1. This algorithm can also be summarized, following the approach in [**11**], as the unitary transformation

$$\left((\mathbf{F}_M^{-1} \otimes \mathbf{I}) \, \Lambda_M(\mathbf{Q}) \, (\mathbf{F}_M \otimes \mathbf{I})\right)$$

applied on state $|0\rangle\mathcal{A}|0\rangle$, followed by a measurement of the first register and classical post-processing of the outcome. In practice, we could choose M to be a power of 2, which would allow us to use a Walsh–Hadamard transform instead of a Fourier transform in step 2.

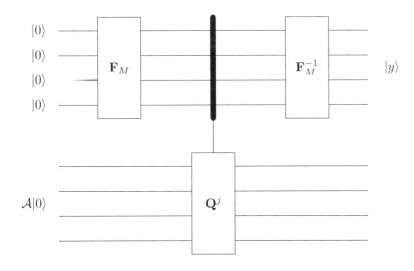

FIGURE 1. Quantum circuit for amplitude estimation.

THEOREM 12 (Amplitude Estimation). *For any positive integer k, the algorithm* **Est_Amp**$(\mathcal{A}, \chi, M)$ *outputs $\tilde{a}$ ($0 \leq \tilde{a} \leq 1$) such that*

$$|\tilde{a} - a| \leq 2\pi k \frac{\sqrt{a(1-a)}}{M} + k^2 \frac{\pi^2}{M^2}$$

with probability at least $\frac{8}{\pi^2}$ when $k = 1$ and with probability greater than $1 - \frac{1}{2(k-1)}$ for $k \geq 2$. It uses exactly M evaluations of f. If $a = 0$ then $\tilde{a} = 0$ with certainty, and if $a = 1$ and M is even, then $\tilde{a} = 1$ with certainty.

PROOF. After step 1, by Equation 6, we have state

$$|0\rangle \mathcal{A}|0\rangle = \frac{-\imath}{\sqrt{2}}|0\rangle \left(e^{\imath\theta_a}|\Psi_+\rangle - e^{-\imath\theta_a}|\Psi_-\rangle\right).$$

After step 2, ignoring global phase, we have

$$\frac{1}{\sqrt{2M}} \sum_{j=0}^{M-1} |j\rangle \left(e^{\imath\theta_a}|\Psi_+\rangle - e^{-\imath\theta_a}|\Psi_-\rangle\right)$$

and after applying $\Lambda_M(\mathbf{Q})$ we have

$$\frac{1}{\sqrt{2M}} \sum_{j=0}^{M-1} |j\rangle \left(e^{\imath\theta_a} e^{2\imath j\theta_a}|\Psi_+\rangle - e^{-\imath\theta_a} e^{-2\imath j\theta_a}|\Psi_-\rangle\right)$$

$$= \frac{e^{\imath\theta_a}}{\sqrt{2M}} \sum_{j=0}^{M-1} e^{2\imath j\theta_a}|j\rangle|\Psi_+\rangle - \frac{e^{-\imath\theta_a}}{\sqrt{2M}} \sum_{j=0}^{M-1} e^{-2\imath j\theta_a}|j\rangle|\Psi_-\rangle$$

$$= \frac{e^{\imath\theta_a}}{\sqrt{2}}|\mathcal{S}_M(\tfrac{\theta_a}{\pi})\rangle|\Psi_+\rangle - \frac{e^{-\imath\theta_a}}{\sqrt{2}}|\mathcal{S}_M(1 - \tfrac{\theta_a}{\pi})\rangle|\Psi_-\rangle.$$

We then apply $\mathbf{F}_M^{-1}$ to the first register and measure it in the computational basis.

The rest of the proof follows from Theorem 11. Tracing out the second register in the eigenvector basis, we see that the first register is in an equally weighted mixture of $\mathbf{F}_M^{-1}|\mathcal{S}_M(\tfrac{\theta_a}{\pi})\rangle$ and $\mathbf{F}_M^{-1}|\mathcal{S}_M(1 - \tfrac{\theta_a}{\pi})\rangle$. Thus the measured value $|y\rangle$ is the result of measuring either the state $\mathbf{F}_M^{-1}|\mathcal{S}_M(\tfrac{\theta_a}{\pi})\rangle$ or the state $\mathbf{F}_M^{-1}|\mathcal{S}_M(1 - \tfrac{\theta_a}{\pi})\rangle$. The probability of measuring $|y\rangle$ given the state $\mathbf{F}_M^{-1}|\mathcal{S}_M(1 - \tfrac{\theta_a}{\pi})\rangle$ is equal to the probability of measuring $|M - y\rangle$ given the state $\mathbf{F}_M^{-1}|\mathcal{S}_M(\tfrac{\theta_a}{\pi})\rangle$. Since $\sin^2\left(\pi \tfrac{(M-y)}{M}\right) = \sin^2\left(\pi\tfrac{y}{M}\right)$, we can assume we measured $|y\rangle$ given the state $\mathbf{F}_M^{-1}|\mathcal{S}_M(\tfrac{\theta_a}{\pi})\rangle$ and $\tilde{\theta}_a = \pi\tfrac{y}{M}$ estimates θ_a as described in Theorem 11. Thus we obtain bounds on $d(\tilde{\theta}_a, \theta_a)$ that translate, using Lemma 7, into the appropriate bounds on $|\tilde{a} - a|$. □

A straightforward application of this algorithm is to approximately count the number of solutions t to $f(x) = 1$. To do this we simply set $\mathcal{A} = \mathbf{W}$ if N is a power of 2, or in general $\mathcal{A} = \mathbf{F}_N$ or any other transformation that maps $|0\rangle$ to $\frac{1}{\sqrt{N}} \sum_{j=0}^{N-1} |j\rangle$. Setting $\chi = f$, we then have $a = \langle \Psi_1 | \Psi_1 \rangle = t/N$, which suggests the following algorithm.

Algorithm(Count(f, M))

1. Output $t' = N \times$ **Est_Amp**$(\mathbf{F}_N, f, M)$.

By Theorem 12, we obtain the following.

THEOREM 13 (Counting). *For any positive integers M and k, and any Boolean function $f : \{0, 1, \ldots, N - 1\} \to \{0, 1\}$, the algorithm $\mathbf{Count}(f, M)$ outputs an estimate t' to $t = |f^{-1}(1)|$ such that*

$$|t' - t| \leq 2\pi k \frac{\sqrt{t(N-t)}}{M} + \pi^2 k^2 \frac{N}{M^2}$$

with probability at least $8/\pi^2$ when $k = 1$, and with probability greater than $1 - \frac{1}{2(k-1)}$ for $k \geq 2$. If $t = 0$ then $t' = 0$ with certainty, and if $t = N$ and M is even, then $t' = N$ with certainty.

Note that $\mathbf{Count}(f, M)$ outputs a real number. In the following counting algorithms we will wish to output an integer, and therefore we will round off the output of $\mathbf{Count}$ to an integer. To assure that the rounding off can be done efficiently[1] we will round off to an integer $\tilde{t}$ satisfying $|\tilde{t} - \mathbf{Count}(f, M)| \leq \frac{2}{3}$.

If we want to estimate t within a few standard deviations, we can apply algorithm $\mathbf{Count}$ with $M = \lceil \sqrt{N} \rceil$.

COROLLARY 14. *Given a Boolean function $f : \{0, 1, \ldots, N - 1\} \to \{0, 1\}$ with t defined as above, rounding off the output of $\mathbf{Count}(f, \lceil \sqrt{N} \rceil)$ gives an estimate $\tilde{t}$ such that*

(15) $$|\tilde{t} - t| < 2\pi \sqrt{\frac{t(N-t)}{N}} + 11$$

with probability at least $8/\pi^2$ and requires exactly $\lceil \sqrt{N} \rceil$ evaluations of f.

We now look at the case of estimating t with some relative error, also referred to as *approximately counting t with accuracy ε*. For this we require the following crucial observation about the output t' of algorithm $\mathbf{Count}(f, L)$. Namely t' is likely to be equal to zero if and only if $L \in o(\sqrt{N/t})$. Thus, we can find a rough estimate of $\sqrt{N/t}$ simply by running algorithm $\mathbf{Count}(f, L)$ with exponentially increasing values of L until we obtain a non-zero output. Having this rough estimate L of $\sqrt{N/t}$ we can then apply Theorem 13 with M in the order of $\frac{1}{\varepsilon} L$ to find an estimate $\tilde{t}$ of t with the required accuracy. The precise algorithm is as follows.

Algorithm(Basic_Approx_Count(f, ε))
1. Start with $\ell = 0$.
2. Increase ℓ by 1.
3. Set $t' = \mathbf{Count}(f, 2^\ell)$.
4. If $t' = 0$ and $2^\ell < 2\sqrt{N}$ then go to step 2.
5. Set $M = \lceil \frac{20\pi^2}{\varepsilon} 2^\ell \rceil$.
6. Set $t' = \mathbf{Count}(f, M)$.
7. Output an integer $\tilde{t}$ satisfying $|\tilde{t} - t'| \leq \frac{2}{3}$.

THEOREM 15. *Given a Boolean function f with N and t defined as above, and any $0 < \varepsilon \leq 1$, $\mathbf{Basic\_Approx\_Count}(f, \varepsilon)$ outputs an estimate $\tilde{t}$ such that*

$$|\tilde{t} - t| \leq \varepsilon t$$

---

[1] For example, if $t' + \frac{1}{2}$ is super-exponentially close to an integer n we may not be able to decide efficiently if t' is closer to n or $n - 1$.

with probability at least $\frac{2}{3}$, using an expected number of evaluations of f which is in $\Theta\left(\frac{1}{\varepsilon}\sqrt{N/t}\right)$. If $t=0$, the algorithm outputs $\tilde{t}=t$ with certainty and f is evaluated a number of times in $\Theta(\sqrt{N})$.

PROOF. When $t=0$, the analysis is straightforward. For $t>0$, let θ denote $\theta_{t/N}$ and $m = \left\lfloor \log_2(\frac{1}{5\theta}) \right\rfloor$. From Theorem 11 we have that the probability that step 3 outputs $\mathbf{Count}(f, 2^\ell) = 0$ for $\ell = 1, 2, \ldots, m$ is

$$\prod_{\ell=1}^{m} \frac{\sin^2(2^\ell \theta)}{2^{2\ell} \sin^2(\theta)} \geq \prod_{\ell=1}^{m} \cos^2(2^\ell \theta) = \frac{\sin^2(2^{m+1}\theta)}{2^{2m} \sin^2(2\theta)} \geq \cos^2\left(\tfrac{2}{5}\right).$$

The previous inequalities are obtained by using the fact that $\sin(M\theta) \geq M \sin(\theta) \cos(M\theta)$ for any $M \geq 0$ and $0 \leq M\theta < \frac{\pi}{2}$, which can be readily seen by considering the Taylor expansion of $\tan(x)$ at $x = M\theta$.

Now assuming step 3 has outputted 0 at least m times (note that $2^m \leq \frac{1}{5\theta} \leq \frac{1}{5}\sqrt{N/t} < 2\sqrt{N}$), after step 5 we have $M \geq \frac{20\pi^2}{\varepsilon} 2^{m+1} \geq \frac{4\pi^2}{\varepsilon\theta}$ and by Theorem 13 (and the fact that $\theta \leq \frac{\pi}{2} \sin(\theta) = \frac{\pi}{2}\sqrt{t/N}$) the probability that $\mathbf{Count}(f, M)$ outputs an integer t' satisfying $|t' - t| \leq \frac{\varepsilon}{4}t + \frac{\varepsilon^2}{64}t$ is at least $8/\pi^2$. Let us suppose this is the case. If $\varepsilon t < 1$, then $|\tilde{t} - t| < 1$ and, since $\tilde{t}$ and t are both integers, we must have $t = \tilde{t}$. If $\varepsilon t \geq 1$, then rounding off t' to $\tilde{t}$ introduces an error of at most $\frac{2}{3} \leq \frac{2\varepsilon}{3}t$, making the total error at most $\frac{\varepsilon}{4}t + \frac{\varepsilon^2}{64}t + \frac{2\varepsilon}{3}t < \varepsilon t$. Therefore the overall probability of outputting an estimate with error at most εt is at least $\cos^2\left(\tfrac{2}{5}\right) \times (8/\pi^2) > \tfrac{2}{3}$.

To upper bound the number of applications of f, note that by Theorem 13, for any integer $L \geq 18\pi\sqrt{N/t}$, the probability that $\mathbf{Count}(f, L)$ outputs 0 is less than $1/4$. Thus the expected value of M at step 6 is in $\Theta(\frac{1}{\varepsilon}\sqrt{N/t})$. □

We remark that in algorithm **Basic_Approx_Count**, we could alternatively to steps 1 to 4 use algorithm **QSearch** of Section 2, provided we have **QSearch** also output its final value of M. In this case, we would use (a multiple of) that value as our rough estimate of $\sqrt{N/t}$, instead of using the final value of 2^ℓ found in step 4 of **Basic_Approx_Count**.

Algorithm **Basic_Approx_Count** is optimal for any fixed ε, but not in general. In Appendix A we give an optimal algorithm, while we now present two simple optimal algorithms for counting the number of solutions exactly. That is, we now consider the problem of determining the exact value of $t = |f^{-1}(-1)|$. In the special case that we are given a nonzero integer t_0 and promised that either $t = 0$ or $t = t_0$, then we can determine which is the case with certainty using a number of evaluations of f in $O(\sqrt{N/t_0})$. This is an easy corollary of Theorem 4 and we state it without proof.

THEOREM 16. *Let $f : \{0, 1, \ldots, N-1\} \to \{0, 1\}$ be a given Boolean function such that the cardinality of the preimage of 1 is either 0 or t_0. Then there exists a quantum algorithm that determines with certainty which is the case using a number of evaluations of f which is in $\Theta(\sqrt{N/t_0})$, and in the latter case, also outputs a random element of $f^{-1}(1)$.*

For the general case in which we do not have any prior knowledge about t, we offer the following algorithm.

Algorithm(Exact_Count(f))

1. Set $t'_1 = \mathbf{Count}(f, \lceil 14\pi\sqrt{N}\rceil)$ and $t'_2 = \mathbf{Count}(f, \lceil 14\pi\sqrt{N}\rceil)$.
2. Let $M_i = \lceil 30\sqrt{(t'_i+1)(N-t'_i+1)}\rceil$ for $i = 1, 2$.
3. Set $M = \min\{M_1, M_2\}$.
4. Set $t' = \mathbf{Count}(f, M)$.
5. Output an integer $\tilde{t}$ satisfying $|\tilde{t} - t'| \leq \frac{2}{3}$.

The main idea of this algorithm is the same as that of algorithm **Basic_Approx_Count**. First we find a rough estimate t'_r of t, and then we run algorithm $\mathbf{Count}(f, M)$ with a value of M that depends on t'_r. By Theorem 13, if we set M to be in the order of $\sqrt{t'_r(N - t'_r)}$, then the output $t' = \mathbf{Count}(f, M)$ is likely to be so that $|t' - t| < \frac{1}{3}$, in which case $\tilde{t} = t$.

THEOREM 17. *Given a Boolean function f with N and t defined as above, algorithm* **Exact_Count** *requires an expected number of evaluations of f which is in $\Theta(\sqrt{(t+1)(N-t+1)})$ and outputs an estimate $\tilde{t}$ which equals t with probability at least $\frac{2}{3}$ using space only linear in $\log(N)$.*

PROOF. Apply Theorem 13 with $k = 7$. For each $i = 1, 2$, with probability greater than $\frac{11}{12}$, outcome t'_i satisfies $|t'_i - t| < \sqrt{\frac{t(N-t)}{N}} + 1/4$, in which case we also have that $\sqrt{t(N-t)} \leq \frac{\sqrt{2}}{30} M_i$. Thus, with probability greater than $\left(\frac{11}{12}\right)^2$, we have

$$\frac{\sqrt{t(N-t)}}{M} \leq \frac{\sqrt{2}}{30}.$$

Suppose this is the case. Then by Theorem 13, with probability at least $8/\pi^2$,

$$|t' - t| \leq \frac{2\pi\sqrt{2}}{30} + \frac{4\pi^2}{30^2} < \frac{1}{3}$$

and consequently

$$|\tilde{t} - t| < 1.$$

Hence, with probability at least $\left(\frac{11}{12}\right)^2 \times 8/\pi^2 > \frac{2}{3}$, we have $\tilde{t} = t$.

The number of applications of f is $2\lceil 14\pi\sqrt{N}\rceil + M$. Consider the expected value of M_i for $i = 1, 2$. Since

$$\sqrt{(t'_i+1)(N-t'_i+1)} \leq \sqrt{(t+1)(N-t+1)} + \sqrt{N|t'_i - t|}$$

for any $0 \leq t'_i, t \leq N$, we just need to upper bound the expected value of $\sqrt{N|t'_i - t|}$. By Theorem 13, for any $k \geq 2$,

$$|t'_i - t| \leq k\sqrt{\frac{t(N-t)}{N}} + k^2$$

with probability at least $1 - \frac{1}{k}$. Hence M_i is less than

(16) $$30(1+k)\left(\sqrt{(t+1)(N-t+1)} + \sqrt{N}\right) + 1$$

with probability at least $1 - \frac{1}{k}$.

In particular, the minimum of M_1 and M_2 is greater than the expression given in Equation 16 with probability at most $\frac{1}{k^2}$. Since any positive random variable Z satisfying $\mathrm{Prob}(Z > k) \leq \frac{1}{k^2}$ has expectation upper bounded by a constant, the expected value of M is in $O(\sqrt{(t+1)(N-t+1)})$. $\square$

It follows from Theorem 4.10 of [1] that any quantum algorithm capable of deciding with high probability whether or not a function $f : \{0, 1, \ldots, N-1\} \to \{0, 1\}$ is such that $|f^{-1}(1)| \leq t$, given some $0 < t < N$, must query f a number of times which is at least in $\Omega\big(\sqrt{(t+1)(N-t+1)}\,\big)$ times. Therefore, our exact counting algorithm is optimal up to a constant factor.

Note also that successive applications of Grover's algorithm in which we strike out the solutions as they are found will also provide an algorithm to perform exact counting. In order to obtain a constant probability of success, if the algorithm fails to return a new element, one must do more than a constant number of trials. In particular, repeating until we get $\log(N)$ failures will provide an overall constant probability of success. Unfortunately, the number of applications of f is then in $O\big(\sqrt{tN} + \log(N)\sqrt{N/t}\,\big)$ and the cost in terms of additional quantum memory is prohibitive, that is in $\Theta(t)$.

5. Concluding remarks

Let $f : \{0, 1, \ldots, N-1\} \to \{0, 1\}$ be a function provided as a black box, in the sense that the only knowledge available about f is given by evaluating it on arbitrary points in its domain. We are interested in the number of times that f must be evaluated to achieve certain goals, and this number is our measure of efficiency. Grover's algorithm can find the x_0 such that $f(x_0) = 1$ quadratically faster in the expected sense than the best possible classical algorithm provided the solution is known to be unique [8, 9]. We have generalized Grover's algorithm in several directions.

- ⋄ The quadratic speedup remains when the solution is not unique, even if the number of solutions is not known ahead of time.
- ⋄ If the number of solutions is known (and nonzero), we can find one quadratically faster in the worst case than would be possible classically even in the expected case.
- ⋄ If the number t of solutions is known to be either 0 or t_0, we can tell which is the case with certainty, and exhibit a solution if $t > 0$, in a time in $O(\sqrt{N/t_0}\,)$ in the worst case. By contrast, the best classical algorithm would need $N - t_0 + 1$ queries in the worst case. This is much better than a quadratic speedup when t_0 is large.
- ⋄ The quadratic speedup remains in a variety of settings that are not constrained to the black-box model: even if additional information about f can be used to design efficient classical heuristics, we can still find solutions quadratically faster on a quantum computer, provided the heuristic falls under the broad scope of our technique.
- ⋄ We give efficient quantum algorithms to estimate the number of solutions in a variety of error models. In all cases, our quantum algorithms are proven optimal, up to a multiplicative constant, among all possible quantum algorithms. In most cases, our quantum algorithms are known to be quadratically faster than the best possible classical algorithm. In the case of counting the number of solutions up to relative error ε, our optimal quantum algorithm is quadratically faster than the best known classical algorithm for fixed ε, but in fact it is better than that when ε is not a constant. Since we do not believe that a super-quadratic quantum improvement for a non-promise black-box problem is possible, we conjecture that there exists a

classical algorithm that uses a number of queries in $O(\min\{M^2, N\})$, where $M = \sqrt{\frac{N}{\lfloor \varepsilon t \rfloor + 1}} + \frac{\sqrt{t(N-t)}}{\lfloor \varepsilon t \rfloor + 1}$ is proportional to the number of queries required by our optimal quantum algorithm. This conjecture is further supported by the fact that we can easily find a good estimate for M^2, without prior knowledge of t, using a number of classical queries in $O(\frac{1}{\varepsilon} + \frac{N}{t+1})$.

⋄ We can amplify efficiently the success probability not only of classical search algorithms, but also of quantum algorithms. More precisely, if a quantum algorithm can output an x that has probability $a > 0$ of being such that $f(x) = 1$, then a solution can be found after evaluating f an expected number of time in $O(1/\sqrt{a})$. If the value of a is known, a solution can be found after evaluating f a number of time in $O(1/\sqrt{a})$ even in the worst case. We call this process *amplitude amplification*. Again, this is quadratically faster than would be possible if the quantum search algorithm were available as a black box to a classical algorithm.

⋄ Finally, we provide a general technique, known as *amplitude estimation*, to estimate efficiently the success probability a of quantum search algorithms. This is the natural quantum generalization of the above-mentioned technique to estimate the number of classical solutions to the equation $f(x) = 1$.

The following table summarizes the number of applications of the given function f in the quantum algorithms presented in this paper. The table also compares the quantum complexities with the classical complexities of these problems, when the latter are known. Any lower bounds indicated (implicit in the use of the "Θ" notation) correspond to those in the black-box model of computation. In the case of the efficiency of quantum counting with accuracy ε, we refer to the algorithm given below in the Appendix.

Problem	Quantum Complexity	Classical Complexity
Decision	$\Theta(\sqrt{N/(t+1)})$	$\Theta(N/(t+1))$
Searching	$\Theta(\sqrt{N/(t+1)})$	$\Theta(N/(t+1))$
Counting with error $\sqrt{t}$	$\Theta(\sqrt{N})$	
Counting with accuracy ε	$\Theta\left(\sqrt{\frac{N}{\lfloor \varepsilon t \rfloor + 1}} + \frac{\sqrt{t(N-t)}}{\lfloor \varepsilon t \rfloor + 1}\right)$	$O(\frac{1}{\varepsilon^2} N/(t+1))$
Exact counting	$\Theta(\sqrt{(t+1)(N-t+1)})$	$\Theta(N)$

We leave as open the problem of finding a quantum algorithm that exploits the structure of some searching or counting problem in a genuinely quantum way. By this, we mean in a way that is not equivalent to applying amplitude amplification or amplitude estimation to a classical heuristic. Note that Shor's factoring algorithm does this in the different context of integer factorization.

Acknowledgements

We are grateful to Joan Boyar, Harry Buhrman, Artur Ekert, Ashwin Nayak, Jeff Shallitt, Barbara Terhal and Ronald de Wolf for helpful discussions.

Appendix A. Tight Algorithm for Approximate

Counting

Here we combine the ideas of algorithms **Basic_Approx_Count** and **Exact_Count** to obtain an optimal algorithm for approximately counting. That this algorithm is optimal follows readily from Corollary 1.2 and Theorem 1.13 of Nayak and Wu [14].

THEOREM 18. *Given a Boolean function f with N and t defined as above, and any ε such that $\frac{1}{3N} < \varepsilon \leq 1$, the following algorithm $\mathbf{Approx\_Count}(f, \varepsilon)$ outputs an estimate $\tilde{t}$ such that*

$$|\tilde{t} - t| \leq \varepsilon t$$

with probability at least $\frac{2}{3}$, using an expected number of evaluations of f in the order of

$$S = \sqrt{\frac{N}{\lfloor \varepsilon t \rfloor + 1}} + \frac{\sqrt{t(N-t)}}{\lfloor \varepsilon t \rfloor + 1}.$$

If $t = 0$ or $t = N$, the algorithm outputs $\tilde{t} = t$ with certainty.

We assume that $\varepsilon N > 1/3$, since otherwise approximately counting with accuracy ε reduces to exact counting. Set

$$(17) \qquad S' = \min\left\{\frac{1}{\sqrt{\varepsilon}}\sqrt{\frac{N}{t}}\left(1 + \sqrt{\frac{N-t}{\varepsilon N}}\right), \sqrt{(t+1)(N-t+1)}\right\}$$

and note that $S' \in \Theta(S)$ where S is defined as in Theorem 18. The algorithm works by finding approximate values for each of the different terms in Equation 17. The general outline of the algorithm is as follows.

Algorithm(Approx_Count(f, ε))
1. Find integer L_1 approximating $\sqrt{N/(t+1)}$.
2. Find integer L_2 approximating $\sqrt{(N-t)/(\varepsilon N)}$.
3. Set $M_1 = \frac{1}{\sqrt{\varepsilon}} L_1(1 + L_2)$.
4. If $M_1 > \sqrt{N}$ then find integer M_2 approximating $\sqrt{(t+1)(N-t+1)}$. If $M_1 \leq \sqrt{N}$ then set $M_2 = \infty$.
5. Set $M = \min\{M_1, M_2\}$.
6. Set $t' = \mathbf{Count}(f, \lceil 10\pi M \rceil)$.
7. Output an integer $\tilde{t}$ satisfying $|\tilde{t} - t'| \leq \frac{2}{3}$.

PROOF. To find L_1, we run steps 1 to 4 of algorithm **Basic_Approx_Count** and then set $L_1 = \lceil 9\pi \times 2^l \rceil$. A proof analogous to that of Theorem 15 gives that
- $L_1 > \sqrt{N/(t+1)}$ with probability at least 0.95, and
- the expected value of L_1 is in $\Theta(\sqrt{N/(t+1)})$.

This requires a number of evaluations of f which is in $\Theta(L_1)$, and thus, the expected number of evaluations of f so far is in $O(S')$.

In step 2, for some constant c to be determined below, we use $2\lceil \frac{c}{\sqrt{\varepsilon}} \rceil$ evaluations of f to find integer L_2 satisfying
- $L_2 > \sqrt{(N-t)/(\varepsilon N)}$ with probability at least 0.95, and
- the expected value of L_2 is in $O(\sqrt{(N-t+1)/(\varepsilon N)})$.

Since $N - t = |f^{-1}(0)|$, finding such L_2 boils down to estimating, with accuracy in $\Theta(\sqrt{\varepsilon})$, the square root of the probability that f takes the value 0 on a random point in its domain. Or equivalently, the probability that $\neg f$ takes the value 1, where $\neg f = 1 - f$. Suppose for some constant c, we run **Count**$(\neg f, \lceil \frac{c}{\sqrt{\varepsilon}} \rceil)$ twice with outputs $\tilde{r}_1$ and $\tilde{r}_2$. By Theorem 13, each output $\tilde{r}_i$ ($i = 1, 2$) satisfies that

$$\left| \sqrt{\frac{\tilde{r}_i}{\varepsilon N}} - \sqrt{\frac{N-t}{\varepsilon N}} \right| \leq \sqrt{\frac{2\pi k}{c}} \sqrt[4]{\frac{N-t}{\varepsilon N}} + \frac{\pi k}{c}$$

with probability at least $1 - \frac{1}{2(k-1)}$ for every $k \geq 2$. It follows that $\tilde{r} = \min\left\{ \sqrt{\tilde{r}_1/(\varepsilon N)}, \sqrt{\tilde{r}_2/(\varepsilon N)} \right\}$ has expected value in $O(\sqrt{(N-t+1)/(\varepsilon N)})$. Setting $k = 21$, $c = 8\pi k$, and $L_2 = \lceil 2\tilde{r} \rceil + 1$, ensures that L_2 satisfies the two properties mentioned above. The number of evaluations of f in step 2 is in $\Theta(\frac{1}{\sqrt{\varepsilon}})$ which is in $O(S')$.

In step 3, we set $M_1 = \frac{1}{\sqrt{\varepsilon}} L_1(1 + L_2)$. Note that

- $M_1 > \frac{1}{\sqrt{\varepsilon}} \sqrt{\frac{N}{t+1}} \left(1 + \sqrt{\frac{N-t}{\varepsilon N}}\right)$ with probability at least 0.95^2, and
- the expected value of M_1 is in the order of $\frac{1}{\sqrt{\varepsilon}} \sqrt{\frac{N}{t+1}} \left(1 + \sqrt{\frac{N-t+1}{\varepsilon N}}\right)$.

In step 4, analogously to algorithm **Exact_Count**, a number of evaluations of f in $\Theta(\sqrt{N})$ suffices to find an integer M_2 such that

- $M_2 > \sqrt{(t+1)(N-t+1)}$ with probability at least 0.95, and
- the expected value of M_2 is in $\Theta(\sqrt{(t+1)(N-t+1)})$.

Fortunately, since $\sqrt{(t+1)(N-t+1)} \geq \sqrt{N}$, we shall only need M_2 if $M_1 > \sqrt{N}$. We obtain that, after step 5,

- M is greater than

$$\min\left\{ \frac{1}{\sqrt{\varepsilon}} \sqrt{\frac{N}{t+1}} \left(1 + \sqrt{\frac{N-t}{\varepsilon N}}\right), \sqrt{(t+1)(N-t+1)} \right\}$$

  with probability at least $0.95^3 > 0.85$, and
- the expected value of M is in $O(S')$.

To derive this latter statement, we use the fact that the expected value of the minimum of two random variables is at most the minimum of their expectation.

Finally, by Theorem 13, applying algorithm **Count**$(f, \lceil 10\pi M \rceil)$ given such an M, produces an estimate t' of t such that $|t' - t| \leq \frac{\varepsilon t}{3}$ (which implies that $|\tilde{t} - t| \leq \varepsilon t$) with probability at least $8/\pi^2$. Hence our overall success probability is at least $0.85 \times 8/\pi^2 > 2/3$, and the expected number of evaluations of f is in $O(S')$. $\square$

References

[1] BEALS, Robert, Harry BUHRMAN, Richard CLEVE, Michele MOSCA and Ronald DE WOLF, "Quantum lower bounds by polynomials", *Proceedings of 39th Annual Symposium on Foundations of Computer Science*, November 1998, pp. 352–361.

[2] BENNETT, Charles H., "Notes on the history of reversible computation", *IBM Journal of Research and Development*, 1988, Vol. 32, pp. 16–23.

[3] BOYER, Michel, Gilles BRASSARD, Peter HØYER and Alain TAPP, "Tight bounds on quantum searching", *Fortschritte Der Physik*, special issue on quantum computing and quantum cryptography, 1998, Vol. 46, pp. 493–505.

[4] BRASSARD, Gilles and Peter HØYER, "An exact quantum polynomial-time algorithm for Simon's problem", *Proceedings of Fifth Israeli Symposium on Theory of Computing and Systems*, IEEE Computer Society Press, June 1997, pp. 12–23.

[5] BRASSARD, Gilles, Peter HØYER and Alain TAPP, "Quantum counting", *Proceedings of 25th International Colloquium on Automata, Languages, and Programming*, Lecture Notes in Computer Science, Vol. 1443, Springer-Verlag, July 1998, pp. 820–831.

[6] CHI, Dong–Pyo and Jinsoo KIM, "Quantum database searching by a single query", Lecture at *First NASA International Conference on Quantum Computing and Quantum Communications*, Palm Springs, February 1998.

[7] CLEVE, Richard, Artur EKERT, Chiara MACCHIAVELLO and Michele MOSCA, "Quantum algorithms revisited", *Proceedings of the Royal Society, London*, Vol. A354, 1998, pp. 339–354.

[8] GROVER, Lov K., "A fast quantum mechanical algorithm for database search", *Proceedings of 28th Annual ACM Symposium on Theory of Computing*, May 1996, pp. 212–219.

[9] GROVER, Lov K., "Quantum mechanics helps in searching for a needle in a haystack", *Physical Review Letters*, Vol. 79, July 1997, pp. 325–328.

[10] GROVER, Lov K., "Quantum computers can search rapidly by using almost any transformation", *Physical Review Letters*, Vol. 80, May 1998, pp. 4329–4332.

[11] HØYER, Peter, "Conjugated operators in quantum algorithms", *Physical Review A*, Vol. 59, May 1999, pp. 3280–3289.

[12] KITAEV, A. Yu., "Quantum measurements and the Abelian stabilizer problem", November 1995. Available at Los Alamos e-Print archive as <http://arXiv.org/abs/quant-ph/9511026>.

[13] MOSCA, Michele, "Quantum searching and counting by eigenvector analysis", *Proceedings of Randomized Algorithms*, Satellite Workshop of 23rd International Symposium on Mathematical Foundations of Computer Science, Brno, Czech Republic, August 1998, pp. 90–100.

[14] NAYAK, Ashwin and Felix WU, "The quantum query complexity of approximating the median and related statistics", *Proceedings of 31st Annual ACM Symposium on Theory of Computing*, May 1999, pp. 384–393.

[15] SHOR, Peter W., "Polynomial-time algorithms for prime factorization and discrete logarithms on a quantum computer", *SIAM Journal on Computing*, Vol. 26, October 1997, pp. 1484–1509.

(Gilles Brassard) DÉPARTEMENT IRO, UNIVERSITÉ DE MONTRÉAL, C.P. 6128, SUCCURSALE CENTRE-VILLE, MONTRÉAL (QUÉBEC), CANADA H3C 3J7.
E-mail address, Gilles Brassard: `brassard@iro.umontreal.ca`.

(Peter Høyer) BRICS, DEPARTMENT OF COMPUTER SCIENCE, UNIVERSITY OF AARHUS, NY MUNKEGADE, BLDG. 540, DK-8000 AARHUS C, DENMARK.
E-mail address, Peter Høyer: `hoyer@brics.dk`.

(Michele Mosca) CACR, DEPARTMENT OF C&O, FACULTY OF MATHEMATICS, UNIVERSITY OF WATERLOO, WATERLOO, ONTARIO, CANADA N2L 3G1.
E-mail address, Michele Mosca: `mmosca@cacr.math.uwaterloo.ca`.

(Alain Tapp) DÉPARTEMENT IRO, UNIVERSITÉ DE MONTRÉAL, C.P. 6128, SUCCURSALE CENTRE-VILLE, MONTRÉAL (QUÉBEC), CANADA H3C 3J7.
E-mail address, Alain Tapp: `tappa@iro.umontreal.ca`.

Manipulating the Entanglement of One Copy of a Two-Particle Pure Entangled State

Lucien Hardy

ABSTRACT. One copy of a general bipartite pure entangled quantum state is considered. We show how it is possible to manipulate the entanglement by local operations and classical communication in order to produce to produce maximally entangled states. It is shown that this problem is equivalent to a certain colouring problem.

There has been much work recently concerning the manipulation of the entanglement of one copy of a pure bipartite entangled state by local operations and classical communication (LOCC). Lo and Popescu [1] obtained some important results in 1997 which were taken up by other researchers more recently. Nielsen [2] obtained necessary and sufficient conditions for it to be possible to transform one state by LOCC into another. Vidal [3] developed entanglement monotones (quantities which cannot increase under LOCC transformations) and studied probabilistic transformations of one state to another. Jonathan and Plenio [4] showed that some transformations which cannot be performed directly can nevertheless be performed if the parties are allowed to borrow entanglement which they later give back.

In this paper we will consider the problem of how to manipulate the entanglement of a general bipartite pure state by local operations and classical communication (LOCC) in order to produce maximally entangled states. This problem was originally studied in a limited context by Lo and Popescu [1]. In Hardy [5] and Jonathan and Plenio [6] the problem was solved in full generality. The purpose of this paper is to present the main points in [5]. The details of some of the more involved proofs are not given here and for those the reader is referred to [5].

We start with two systems A and B prepared in a general pure state $|\psi\rangle$. It follows from the Schmidt decomposition theorem that we can write

(1) $$|\psi\rangle = \sum_{j=1}^{I} \sqrt{\lambda_j} |j\rangle_A |j\rangle_B$$

where by appropriate labeling we choose $\lambda_j \geq \lambda_{j+1}$ and where the states $|j\rangle_{A,B}$ are orthonormal. We have $\sum_i \lambda_i = 1$ by normalisation. Let system A be in Alice's possession and let system B be in Bob's possession. We will assume that Alice and Bob are allowed to do anything which does not involve their sending quantum

---

2000 *Mathematics Subject Classification.* Primary 81P68.

©2002 American Mathematical Society

systems to one another (for if they could do this then they could generate new entanglement between themselves - for example, Alice could send one particle of an entangled pair to Bob and keep the other particle in her own hands). This means that Alice and Bob can do anything locally that they want, in particular they can make measurements, introduce new local quantum systems, and discard quantum systems. In addition, they can communicate classically with one another, for example by talking down a telephone wire. Their task here is to obtain states like

$$|\varphi_m\rangle = \frac{1}{\sqrt{m}} \sum_{k=1}^{m} |k\rangle_A |k\rangle_B \qquad (2)$$

We will call this an m-state. This is a maximally entangled state involving m levels of systems A and B. We say it is maximally entangled because modulus of the amplitude of each term when written in this Schmidt form is equal. The bigger m is the more entanglement there is in an m-state. The usual measure that is used is that one m-state is equivalent to $\log_2 m$ 2-states. Because of their simplicity, 2-states, such as the state

$$\frac{1}{\sqrt{2}}(|1\rangle_A|1\rangle_B + |2\rangle_A|2\rangle_B),$$

are a useful unit for measuring quantum entanglement. The $\log_2 m$ relation makes most sense when $m = 2^r$ for integer r. Consider $m = 4$. Since $\log_2 4 = 2$ a 4-state should be equivalent to two 2-states. We can see this is true. Take two 2-states

$$\frac{1}{2}(|1\rangle_{A'}|1\rangle_{B'} + |2\rangle_{A'}|2\rangle_{B'})(|1\rangle_{A''}|1\rangle_{B''} + |2\rangle_{A''}|2\rangle_{B''})$$

We can expand this:

$$\frac{1}{\sqrt{4}}(|11\rangle_{A'A''}|11\rangle_{B'B''} + |12\rangle_{A'A''}|12\rangle_{B'B''}|21\rangle_{A'A''}|21\rangle_{B'B''} + |22\rangle_{A'A''}|22\rangle_{B'B''})$$

where we have used the notation $|1\rangle_{A'}|1\rangle_{A''} = |11\rangle_{A'A''}$. Now we can think of the system $A'A''$ as a single system A and similarly for B. Then we can rename the states in this expansion as follows

$$|11\rangle_{A'A''} = |1\rangle_A$$

$$|12\rangle_{A'A''} = |2\rangle_A$$

$$|21\rangle_{A'A''} = |3\rangle_A$$

$$|22\rangle_{A'A''} = |4\rangle_A$$

and similarly for B. Hence the above state becomes

$$\frac{1}{\sqrt{4}} \sum_{k=1}^{4} |k\rangle_A |k\rangle_B$$

and this is just a 4-state. It is clear that the same can be done for any other $m = 2^r$. This proves the $\log_2 m$ relation for values of m which are powers of 2. Although this relation works strictly only for these values, it is reasonable to use this as a measure for other values of m also (especially when we consider that if Alice and Bob share an m-state and an m'-state then this is equivalent to their sharing a state with mm' terms, namely an mm'-state and therefore the measure is additive because $\log_2 m + \log_2 m' = \log_2 mm'$).

Starting with the state $|\psi\rangle$ Alice and Bob will end up with a state $|\varphi_m\rangle$ with some probability p_m where, at the end of the procedure, they know which $|\varphi_m\rangle$ they have (so they know what m is). The expected entanglement, using the $\log_2 m$ measure is

$$E = \sum_m p_m \log_2 m \tag{3}$$

We will describe a method of manipulating the state by LOCC which can yield the most general transformation of the type $|\psi\rangle \to \{|\varphi_m\rangle; p_m\}$. We will not prove here that this is the most general transformation (for a proof of this the reader is referred to [5] (see also [3,6])).

Most of the work is done by Alice. First she introduces an additional quantum system R which she has prepared in the state $|1\rangle_R$. The system R has orthonormal basis states $|n\rangle_R$ where $n = 1$ to N. The initial state is

$$|1\rangle_R|\psi\rangle = \sum_{j=1}^{I} \sqrt{\lambda_j}(|1\rangle_R|j\rangle_A)|j\rangle_B \tag{4}$$

We define the numbers $N_j = N\lambda_j$. We will take $N \to \infty$ so that, at least in this limit, the numbers N_j can be taken to be integers. Alice has the systems R and A under her control. She performs the unitary transformation described by

$$|1\rangle_R|i\rangle_A \to \frac{1}{\sqrt{N_j}}\Big(\sum_{n=1}^{N_j} |n\rangle_R\Big)|i\rangle_A \tag{5}$$

for all j (this transformation is unitary since orthogonal states are mapped onto orthogonal states). After this transformation the state becomes

$$|\Psi_{\text{start}}\rangle = \frac{1}{\sqrt{N}} \sum_j \sum_{n=1}^{N_j} |n\rangle_R|j\rangle_A|j\rangle_B \tag{6}$$

We will call this the *start state*. It is possible to go back to the original state from the start state by reversing the above transformation so there is no change in the entanglement properties at this stage. The start state has the property that each term in it has the same amplitude. It is this property that will make it possible to describe a graphical method for manipulating the entanglement. The terms in the start state are of the form $\frac{1}{\sqrt{N}}|n\rangle_R|j\rangle_A|i\rangle_B$. (Initially $j = i$, however this will not remain true as Alice performs some basic manipulations.) We associate with each such term a rectangle of width 1, height $\frac{1}{N}$, and colour j where this rectangle is placed on a graph with its top right hand corner at position $(\frac{n}{N}, i)$. Each rectangle for all the terms in the start state is plotted on the graph and the resulting graph looks like that shown in figure 1. The colour is not shown, but since initially $j = i$ each column will be of one colour which will be different to the colour of all the other columns. The height of the ith column is λ_i. The area of each rectangle is $\frac{1}{N}$ which is equal to the probability associated with the corresponding term in the start state. The total area is equal to 1.

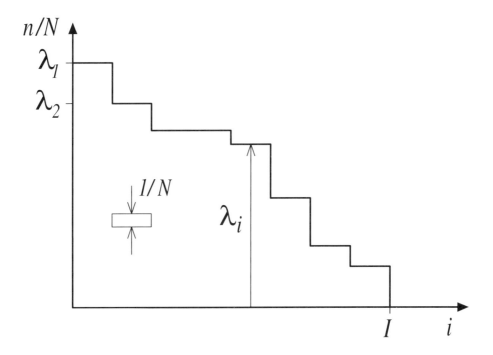

FIGURE 1. The start state can be plotted on a graph. Such a graph is called an area diagram. This area diagram has step structure in which the steps go down towards the right.

The basic unitary operation, $U(n, i \leftrightarrow n', i')$, employed by Alice is defined by the transformation equations

(7) $$|n\rangle_R |i\rangle_A \to |n'\rangle_R |i'\rangle_A$$

(8) $$|n'\rangle_R |i'\rangle_A \to |n\rangle_R |i\rangle_A$$

with no change for all other $|n''\rangle_R |i''\rangle_A$. We will call this the swap operation. The effect of this operation is to move elements around on the area diagram. If there are elements at both the (n, i) and (n', i') positions then they will have their positions swapped. If, there is only an element at one of the two positions then it will be moved to the other position while the original position will become vacant. These moves will not effect Bob's part of the state which means that the value of j and hence the colour of each rectangle will not be changed as it is moved to its new position. Hence, by using this swap operation it is possible to redistribute the colour on the graph. Note the following: (i) Alice can move as many rectangles around a once as she likes; (2) As $N \to \infty$ the height of the rectangles tends to zero and hence we can cut along any horizontal line; (3) we can temporarily store pieces in the empty space above the graph. Hence, by local operations Alice can cut and paste the area on the graph in any way she likes as long as the cuts consist of horizontal cuts and vertical cuts between the columns.

Before considering the effect of moving area around, consider what happens if Alice makes a measurement of the value of n by measuring the ancilla R on the basis $|n\rangle_R$. Since we initially have $i = j$ the effect will be to obtain the state (written in unnormalised form)

$$\sum_j |j\rangle_A |j\rangle_B \tag{9}$$

This is an m-state where the value of m is equal to the width of the graph at that value of n. In fact what we have done is pick out a row of rectangles corresponding to a particular value of n. The rectangles in this row all have different values of i (necessarily since they have different horizontal locations on the graph) and, in this case, they also all are of different colour. This second property is essential if the resulting state is to be an m-state. If it were the case that some of the rectangles had the same colour then these terms would not be bi-orthogonal (orthogonal at both Alice's and at Bob's end) and hence the state would not be an m-state.

Now consider moving the area around on the graph. After moving the area around we will make a measurement of n. This will yield a state between Alice and Bob. We want such states to be m-states. Hence we must impose the constraint that after the re-colouring the same colour is not repeated as we go across any given row (for any given n). We can impose a second constraint - that the new graph also has a step structure in which the steps go down as we go towards the right (i.e. that there are no gaps in the columns and that the height of the columns decreases with i). Hence, the new graph will look like figure 1 except but the columns will have different heights λ'_i (where $\lambda'_i \geq \lambda'_{i+1}$) and the columns can be multicoloured rather than only of one colour. This second constraint (imposing step structure) is not necessary, but it does not lead to any loss of generality and so we can impose it to simplify matters. To see that it does not lead to any loss of generality consider a graph which does not have this property. Nevertheless, a measurement onto n will yield a distribution of m-states (assuming the colours are different across rows). The rows corresponding to these m-states could be contracted by moving all pieces to the leftmost empty position in the row and then these rows could be stacked on top of one another, the widest going at the bottom. In this way we would arrive at a new graph with the same distribution of m-states but now respecting the second assumption. Thus, we want to find the most general way of recolouring the graph in which (a) the new graph has the property that the colours across any row are all different, and (b) the new graph has step structure.

We can see immediately that any net movement of area down is not possible since this will lead to the same colour being in the same row in at least two positions. This means that any successful strategy must involve a net movement of area upwards or not at all. What is surprising is that any new graph having having a net movement of area upwards (or at least not downwards) can be recoloured in accordance with (a) and (b). For a proof of this see [5]. Since the new graph must have step structure, a net movement of area upwards (or at least not downwards) is equivalent to a net movement of area to the left. Hence, the new graph must satisfy the algebraic constraints that

$$\sum_{i=p}^{I} \lambda'_i \leq \sum_{i=p}^{I} \lambda_i \tag{10}$$

for all $r = 1$ to I with equality holding when $r = 1$. These constraints follow because the area to the right of any vertical line drawn between two columns must have less area to its left afterwards

After performing swap transformations corresponding to an appropriate recolouring, Alice will make a measurement of n and communicate the result to Bob. For this value of n Alice and Bob can read from the final graph what state they have. This state will always be an m-state. If they want to put their state into the standard form in (2) then Bob can perform an appropriate unitary transformation. For example, they might end up with the 2-state $|1\rangle_A|3\rangle_B + |2\rangle_A|4\rangle_B$. Bob can perform the unitary transformation defined by $|3\rangle_B \leftrightarrow |1\rangle_B$ and $|4\rangle_B \leftrightarrow |2\rangle_B$ which will put the state in the standard form $|1\rangle_A|1\rangle_B + |2\rangle_A|2\rangle_B$.

The probability of obtaining some particular value of m for the m-state is equal to the area of rectangle obtained by extending back horizontal lines from the top and bottom of the step having width m, i.e. $p_m = (\lambda'_m - \lambda'_{m+1})m$. Hence, the expected entanglement using the $\log_2 m$ measure is

$$E = \sum_i (\lambda'_i - \lambda'_{i+1}) i \log_2 i \tag{11}$$

In fact any net movement of area to the left will decrease E (this is not obvious but follows from the convex nature of the log function) and hence the maximum expected entanglement is

$$E_{\max} = \sum_i (\lambda_i - \lambda_{i+1}) i \log_2 i \tag{12}$$

corresponding to the original graph. For a full proof that this is the maximum see [5] (see also [6]).

The main advantage of the approach described above is that the transformations required can be pictured as graph transformations. This method can be applied to other problems concerning the manipulation of the entanglement of a single copy of a pure state. For example, it can be used to perform the transformation, by LOCC, of one entangled state to another in accordance with Nielsen's theorem. This method does not, however, generalise easily multipartite pure states. The principle reason for this is that there is no Schmidt decomposition for such states in general. This leaves open multipartite entangled states as an area where exciting developments are likely to happen in the near future.

References

[1] H-K Lo and S. Popescu, Concentrating entanglement by local actions - beyond mean values, Report No. quant-ph/9707038
[2] M. A. Nielsen, Phys. Rev. Lett. **83**, 436 (1999), report No. quant-ph/9811053
[3] G. Vidal, Phys. Rev. Lett. **83**, 1046 (1999), report No. quant-ph/9902033
[4] D. Jonathan and M. Plenio, Phys. Rev. Lett. **81**, 2839 (1999), report No. quant-ph/9905071.
[5] L. Hardy, Phys. Rev. A **60**, 1912 (1999), report no. quant-ph/9903001.
[6] D. Jonathan and M. Plenio, Minimal conditions for local pure-state entanglement manipulation, Report No. quant-ph/9903054

(Lucien Hardy) CENTRE FOR QUANTUM COMPUTATION, THE CLARENDON LABORATORY, PARKS ROAD, OXFORD OX1 3PU, UK

Geometric Algebra in Quantum Information Processing

Timothy F. Havel and Chris J. L. Doran

ABSTRACT. This paper develops a geometric model for coupled two-state quantum systems (qubits) using geometric (aka Clifford) algebra. It begins by showing how Euclidean spinors can be interpreted as entities in the geometric algebra of a Euclidean vector space. This algebra is then lifted to Minkowski space-time and its associated geometric algebra, and the insights this provides into how density operators and entanglement behave under Lorentz transformations are discussed. The direct sum of multiple copies of space-time induces a tensor product structure on the associated algebra, in which a suitable quotient is isomorphic to the matrix algebra conventionally used in multi-qubit quantum mechanics. Finally, the utility of geometric algebra in understanding both unitary and nonunitary quantum operations is demonstrated on several examples of interest in quantum information processing.

1. Introduction

Quantum mechanics attaches physical significance to representations of the rotation group which differ substantially from those studied in classical geometry. Much of the mystery surrounding it is due to this fact. The enormous interest recently generated by proposals to build a *quantum computer* [**Llo95, EJ96, Ste98, WC98, Bro99, BD00**] has focussed attention on the simplest possible quantum system: a two-state system or *qubit*. Our understanding of qubits is based on two distinct geometric models of their states and transformations:

- A complex projective line under the action of $SU(2)$.
- A Euclidean unit 2-sphere under the action of $SO(3)$.

The first is used almost exclusively in fundamental quantum physics, while the second ("classical") model is used extensively in certain applications, e.g. nuclear magnetic resonance (NMR) spectroscopy [**HSTC00**]. In particular, in quantum computing a qubit represents a binary 0 or 1 as its state corresponds to one of a pair of conjugate (orthogonal) projective points $|0\rangle$ or $|1\rangle$. These in turn correspond

---

2000 *Mathematics Subject Classification.* Primary 81R99; Secondary 15A66, 94B27.

TH was support during this work by ARO grants DAAD19-01-1-0519 & DAAD19-01-1-0678, DAAG55-97-1-0342 from DARPA/MTO, MDA972-01-1-0003 from DARPA/DSO, and by the Cambridge-MIT Institute, Ltd.

CD was supported by an EPSRC Advanced Fellowship.

©2002 American Mathematical Society

to a pair of diametrical points on the unit sphere, which determine the alignment of the qubit with or against the corresponding axis of quantization.

Formally, these two models are related by stereographic projection of the Riemann (unit) sphere onto the Argand plane, the points of which are the ratios of the homogeneous coordinates of points on the projective line (see e.g. [**Alt86, FH81**]). While this elegant construction describes the mapping between the two representations in a geometric fashion, it does not unite them in a single mathematical structure. This paper provides an informal account of how this is done by geometric (aka Clifford) algebra; in addition, it describes an extension of this formalism to multi-qubit systems, and shows that it provides a concise and lucid means of describing the operations of quantum information processing [**SCH98, HSTC00**]. Significantly, this extension is most naturally derived via the geometric algebra of Minkowski space-time [**DLG93**], which has also been shown to be an efficient formalism within which to study a very wide range of problems in classical [**Hes99, Jan89**], relativistic [**Hes66, Bay96**] and fundamental quantum [**DLG**$^+$**96**] physics. More complete and rigorous accounts may be found in these references, and in [**HCST00, Hav01**].

2. Euclidean Geometry and Spinors

Let $\mathfrak{R}_3$ be a three-dimensional Euclidean vector space whose inner product is denoted by $(\boldsymbol{a}, \boldsymbol{b}) \mapsto \boldsymbol{a} \cdot \boldsymbol{b}$. The Clifford or *geometric algebra* of $\mathfrak{R}_3$ is the associative algebra generated by $\mathfrak{R}_3$ over $\mathbb{R}$ such that $\boldsymbol{a}^2 = \|\boldsymbol{a}\|^2 \equiv \boldsymbol{a} \cdot \boldsymbol{a}$ for all $\boldsymbol{a} \in \mathfrak{R}_3$. This algebra will be referred to in the following as the *Pauli algebra*, and denoted by $\mathcal{G}_3$. The interesting thing about this algebra is its geometric interpretation, which will now be described.

To begin, note that every nonzero vector $\boldsymbol{a} \in \mathfrak{R}_3$ has an inverse $\boldsymbol{a}/\|\boldsymbol{a}\|^2$. In addition, a simple application of the law of cosines shows that the inner product of $\boldsymbol{a}$ with any other vector $\boldsymbol{b} \in \mathfrak{R}_3$ is given by the symmetric part of the geometric product:

$$\begin{aligned}(2.1)\quad \tfrac{1}{2}(\boldsymbol{a}\boldsymbol{b} + \boldsymbol{b}\boldsymbol{a}) &= \tfrac{1}{2}((\boldsymbol{a}+\boldsymbol{b})^2 - \boldsymbol{a}^2 - \boldsymbol{b}^2) \\ &= \tfrac{1}{2}(\|\boldsymbol{a}+\boldsymbol{b}\|^2 - \|\boldsymbol{a}\|^2 - \|\boldsymbol{b}\|^2) = \boldsymbol{a} \cdot \boldsymbol{b}\end{aligned}$$

The antisymmetric part, by way of contrast, is called the *outer product*, and denoted by $(\boldsymbol{a}, \boldsymbol{b}) \mapsto \boldsymbol{a} \wedge \boldsymbol{b} \equiv (\boldsymbol{a}\boldsymbol{b} - \boldsymbol{b}\boldsymbol{a})/2$. Since the outer product of two vectors $\boldsymbol{a} \wedge \boldsymbol{b}$ is invariant under inversion in the origin, it cannot itself be a vector. The space $\langle \boldsymbol{a} \wedge \boldsymbol{b} \mid \boldsymbol{a}, \boldsymbol{b} \in \mathfrak{R}_3 \rangle$ therefore carries an inequivalent representation of the orthogonal group $\mathsf{O}(3)$, and its elements are accordingly called *bivectors*. These are most naturally interpreted as oriented plane segments, instead of oriented line segments like vectors in $\mathfrak{R}_3$. If we similarly define the outer product of a vector with a bivector and require it to be associative, i.e.

$$(2.2)\quad \boldsymbol{a} \wedge (\boldsymbol{b} \wedge \boldsymbol{c}) \equiv \tfrac{1}{2}(\boldsymbol{a}\boldsymbol{b}\boldsymbol{c} - \boldsymbol{c}\boldsymbol{b}\boldsymbol{a}) \equiv (\boldsymbol{a} \wedge \boldsymbol{b}) \wedge \boldsymbol{c}$$

($\boldsymbol{a}, \boldsymbol{b}, \boldsymbol{c} \in \mathfrak{R}_3$), then it can be shown via straightforward though somewhat lengthy calculations that this product of three vectors is totally antisymmetric, meaning that the outer product generates the well-known exterior algebra $\bigwedge \mathfrak{R}_3$ (cf. [**Hes99, Rie58**]). The outer product of three vectors is called a *trivector*, and (since it changes sign under inversion) is most appropriately interpreted as an oriented space segment or volume element.

The general properties of inner and outer products in the geometric algebras of arbitrary metric vector spaces can be worked out along these lines in a coordinate-free fashion [**HS84**]. The remainder of this section will focus on how the Pauli algebra is used to describe the quantum mechanics of qubits. In this application it is more common to work with a fixed orthonormal basis $\sigma_x, \sigma_y, \sigma_z \in \Re_3$. Quantum mechanics, however, views these basis vectors in a very different way from that taken above, in that they are regarded as *operators* on a two-dimensional Hilbert space $\mathfrak{H} \approx \mathbb{C}^2$ (see e.g. [**Sak94**]). These operators, in turn, are usually identified with the Pauli matrices

$$(2.3) \quad \sigma_x \leftrightarrow \underline{\sigma}_x \equiv \begin{bmatrix} 0 & 1 \\ 1 & 0 \end{bmatrix}, \; \sigma_y \leftrightarrow \underline{\sigma}_y \equiv \begin{bmatrix} 0 & -\imath \\ \imath & 0 \end{bmatrix}, \; \sigma_z \leftrightarrow \underline{\sigma}_z \equiv \begin{bmatrix} 1 & 0 \\ 0 & -1 \end{bmatrix},$$

where $\imath$ is an imaginary unit ($\imath^2 = -1$), the underline signifies that the associated symbol is a matrix, and throughout this paper the symbol "$\leftrightarrow$" should be read as "is represented by" or "is equivalent to". The connection between the two viewpoints lies the fact that these matrices satisfy the defining relations of the *abstract* Pauli algebra $\mathcal{G}_3$, namely

$$(2.4) \quad (\sigma_\mu)^2 = \mathbf{1} \leftrightarrow \underline{1}, \quad \sigma_\mu \sigma_\nu = -\sigma_\nu \sigma_\mu \quad (\mu, \nu \in \{x, y, z\}, \mu \neq \nu),$$

and hence constitute a faithful matrix representation of it. This shows, in particular, that $\mathcal{G}_3$ is 8-dimensional as a *real* linear space.[1]

In most physical situations, these operators (times $\hbar$) represent measurements of the intrinsic angular momentum of the qubits, and hence are regarded as generators of rotations in the Lie algebra so(3) over $\mathbb{C}$ satisfying the commutator relation

$$(2.5) \quad \tfrac{1}{2}[\sigma_x, \sigma_y] = \imath \sigma_z,$$

and its cyclic permutations. In terms of geometric algebra, the left-hand side is just the outer product of the vectors. The right-hand side is somewhat harder to interpret, because the Pauli algebra is defined over the real numbers. The trick is to observe that, in terms of the matrix representation, $\underline{\sigma}_x \underline{\sigma}_y \underline{\sigma}_z = \imath \underline{1}$. Thus by interpreting the abstract imaginary $\imath$ as the trivector $\iota \equiv \sigma_x \sigma_y \sigma_z = \sigma_x \wedge \sigma_y \wedge \sigma_z$, the angular momentum relations become a triviality:

$$(2.6) \quad \sigma_x \wedge \sigma_y = \sigma_x \sigma_y = \sigma_x \sigma_y (\sigma_z)^2 = \iota \sigma_z$$

More generally, the vector cross product is related to the outer product by

$$(2.7) \quad \boldsymbol{a} \times \boldsymbol{b} = -\tfrac{\iota}{2}(\boldsymbol{ab} - \boldsymbol{ba}) = -\iota(\boldsymbol{a} \wedge \boldsymbol{b}),$$

from which it may be seen that multiplication by the unit trivector ι maps vectors to orthogonal bivectors and vice versa. Since they span a one-dimensional space but change sign under inversion in the origin, trivectors can also be regarded as *pseudoscalars*. Perhaps the most important thing which geometric algebra contributes to physics are geometric interpretations for the imaginary units which it otherwise uses blindly.

---

[1] In the quantum mechanics literature, the notation $\underline{\boldsymbol{a}} \cdot \underline{\vec{\sigma}}$ is often used for $\sum_\mu a_\mu \underline{\sigma}_\mu$. Because $\boldsymbol{a}$ is geometrically just a vector in $\Re_3$ (*not* a matrix for it in a basis-dependent representation of $\mathcal{G}_3$), this is an abuse of the dot-notation for the Euclidean inner product, which is otherwise perhaps the most consistently used notation in all of science. This abuse of notation will not be perpetrated in this paper.

If we denote the induced bivector basis by

$$\text{(2.8)} \qquad \boldsymbol{I} \equiv \boldsymbol{\sigma}_y \wedge \boldsymbol{\sigma}_z, \quad \boldsymbol{J} \equiv \boldsymbol{\sigma}_z \wedge \boldsymbol{\sigma}_x, \quad \boldsymbol{K} \equiv \boldsymbol{\sigma}_x \wedge \boldsymbol{\sigma}_y,$$

it is readily seen that these basis bivectors likewise square to -1. On multiplying the angular momentum generating relations through by $-1 = \iota^2$, we obtain

$$\text{(2.9)} \qquad \boldsymbol{JI} = \boldsymbol{K}, \quad \boldsymbol{IK} = \boldsymbol{J}, \quad \boldsymbol{KJ} = \boldsymbol{I}, \quad \text{and} \quad \boldsymbol{KJI} = -1.$$

This shows that these basis bivectors generate a subalgebra of $\mathcal{G}_3$ isomorphic to Hamilton's quaternions [**Alt86, Alt89**], which is also known as the *even subalgebra* $\mathcal{G}_3^+$ (since it is generated by the products of even numbers of vectors). It is well-known that the quaternions' multiplicative group is $\mathbb{R}^* \oplus \mathsf{SU}(2)$, which implies that the even subalgebra should be closely related to rotations. This relationship will now be worked out explicitly.

Consider the result of conjugating a vector $\boldsymbol{x}$ by a vector $\boldsymbol{a}$, i.e.

$$\text{(2.10)} \qquad \boldsymbol{axa}^{-1} = \boldsymbol{a}(\boldsymbol{x}_{\|} + \boldsymbol{x}_{\perp})\boldsymbol{a}^{-1} = \boldsymbol{aa}^{-1}\boldsymbol{x}_{\|} - \boldsymbol{aa}^{-1}\boldsymbol{x}_{\perp},$$

where we have split $\boldsymbol{x} = (\boldsymbol{x} \cdot \boldsymbol{a}^{-1} + \boldsymbol{x} \wedge \boldsymbol{a}^{-1})\boldsymbol{a} \equiv \boldsymbol{x}_{\|} + \boldsymbol{x}_{\perp}$ into its parts parallel and perpendicular to $\boldsymbol{a}$. This shows that $-\boldsymbol{axa}^{-1}$ is the *reflection* of $\boldsymbol{x}$ in the plane orthogonal to $\boldsymbol{a}$. ¿From the well-known fact that the composition of two reflections is a rotation by *twice* the lessor angle between their planes and about these planes' line of intersection, it follows that conjugating a vector by an element of the even subalgebra just rotates it accordingly:

$$\text{(2.11)} \qquad (\boldsymbol{ba})\,\boldsymbol{x}\,(\boldsymbol{ba})^{-1} = \boldsymbol{ba}\,\boldsymbol{x}\,\boldsymbol{a}^{-1}\boldsymbol{b}^{-1} = \frac{\boldsymbol{ba}\,\boldsymbol{x}\,\boldsymbol{ab}}{\|\boldsymbol{a}\|^2 \|\boldsymbol{b}\|^2}$$

Let $\boldsymbol{u} \equiv \boldsymbol{a}/\|\boldsymbol{a}\|$, $\boldsymbol{v} \equiv \boldsymbol{b}/\|\boldsymbol{b}\|$ and $\boldsymbol{R} \equiv \boldsymbol{vu}$ be the corresponding *unit* quaternion. Then $\boldsymbol{R} = \cos(\theta/2) - \iota\boldsymbol{r}\sin(\theta/2)$ where $\cos(\theta/2) = \boldsymbol{u} \cdot \boldsymbol{v}$ and $\iota\boldsymbol{r} = \boldsymbol{u} \wedge \boldsymbol{v}/\|\boldsymbol{u} \wedge \boldsymbol{v}\|$. Moreover, the inverse $(\boldsymbol{vu})^{-1}$ is now simply the *reverse* $\boldsymbol{uv} \equiv (\boldsymbol{vu})^\dagger$, which in turn corresponds to the *conjugate* quaternion $\boldsymbol{R}^\dagger \equiv \cos(\theta/2) + \iota\boldsymbol{r}\sin(\theta/2)$. This reversal operation on $\mathcal{G}_3^+$ extends to a well-defined anti-automorphism of $\mathcal{G}_3$, which corresponds to Hermitian conjugation in its representation by Pauli matrices. On splitting $\boldsymbol{x}$ into its parts parallel $\boldsymbol{x}_{\|}$ and perpendicular $\boldsymbol{x}_{\perp}$ to $\boldsymbol{r}$ as above, the rotation may now be written as $\boldsymbol{R}\,\boldsymbol{x}\,\boldsymbol{R}^\dagger = \boldsymbol{R}(\boldsymbol{x}_{\|} + \boldsymbol{x}_{\perp})\boldsymbol{R}^\dagger = \boldsymbol{x}_{\|} + \boldsymbol{x}_{\perp}(\boldsymbol{R}^\dagger)^2$

$$\text{(2.12)} \qquad \begin{aligned} &= \boldsymbol{x}_{\|} + \boldsymbol{x}_{\perp}(\cos^2(\theta/2) - \sin^2(\theta/2) + 2\iota\boldsymbol{r}\cos(\theta/2)\sin(\theta/2)) \\ &= \boldsymbol{x}_{\|} + \boldsymbol{x}_{\perp}(\cos(\theta) + \iota\boldsymbol{r}\sin(\theta)), \end{aligned}$$

and so may be viewed as multiplication of $\boldsymbol{x}_{\perp}$ by the "complex number" $\cos(\theta) + \iota\boldsymbol{r}\sin(\theta)$ in the Argand plane defined by the bivector $\iota\boldsymbol{r}$.

By collecting even and odd powers in its Taylor series, it may be seen that any unit quaternion can be written as the exponential of a bivector orthogonal to the axis of rotation $\boldsymbol{r}$:

$$\text{(2.13)} \qquad e^{-\iota\boldsymbol{r}\theta/2} = 1 - \iota\boldsymbol{r}\tfrac{\theta}{2} - \tfrac{1}{2}\left(\tfrac{\theta}{2}\right)^2 + \cdots = \cos(\theta/2) - \iota\boldsymbol{r}\sin(\theta/2)$$

This is formally analogous to a complex exponential, and is also in accord with our previous observation that the space of bivectors is isomorphic to the Lie algebra $\mathsf{su}(2) \approx \mathsf{so}(3)$ under the commutator product. The pair $[\cos(\theta/2); \sin(\theta/2)\boldsymbol{r}]$ are known as *Euler-Rodrigues* parameters for the rotation group $\mathsf{SO}(3)$; since

$[-\cos(\theta/2); -\sin(\theta/2)\boldsymbol{r}]$ determines the same rotation, this parametrization is two-to-one. A one-to-one parametrization is obtained from the *outer exponential*, i.e.

$$(2.14) \qquad \wedge^{-\iota \boldsymbol{r}\tau} = 1 - \iota\boldsymbol{r}\tau - \tfrac{1}{2}\boldsymbol{r}\wedge \boldsymbol{r}\tau^2 + \cdots = 1 - \iota\boldsymbol{r}\tau \quad (\text{since } \boldsymbol{r}\wedge\boldsymbol{r} = 0) \ .$$

The squared norm of this outer exponential is $1 + \tau^2$, so that the normalized outer exponential equals the usual exponential if we set $\tau = \tan(\theta/2)$. Because $\boldsymbol{t} \equiv \tan(\theta/2)\boldsymbol{r}$ is the four-dimensional stereographic projection of $[\cos(\theta); \sin(\theta)\boldsymbol{r}]$ from $\theta = \pi$, it has been called the *stereographic* parameter for $\mathsf{SO}(3)$. Note, however, that this parametrization does not include rotations by π.

Another two-to-one parametrization of rotations is given by the *Cayley-Klein* parameters $[\psi_1; \psi_2] \in \mathbb{C}^2$, where

$$(2.15) \qquad \begin{aligned} \psi_1 &\equiv \cos(\theta/2) - \iota \sin(\theta/2)(\boldsymbol{r}\cdot\boldsymbol{\sigma}_{\mathsf{z}}) \ , \\ \psi_2 &\equiv \sin(\theta/2)(\boldsymbol{r}\cdot\boldsymbol{\sigma}_{\mathsf{x}}) + \iota \sin(\theta/2)(\boldsymbol{r}\cdot\boldsymbol{\sigma}_{\mathsf{y}}) \ . \end{aligned}$$

The corresponding $\mathsf{SU}(2)$ matrix is simply

$$(2.16) \qquad \underline{\boldsymbol{\Psi}} = \begin{bmatrix} \psi_1 & -\psi_2^* \\ \psi_2 & \psi_1^* \end{bmatrix} \ .$$

It follows that the complex column vector $|\psi\rangle \equiv [\psi_1; \psi_2]$ itself transforms under left-multiplication with matrices in $\mathsf{SU}(2)$, which is commonly described in quantum mechanics by calling it a *spinor*. In particular, the spinors $|0\rangle \equiv [1; 0]$ and $|1\rangle \equiv [0; 1]$ are those commonly used in quantum computing to store binary information. Since the Cayley-Klein parameters uniquely determine the $\mathsf{SU}(2)$ matrix, however, we can just as well regard spinors as entities *in* $\mathsf{SU}(2)$, e.g. $|0\rangle \leftrightarrow \boldsymbol{1}$ and $|1\rangle \leftrightarrow -\iota\boldsymbol{\sigma}_{\mathsf{y}}$. The usual action of $\mathsf{SU}(2)$ on spinors then becomes the left-regular action of $\mathsf{SU}(2)$ on itself.

The representation of $\mathsf{SU}(2)$ used above depends upon the choice of coordinate system: Changing to a different the coordinate system gives a different (though equivalent) representation. Recalling that $\mathsf{SU}(2)$ is isomorphic to the multiplicative group of unit elements (quaternions) in the even subalgebra $\mathcal{G}_3^+$, a coordinate-free or *geometric* interpretation of spinors is obtained by regarding them as elements of $\mathcal{G}_3^+$ itself. This interpretation of spinors as entities in ordinary Euclidean geometry was first pointed out by Hestenes over thirty years ago [**Hes66**], but physicists persist in putting operators and operands into separate spaces, and in working with a matrix representation instead of directly with the geometric entities themselves. The perceived nonintuitive nature of quantum mechanics is due in large part to the resulting confusion over the geometric meaning of the objects with which it deals, which is spelled out explicitly in geometric algebra.

As another example, consider how the *density operator* of an "ensemble" of qubits can be interpreted in geometric algebra. This operator $\boldsymbol{\rho}$ is usually defined via a matrix representation as $\underline{\boldsymbol{\rho}} \equiv \overline{|\psi\rangle\langle\psi|}$, where the overline denotes the average over the ensemble. As first observed by von Neumann, this matrix contains all the information needed to compute the ensemble average expectation values of the qubit observables, since

$$(2.17) \qquad \overline{\langle\psi|\underline{\boldsymbol{\sigma}}_\mu|\psi\rangle} = \overline{\mathrm{tr}(\underline{\boldsymbol{\sigma}}_\mu|\psi\rangle\langle\psi|)} = \mathrm{tr}(\underline{\boldsymbol{\sigma}}_\mu\overline{|\psi\rangle\langle\psi|}) = \mathrm{tr}(\underline{\boldsymbol{\sigma}}_\mu\underline{\boldsymbol{\rho}})$$

($\mu \in \{\mathsf{x}, \mathsf{y}, \mathsf{z}\}$). To translate this into geometric algebra, we set the second column of $\underline{\boldsymbol{\Psi}}$ to zero by right-multiplying it by the idempotent matrix $\underline{\boldsymbol{E}}_+ \equiv (\boldsymbol{1}+\underline{\boldsymbol{\sigma}}_{\mathsf{z}})/2$, i.e.

(2.18) $$\underline{\boldsymbol{\Psi}}\,\underline{\boldsymbol{E}}_+ \;\equiv\; \begin{bmatrix} \psi_1 & -\psi_2^* \\ \psi_2 & \psi_1^* \end{bmatrix} \begin{bmatrix} 1 & 0 \\ 0 & 0 \end{bmatrix} \;=\; \begin{bmatrix} \psi_1 & 0 \\ \psi_2 & 0 \end{bmatrix}.$$

This corresponds to projecting $\boldsymbol{\Psi} \in \mathcal{G}_3^+$ onto a *left-ideal* in $\mathcal{G}_3$, and allows the dyadic product $|\psi\rangle\langle\psi|$ in Eq. (2.17) to be written as:

(2.19) $$|\psi\rangle\langle\psi| \;=\; \begin{bmatrix} \psi_1 & 0 \\ \psi_2 & 0 \end{bmatrix} \begin{bmatrix} \psi_1^* & \psi_2^* \\ 0 & 0 \end{bmatrix} \;\equiv\; (\underline{\boldsymbol{\Psi}}\,\underline{\boldsymbol{E}}_+)(\underline{\boldsymbol{\Psi}}\,\underline{\boldsymbol{E}}_+)^\dagger$$

Thus the interpretation of the density operator in geometric algebra is

(2.20) $$\rho \;=\; \overline{(\boldsymbol{\Psi}\boldsymbol{E}_+)(\boldsymbol{\Psi}\boldsymbol{E}_+)^\dagger} \;=\; \overline{\boldsymbol{\Psi}\boldsymbol{E}_+\boldsymbol{\Psi}^\dagger} \;=\; \tfrac{1}{2}\left(1 + \overline{\boldsymbol{\Psi}\boldsymbol{\sigma}_z\boldsymbol{\Psi}^\dagger}\right)$$

(cf. [**SLD99**]). The vector part $\boldsymbol{p} \equiv \langle\rho\rangle_1 = \overline{\boldsymbol{\Psi}\boldsymbol{\sigma}_z\boldsymbol{\Psi}^\dagger}$ is called the *polarization vector* (in optics, its components are known as the *Stokes parameters* [**BBDH93**], while in NMR it is known as the *Bloch vector* after the pioneer of NMR who rediscovered it [**Blo46**]). Its length is $\|\boldsymbol{p}\| \leq 1$ with equality if and only if all members of the ensemble are in the same state $\boldsymbol{\Psi}$. In this case the ensemble is said to be in a *pure* state, and the density operator is itself an idempotent $(1+\boldsymbol{p})/2$, where $\boldsymbol{p} \equiv \boldsymbol{\Psi}\boldsymbol{\sigma}_z\boldsymbol{\Psi}^\dagger$. For an ensemble in a general *mixed* state, the length of the ensemble-average polarization vector measures the degree of alignment among the (unit length) polarization vectors of the individual members of the ensemble, and is called the *polarization* of the ensemble.

In many physical situations there is a natural reference direction; for example, in NMR computing the qubits are spin $1/2$ atomic nuclei whose intrinsic magnetic dipoles have been polarized by the application of a strong magnetic field [**HCST00**]. ¿From a geometric perspective, however, the density operator is just the sum of a scalar and a vector, which for a pure state is related to the corresponding "spinor" by rotation of a fixed reference vector (conventionally taken to be $\boldsymbol{\sigma}_z$ as above) by $\boldsymbol{\Psi}$. Since the trace in the standard matrix representation is simply twice the scalar part $\langle \_ \rangle_0$ of the corresponding expression in geometric algebra, the ensemble-average expectation value

(2.21) $$\tfrac{1}{2}\mathrm{tr}(\underline{\boldsymbol{\sigma}}_\mu\,\underline{\rho}) \;\leftrightarrow\; \langle \boldsymbol{\sigma}_\mu\rho\rangle_0 \;=\; \langle \boldsymbol{\sigma}_\mu\boldsymbol{\Psi}\boldsymbol{\sigma}_z\boldsymbol{\Psi}^\dagger\rangle_0 \;=\; \boldsymbol{\sigma}_\mu\cdot(\boldsymbol{\Psi}\boldsymbol{\sigma}_z\boldsymbol{\Psi}^\dagger) \;=\; \boldsymbol{\sigma}_\mu\cdot\boldsymbol{p}$$

is just the component of the polarization vector along the μ-th axis. Unlike the *strong* measurements usually considered in quantum texts, where measurement of $\boldsymbol{\sigma}_\mu$ yields one of the random outcomes ± 1 with probabilities $(1\pm\boldsymbol{\sigma}_\mu\cdot\boldsymbol{p})/2$ and leaves the system in the corresponding state $\boldsymbol{p} = \pm\boldsymbol{\sigma}_\mu$, *weak* measurements of ensemble-average expectation values can be made with only negligible perturbations to the ensemble as a whole [**Per93**]. This is in fact how quantum mechanical systems are usually manifest at the macroscopic level!

To see how all this relates to conventional wisdom, observe that the polarization vector of a pure state may be written in terms of the Cayley-Klein parameters as

(2.22) $$\boldsymbol{p} \;=\; 2\Re(\psi_1^*\psi_2)\boldsymbol{\sigma}_\mathsf{x} + 2\Im(\psi_1^*\psi_2)\boldsymbol{\sigma}_\mathsf{y} + (|\psi_1|^2 - |\psi_2|^2)\boldsymbol{\sigma}_\mathsf{z}.$$

Its stereographic projection from $-\boldsymbol{\sigma}_\mathsf{z}$ onto the $\boldsymbol{\sigma}_\mathsf{x}\boldsymbol{\sigma}_\mathsf{y}$ plane is therefore

(2.23) $$\frac{2\Re(\psi_1^*\psi_2)\boldsymbol{\sigma}_\mathsf{x} + 2\Im(\psi_1^*\psi_2)\boldsymbol{\sigma}_\mathsf{y}}{1 + |\psi_1|^2 - |\psi_2|^2}.$$

Multiplying by $\boldsymbol{\sigma}_x$ and simplifying the denominator using $|\psi_1|^2 + |\psi_2|^2 = 1$ yields

$$\tag{2.24} \frac{\Re(\psi_1^* \psi_2) + \Im(\psi_1^* \psi_2) \boldsymbol{K}}{|\psi_1|^2}$$

where $\boldsymbol{K} = \boldsymbol{\sigma}_x \boldsymbol{\sigma}_y$ is a square-root of -1. This is the same as the ratio ψ_2/ψ_1 save for the use of $\imath$ instead of $\boldsymbol{K}$ as the imaginary unit, which explains formally why $\mathsf{SO}(3)$ acts on the polarization vector in the same way that $\mathsf{SU}(2)$ acts on the ratio of the Cayley-Klein parameters [**Alt86, FH81**].

3. Space-Time Geometry and Multiparticle Spinors

The above interpretations apply only to single qubits (or to ensembles consisting of noninteracting and identical qubits). Extending them to systems of interacting and distinguishable qubits may be done in a physically significant fashion by considering the geometric algebra of *space-time* (or Minkowski space) $\mathfrak{R}_{1,3}$. This algebra, known as the *Dirac* algebra and denoted by $\mathcal{G}_{1,3}$, may be defined by the generating relations among an orthonormal basis analogous to Eq. (2.4):

$$\tag{3.1} \begin{aligned} \gamma_t^2 &= 1, \quad \gamma_\mu^2 = -1 \quad (\mu \in \{x,y,z\}), \\ \gamma_\mu \gamma_\nu &= -\gamma_\nu \gamma_\mu \quad (\mu, \nu \in \{t,x,y,z\}, \mu \neq \nu) \end{aligned}$$

The corresponding geometric algebra separates into five inequivalent representations under the action of the full Lorentz group $\mathsf{O}(1,3)$, i.e.

$$\tag{3.2} \begin{aligned} \langle 1 \rangle &\quad \text{(scalars, 1-dimensional)} \\ \langle \gamma_\mu \rangle &\quad \text{(vectors, 4-dimensional)} \\ \langle \gamma_\mu \gamma_\nu \rangle &\quad \text{(bivectors, 6-dimensional)} \\ \langle \gamma_\mu \gamma_\nu \gamma_\eta \rangle &\quad \text{(trivectors, 4-dimensional)} \\ \langle \gamma_t \gamma_x \gamma_y \gamma_z \rangle &\quad \text{(pseudo-scalars, 1-dimensional)}, \end{aligned}$$

where $\mu, \nu, \eta \in \{t,x,y,z\}$ with $\mu \neq \nu \neq \eta \neq \mu$, for a total dimension of 16.

The important point for our purposes is that the even subalgebra of the Dirac algebra $\mathcal{G}_{1,3}^+$ is isomorphic to the Pauli algebra $\mathcal{G}_3$ [**Hes66**]. This isomorphism may be constructed by choosing bases $\boldsymbol{\gamma}_\mu \in \mathcal{G}_{1,3}$ and $\boldsymbol{\sigma}_\mu \in \mathcal{G}_3$, and defining an invertible linear mapping by

$$\tag{3.3} \boldsymbol{\sigma}_\mu \in \mathcal{G}_3 \;\leftrightarrow\; \boldsymbol{\gamma}_\mu \boldsymbol{\gamma}_t \in \mathcal{G}_{1,3}^+ \quad (\mu \subset \{x,y,z\}).$$

These so-called *relative spatial vectors* $\boldsymbol{\gamma}_\mu \boldsymbol{\gamma}_t$ satisfy the relations in Eq. (2.4), since

$$\tag{3.4} \begin{aligned} (\boldsymbol{\sigma}_\mu)^2 &\leftrightarrow (\boldsymbol{\gamma}_\mu \boldsymbol{\gamma}_t)^2 = \boldsymbol{\gamma}_\mu (\boldsymbol{\gamma}_t)^2 \boldsymbol{\gamma}_\mu = -(\boldsymbol{\gamma}_\mu)^2 - 1 \\ \boldsymbol{\sigma}_\mu \boldsymbol{\sigma}_\nu &\leftrightarrow (\boldsymbol{\gamma}_\mu \boldsymbol{\gamma}_t)(\boldsymbol{\gamma}_\nu \boldsymbol{\gamma}_t) = \boldsymbol{\gamma}_\nu (\boldsymbol{\gamma}_\mu \boldsymbol{\gamma}_t) \boldsymbol{\gamma}_t = -(\boldsymbol{\gamma}_\nu \boldsymbol{\gamma}_t)(\boldsymbol{\gamma}_\mu \boldsymbol{\gamma}_t) \\ &\leftrightarrow -\boldsymbol{\sigma}_\nu \boldsymbol{\sigma}_\mu \quad (\mu, \nu \in \{x,y,z\}, \mu \neq \nu), \end{aligned}$$

and hence generate an algebra isomorphic to $\mathcal{G}_3$. As bivectors in $\mathcal{G}_{1,3}$, however, they also generate $\mathcal{G}_{1,3}^+$, since

$$\tag{3.5} \boldsymbol{\gamma}_\mu \boldsymbol{\gamma}_\nu = \boldsymbol{\gamma}_\mu (\boldsymbol{\gamma}_t)^2 \boldsymbol{\gamma}_\nu = -(\boldsymbol{\gamma}_\mu \boldsymbol{\gamma}_t)(\boldsymbol{\gamma}_\nu \boldsymbol{\gamma}_t) \;\leftrightarrow\; -\boldsymbol{\sigma}_\mu \boldsymbol{\sigma}_\nu$$

($\mu, \nu \in \{x,y,z\}, \mu \neq \nu$), and similarly

$$\tag{3.6} \iota \equiv \boldsymbol{\gamma}_t \boldsymbol{\gamma}_x \boldsymbol{\gamma}_y \boldsymbol{\gamma}_z = (\boldsymbol{\gamma}_x \boldsymbol{\gamma}_t)(\boldsymbol{\gamma}_y \boldsymbol{\gamma}_t)(\boldsymbol{\gamma}_z \boldsymbol{\gamma}_t) \;\leftrightarrow\; \boldsymbol{\sigma}_x \boldsymbol{\sigma}_y \boldsymbol{\sigma}_z.$$

Thus $\boldsymbol{\gamma}_\mu \boldsymbol{\gamma}_t \leftrightarrow \boldsymbol{\sigma}_\mu$ ($\mu \in \{x,y,z\}$) induces an algebra isomorphism as claimed, and when the bases are understood we may identify $\boldsymbol{\sigma}_\mu \equiv \boldsymbol{\gamma}_\mu \boldsymbol{\gamma}_t$.

The choice of time-like vector $\boldsymbol{\gamma}_\mathrm{t} \in \mathfrak{R}_{1,3}$ in fact determines an inertial frame up to spatial rotation, in which the *time t* and *place s* of an event **e** in that frame are given by

$$(3.7) \qquad t + \boldsymbol{s} = \mathbf{e} \cdot \boldsymbol{\gamma}_\mathrm{t} + \mathbf{e} \wedge \boldsymbol{\gamma}_\mathrm{t} = \mathbf{e}\boldsymbol{\gamma}_\mathrm{t}$$

(note that upright case is used for the space-time vector $\mathbf{e} \in \mathfrak{R}_{1,3}$). Thus the invariant interval between events separated by the space-time vector **e** is $\mathbf{e}^2 = \mathbf{e}\boldsymbol{\gamma}_\mathrm{t}^2\mathbf{e} = (t+\boldsymbol{s})(t-\boldsymbol{s}) = t^2 - \boldsymbol{s}^2$ as usual, while the relative velocity between events whose space-time velocities are $\boldsymbol{\gamma}_\mathrm{t}$ and $\mathbf{v} \equiv \partial\mathbf{e}/\partial\tau$ is

$$(3.8) \qquad \boldsymbol{v} \;=\; \frac{\partial \boldsymbol{s}}{\partial t} \;=\; \frac{\partial \boldsymbol{s}}{\partial \tau}\frac{\partial \tau}{\partial t} \;\equiv\; \left(\frac{\partial \mathbf{e}}{\partial \tau}\wedge \boldsymbol{\gamma}_\mathrm{t}\right)\left(\frac{\partial \mathbf{e}}{\partial \tau}\cdot \boldsymbol{\gamma}_\mathrm{t}\right)^{-1} \;=\; \frac{\mathbf{v}\wedge \boldsymbol{\gamma}_\mathrm{t}}{\mathbf{v}\cdot \boldsymbol{\gamma}_\mathrm{t}} \,,$$

so that $\boldsymbol{v}\cdot\boldsymbol{\gamma}_\mathrm{t}$ lies on an affine hyperplane in space-time.

A great deal of physics can be done in a manifestly Lorentz covariant fashion using the Dirac algebra. For example, the electromagnetic field at a given point in space-time corresponds to an arbitrary bivector $\mathbf{F} \in \bigwedge_2 \mathfrak{R}_{1,3}$, called the *Faraday bivector*, and the covariant form of the Lorentz force equation is

$$(3.9) \qquad m\dot{\mathbf{v}} \;=\; q\mathbf{F}\cdot\mathbf{v} \,,$$

where m is the rest mass, q the charge and **v** the space-time velocity. (This is another example of the general rule that, in geometric algebra, the generators of motion are bivectors [**DHSvA93**].) The usual frame-dependent form is recovered by splitting the quantities in this equation by $\boldsymbol{\gamma}_\mathrm{t}$ as above [**Jan89**]; in particular, the Faraday bivector splits into an electric and a magnetic field as $\mathbf{F} \equiv \boldsymbol{E} + \iota\boldsymbol{B}$, where

$$(3.10) \qquad \boldsymbol{E} \;=\; (\mathbf{F}\cdot\boldsymbol{\gamma}_\mathrm{t})\boldsymbol{\gamma}_\mathrm{t} \quad\text{and}\quad \iota\boldsymbol{B} \;=\; (\mathbf{F}\wedge\boldsymbol{\gamma}_\mathrm{t})\boldsymbol{\gamma}_\mathrm{t} \,.$$

The space-time reverse will be denoted by a tilde, e.g. in the present case $\tilde{\mathbf{F}} = -\mathbf{F}$. This is related to the spatial (or Pauli) reverse by $\mathbf{F}^\dagger = \boldsymbol{E} - \iota\boldsymbol{B} = \boldsymbol{\gamma}_\mathrm{t}\tilde{\mathbf{F}}\boldsymbol{\gamma}_\mathrm{t}$. Both operations agree on the Pauli-even subalgebra, but the spatial reverse is *not* Lorentz coveriant since it depends on a particular $\boldsymbol{\gamma}_\mathrm{t}$.

Returning to our previous discussion of the density operator, we observe that the space-time form of the density operator of a single qubit polarized along z can be written as

$$(3.11) \qquad \rho \;=\; \tfrac{1}{2}(1 + \alpha\boldsymbol{\sigma}_z) \;=\; \tfrac{1}{2}(\boldsymbol{\gamma}_\mathrm{t} + \alpha\boldsymbol{\gamma}_z)\boldsymbol{\gamma}_\mathrm{t} \;\equiv\; \varrho\boldsymbol{\gamma}_\mathrm{t} \,,$$

where $-1 \le \alpha \le 1$ is the polarization and $\boldsymbol{\gamma}_\mathrm{t}$ determines the local inertial frame. It follows that the Lorentz covariant form of the density operator is a time-like vector $\varrho \in \mathfrak{R}_{1,3}$. Under a Lorentz boost $\boldsymbol{L} = \exp(-\lambda\boldsymbol{\sigma}_z/2) \in \mathsf{SO}(1,3)$ along $\boldsymbol{\sigma}_z$, therefore, the relativistic density operator ϱ transforms to $\varrho' \equiv \tfrac{1}{2}\boldsymbol{L}(\boldsymbol{\gamma}_\mathrm{t} + \alpha\boldsymbol{\gamma}_z)\tilde{\boldsymbol{L}} =$

$$(3.12) \qquad \tfrac{1}{2}\left(\cosh(\lambda)\boldsymbol{\gamma}_\mathrm{t} - \sinh(\lambda)\boldsymbol{\gamma}_z + \alpha(\cosh(\lambda)\boldsymbol{\gamma}_z - \sinh(\lambda)\boldsymbol{\gamma}_\mathrm{t})\right) \,.$$

This implies that in the unaccelerated frame (with renormalization by $\varrho'\cdot\boldsymbol{\gamma}_\mathrm{t}$),

$$(3.13) \qquad \begin{aligned} \rho' &= \frac{\varrho'\boldsymbol{\gamma}_\mathrm{t}}{\varrho'\cdot\boldsymbol{\gamma}_\mathrm{t}} = \frac{\varrho'\cdot\boldsymbol{\gamma}_\mathrm{t} + \varrho'\wedge\boldsymbol{\gamma}_\mathrm{t}}{\varrho'\cdot\boldsymbol{\gamma}_\mathrm{t}} \\ &= \frac{1}{2}\left(1 + \frac{\alpha\cosh(\lambda) - \sinh(\lambda)}{\cosh(\lambda) - \alpha\sinh(\lambda)}\boldsymbol{\sigma}_z\right) \,. \end{aligned}$$

It follows that the polarization itself transforms as

(3.14) $$\alpha' = \frac{\alpha \cosh(\lambda) - \sinh(\lambda)}{\cosh(\lambda) - \alpha \sinh(\lambda)} \ .$$

If we assume the qubit is at equilibrium with a heat bath, statistical mechanics tells us that $\alpha = \tanh(-\beta\epsilon/2)$ where $\beta = 1/(k_\mathrm{B} T)$ is the inverse temperature and $\epsilon \in \mathbb{R}$ is the energy difference between the $|0\rangle$ and $|1\rangle$ states [**Tol38**]. Then the addition formulae for cosh and sinh give

(3.15) $$\alpha' = \tanh(-\beta\epsilon/2 - \lambda) \ ,$$

so the apparent equilibrium polarization depends on velocity. These results are not to be found in the classic treatise on relativistic thermodynamics [**Tol34**].

We will now construct a Lorentz covariant *multiparticle theory* of qubit systems in the simplest possible way, by taking a direct sum of copies of space-time (regarded as a vector space, rather than an algebra), one for each of the N qubits, i.e.

(3.16) $$\bigoplus_{q=1}^{N} \langle \gamma_\mathsf{t}^q, \gamma_\mathsf{x}^q, \gamma_\mathsf{y}^q, \gamma_\mathsf{z}^q \rangle \ ,$$

and considering the associated geometric algebra $\mathcal{G}_{N,3N}$. Then the even subalgebras of different particle spaces $p \neq q$ *commute*, since (in any given bases)

(3.17) $$\sigma_\mu^p \sigma_\nu^q = \gamma_\mu^p (\gamma_\nu^q \gamma_\mathsf{t}^q) \gamma_\mathsf{t}^p = \sigma_\nu^q \sigma_\mu^p$$

for all $\mu, \nu \in \{\mathsf{x},\mathsf{y},\mathsf{z}\}$, so that the algebra generated by the even subalgebras is isomorphic to a tensor product of these algebras, written as

(3.18) $$(\mathcal{G}_{1,3}^+)^{\otimes N} \approx \mathcal{G}_3^{\otimes N} \equiv (\mathcal{G}_3)^{\otimes N} \ .$$

This construction of the tensor product was first used by Clifford as a means of studying the tensor products of quaternion algebras [**Cli78**]; van der Waerden has in fact called it a Clifford algebra of the second kind [**vdW85**]. As a means of justifying the tensor product of nonrelativistic quantum mechanics in terms of the underlying geometry of space-time, however, it is a much more recent development [**DLG93**].

A key feature of quantum mechanics, which is needed for quantum computers to be able to solve problems more efficiently than their classical counterparts, is an exponential growth in the dimension of the Hilbert space of a multi-qubit system with the number of particles involved. The complex dimension of the Hilbert space $(\mathfrak{H})^{\otimes N}$ of an N-qubit system is in fact 2^N, and the space of operators (linear transformations) on $(\mathfrak{H})^{\otimes N}$ therefore has *real* dimension 2^{2N+1}. The above construction yields a space of "operators" $\mathcal{G}_3^{\otimes N}$ whose real dimension also grows exponentially, but as 2^{3N}. The extra degrees of freedom are due to the presence of a different unit pseudo-scalar ι^q in every particle space. They can easily be removed by multiplying through by an idempotent element called the *correlator*:

(3.19) $$\boldsymbol{C} \equiv \tfrac{1}{2}(1 - \iota^1 \iota^2) \tfrac{1}{2}(1 - \iota^1 \iota^3) \cdots \tfrac{1}{2}(1 - \iota^1 \iota^N)$$

This commutes with everything in $\mathcal{G}_3^{\otimes N}$ and satisfies $\iota^p \iota^q \boldsymbol{C} = -\boldsymbol{C}$ for $1 \leq p, q \leq N$, so that multiplication by it homomorphically maps $\mathcal{G}_3^{\otimes N}$ onto an ideal $\mathcal{G}_3^{\otimes N}/\boldsymbol{C}$ wherein all the unit pseudo-scalars have been identified,[2] and which therefore has

---

[2]The notation $\mathcal{G}_3^{\otimes N}/\boldsymbol{C}$ is justified by the fact that the two-sided principle ideal $\mathcal{G}_3^{\otimes N}(\boldsymbol{C})$ generated by $\boldsymbol{C}$ is isomorphic to the quotient algebra $\mathcal{G}_3^{\otimes N}/\ker(\boldsymbol{C})$, where $\ker(\boldsymbol{C}) \equiv \{\boldsymbol{g} \in \mathcal{G}_3^{\otimes N} \mid \boldsymbol{g}\boldsymbol{C} = 0\}$.

the correct dimension over $\mathbb{R}$. As a subalgebra, this ideal is in fact isomorphic to the algebra of $2^N \times 2^N$ complex matrices, and hence capable of describing all the states and transformations of (ensembles of) N qubit systems. In the following, we shall generally omit $\boldsymbol{C}$ from our expressions altogether, and use a single unit imaginary ι as in conventional quantum mechanics.

On the "even" subalgebra $(\mathcal{G}_3^+)^{\otimes N}$, multiplication by the correlator turns out to be an algebra automorphism; this algebra can thus be written as

$$(3.20) \qquad (\mathcal{G}_3^+)^{\otimes N} \approx (\mathcal{G}_3^+)^{\otimes N}/\boldsymbol{C} \approx (\mathcal{G}_3^{\otimes N}/\boldsymbol{C})^+ \approx \mathcal{SU}(2)^{\otimes N},$$

where the "+" refers throughout to the subalgebra generated by expressions which are invariant under inversion in the origin, and $\mathcal{SU}(2)^{\otimes N}$ to the algebra generated over $\mathbb{R}$ by the Kronecker products of matrices in the group $\mathsf{SU}(2)$. This subalgebra has real dimension 2^{2N}, but is mapped onto a left-ideal of dimension 2^{N+1} by right-multiplication with another idempotent which is given by the tensor product of those considered earlier, namely

$$(3.21) \qquad \boldsymbol{E}_+ \equiv \boldsymbol{E}_+^1 \boldsymbol{E}_+^2 \cdots \boldsymbol{E}_+^N,$$

where $\boldsymbol{E}_\pm^q \equiv (1 \pm \boldsymbol{\sigma}_z^q)/2$ for $q = 1, \ldots, N$. Henceforth, the term "even subalgebra" will refer to $(\mathcal{G}_3^+)^{\otimes N}$ (suitably correlated) unless otherwise stated.

In terms of the usual matrix representation, right-multiplication of an element of the even subalgebra $\boldsymbol{\Psi}$ by $\boldsymbol{E}_+$ likewise sets all but the first column to zero, so that $\boldsymbol{\Psi}\boldsymbol{E}_+$ transforms like a "spinor" in $\mathfrak{H}^{\otimes N}$ under left-multiplication by single particle rotations $\boldsymbol{R}^q \in (\mathcal{G}_3^+)^{\otimes N}$. Unlike the single particle case, however, this one column does not uniquely determine an element of the even subalgebra $(\mathcal{G}_3^+)^{\otimes N}/\boldsymbol{C}$. What has been proposed instead [**DLG93**] is to use the fact that $\boldsymbol{E}_+$ "absorbs" $\boldsymbol{\sigma}_z$'s to distribute copies of the latter across the correlator, converting it to what will here be called the *directional* correlator $\boldsymbol{D}$, i.e.

$$(3.22) \qquad \boldsymbol{\Psi} \boldsymbol{C} \boldsymbol{E}_+ = \boldsymbol{\Psi} \boldsymbol{C} \left((\boldsymbol{\sigma}_z^1)^{N-1} \boldsymbol{\sigma}_z^2 \cdots \boldsymbol{\sigma}_z^N \right) \boldsymbol{E}_+ = \boldsymbol{\Psi} \boldsymbol{D} \boldsymbol{E}_+,$$

where

$$(3.23) \qquad \boldsymbol{D} \equiv \tfrac{1}{2}(1 - \iota^1 \boldsymbol{\sigma}_z^1 \iota^2 \boldsymbol{\sigma}_z^2) \tfrac{1}{2}(1 - \iota^1 \boldsymbol{\sigma}_z^1 \iota^3 \boldsymbol{\sigma}_z^3) \cdots \tfrac{1}{2}(1 - \iota^1 \boldsymbol{\sigma}_z^1 \iota^N \boldsymbol{\sigma}_z^N).$$

It can be shown that right-multiplication by $\boldsymbol{D}$, unlike $\boldsymbol{C}$, reduces the dimensionality to 2^{N+1}, thereby permitting the objects in this *reduced* even subalgebra $(\mathcal{G}_3^+)^{\otimes N}/\boldsymbol{D}$ to be regarded as spinors, analogous to $\mathcal{G}_3^+$ for a single qubit. In the corresponding left-ideal, $\boldsymbol{K} \equiv \iota^1 \boldsymbol{\sigma}_z^1 \boldsymbol{D} \leftrightarrow \cdots \leftrightarrow \iota^N \boldsymbol{\sigma}_z^N \boldsymbol{D}$ serves as the unit imaginary, since $\boldsymbol{K}^2 = -\boldsymbol{D}$, but is required to always operate from the *right*. Henceforth, unless otherwise mentioned, we will regard spinors $(\mathcal{G}_3^+)^{\otimes N}/\boldsymbol{D} = ((\mathcal{G}_3^+)^{\otimes N}/\boldsymbol{C})/(\boldsymbol{C}\boldsymbol{D})$ as a left-ideal in the $\boldsymbol{C}$-correlated even subalgebra, drop both $\boldsymbol{C}$ and the superscripts on the ι's as above, and use $\boldsymbol{D}$ as a short-hand for $\boldsymbol{C}\boldsymbol{D} = \boldsymbol{D}\boldsymbol{C}$.

In the case of two qubits, for example, the identifications are induced by $\boldsymbol{D}$ are

$$(3.24) \qquad \begin{array}{cccccc}
|00\rangle & -1 \stackrel{D}{\longleftrightarrow} \iota\sigma_z^1 \iota\sigma_z^2 & \iota\sigma_z^1 \stackrel{D}{\longleftrightarrow} \iota\sigma_z^2 \\
|01\rangle & \iota\sigma_y^2 \stackrel{D}{\longleftrightarrow} \iota\sigma_z^1 \iota\sigma_x^2 & -\iota\sigma_x^2 \stackrel{D}{\longleftrightarrow} \iota\sigma_z^1 \iota\sigma_y^2 \\
|10\rangle & \iota\sigma_y^1 \stackrel{D}{\longleftrightarrow} \iota\sigma_x^1 \iota\sigma_z^2 & -\iota\sigma_x^1 \stackrel{D}{\longleftrightarrow} \iota\sigma_y^1 \iota\sigma_z^2 \\
|11\rangle & -\iota\sigma_y^1 \iota\sigma_y^2 \stackrel{D}{\longleftrightarrow} \iota\sigma_x^1 \iota\sigma_x^2 & \iota\sigma_x^1 \iota\sigma_y^2 \stackrel{D}{\longleftrightarrow} \iota\sigma_y^1 \iota\sigma_x^2
\end{array}$$

(where the two columns differ by operation with K). ¿From this it may be seen that any "spinor" in $(\mathcal{G}_3^+)^{\otimes 2}/D$ can be written as

$$\Psi = \Big((\alpha_0 + \beta_0 K) - \iota\sigma_y^2(\alpha_1 + \beta_1 K) - $$
(3.25)
$$\iota\sigma_y^1(\alpha_2 + \beta_2 K) + \iota\sigma_y^1\iota\sigma_y^2(\alpha_3 + \beta_3 K)\Big) D$$

(cf. [**SLD99**]). Alternatively, again using Eq. (3.24), a unit norm spinor may be factorized into a product of entities in the *correlated and reduced* even subalgebra, namely $\Psi = R^1 S^2 TPDC$, where

(3.26)
$$R^1 \equiv e^{-\iota\phi\sigma_z^1/2} e^{-\iota\theta\sigma_y^1/2}, \qquad T \equiv \cos(\varsigma/2) - \sin(\varsigma/2)\,\sigma_y^1\sigma_y^2 K,$$
$$S^2 \equiv e^{-\iota\varphi\sigma_z^2/2} e^{-\iota\vartheta\sigma_y^2/2}, \qquad P \equiv e^{-\tau K/2}.$$

Thus when $\varsigma = \pi$, the factor T becomes

(3.27)
$$-\sigma_y^1\sigma_y^2 K = (-\iota\sigma_y^1)(-\iota\sigma_y^2) K = e^{-(\pi/2)\iota\sigma_y^1} e^{-(\pi/2)\iota\sigma_y^2} K,$$

so that the arguments of the exponentials involving $\iota\sigma_y^1$ and $\iota\sigma_y^2$ in the first two factors are shifted by $\pi/2$ while the total phase is shifted by $\tau = \pi$. It follows that T rotates the first two factors in the planes defined by their conjugate spinors $[-r_2^*; r_1^*], [-s_2^*; s_1^*]$. Thus on right-multiplying by E_+ and expanding in the usual basis, we obtain (up to an overall phase)

(3.28)
$$\underline{\Psi} = \cos(\tfrac{\varsigma}{2})\, e^{\iota\tau/2} \begin{bmatrix} \cos(\tfrac{\theta}{2})e^{\iota\phi/2} \\ \sin(\tfrac{\theta}{2})e^{-\iota\phi/2} \end{bmatrix} \otimes \begin{bmatrix} \cos(\tfrac{\vartheta}{2})e^{\iota\varphi/2} \\ \sin(\tfrac{\vartheta}{2})e^{-\iota\varphi/2} \end{bmatrix} +$$
$$\sin(\tfrac{\varsigma}{2})\, e^{-\iota\tau/2} \begin{bmatrix} \sin(\tfrac{\theta}{2})e^{\iota\phi/2} \\ -\cos(\tfrac{\theta}{2})e^{-\iota\phi/2} \end{bmatrix} \otimes \begin{bmatrix} \sin(\tfrac{\vartheta}{2})e^{\iota\varphi/2} \\ -\cos(\tfrac{\vartheta}{2})e^{-\iota\varphi/2} \end{bmatrix}$$

This is known as the *Schmidt decomposition* [**EK95**]. It is useful in studying the *entanglement* of bipartite quantum systems, which (in conventional terms) means that $|\psi\rangle \in \mathfrak{H}^{\otimes 2}$ cannot be written as a product $|\psi^1\rangle \otimes |\psi^2\rangle \equiv |\psi^1\rangle|\psi^2\rangle \equiv |\psi^1\psi^2\rangle$ for any one-particle spinors $|\psi^1\rangle, |\psi^2\rangle \in \mathfrak{H}$. In fact it is just the singular value decomposition in disguise, since (for example) on arranging the entries of a two-qubit spinor $|\psi\rangle = [\psi_1; \cdots; \psi_4]$ in a 2×2 matrix, we can write

(3.29)
$$\underline{\Psi} \equiv \begin{bmatrix} \psi_1 & \psi_3 \\ \psi_2 & \psi_4 \end{bmatrix} = \underline{U}\,\underline{V}\,\underline{W}^\dagger = \underline{u}^1 v^{11} (\underline{w}^1)^\dagger + \underline{u}^2 v^{22} (\underline{w}^2)^\dagger,$$

where $\underline{V}$ is a 2×2 diagonal matrix containing the singular values $v^{11} \geq v^{22} \geq 0$ and $\underline{U}, \underline{W}$ are unitary matrices with columns $\underline{u}^k, \underline{w}^k$, respectively. Since the entries of the dyadic products $\underline{u}^1(\underline{w}^1)^\dagger, \underline{u}^2(\underline{w}^2)^\dagger$ are exactly the same as the Kronecker matrix products $\underline{u}^1 \otimes \underline{w}^1, \underline{u}^2 \otimes \underline{w}^2$, the equivalence with Eq. (3.28) follows with $v^{11} \equiv \cos(\varsigma/2)$, $v^{22} \equiv \sin(\varsigma/2)$, and the Kronecker products of the columns of $\underline{U}$ and $\underline{W}$ identified with conjugate pairs of single qubit spinors whose relative phases are given by $\exp(\pm\iota\tau/2)$.

Clearly a two-qubit spinor is unentangled if and only if $v^{11} = 1$, which is equivalent to $\varsigma = 0$ or $T = 1$. Thus T describes the entanglement of the qubits, and is accordingly called the *tangler*. The geometric algebra approach clearly provides deeper insight into the structure of entanglement than does one based on mechanical matrix algebra. In particular, the fact that $\tilde{\Psi}\Psi$ is even and reversion-symmetric in the Dirac as well as the Pauli algebra implies that it is the sum of a scalar and a four-vector in the two-particle Dirac algebra $\mathcal{G}_{2,6}$. Since Lorentz transformations of the spinors cancel, this entity is in fact a Lorentz invariant, and dividing out the total phase $\boldsymbol{P}$ as $\boldsymbol{P}(\tilde{\boldsymbol{\Psi}}\boldsymbol{\Psi})\tilde{\boldsymbol{P}}$ yields the *square* of the tangler directly. The availability of such powerful methods of manipulating entities in the multiparticle Dirac algebra promises to be useful in finding analogs of the Schmidt decomposition for three or more qubits.

4. Quantum Operations on Density Operators

Quantum computers operate on information stored in the states of quantum systems. The systems are usually assumed to be arrays of distinguishable qubits (two-state subsystems), whose basis states $|0\rangle$ and $|1\rangle$ correspond to the binary digits 0 and 1, respectively, while the operations are usually taken to be unitary. General unitary transformations of the qubits are built up from simpler ones that affect only a few qubits at a time, which are called *quantum logic gates*. The representation of these gates in suitable products of Clifford algebras has been described in Refs. [**SCH98, Vla01**]. The goal here will be to show how gates act upon spinors in the even subalgebra, and how they can be extended to a wider class of nonunitary quantum operations on density operators.

Given the isomorphism between the algebra of $2^N \times 2^N$ matrices over $\mathbb{C}$ and $\mathcal{G}_3^{\otimes N}/C$ relative to a choice of basis in each particle space, it is straightforward to interpret matrices in the former as geometric entities in the latter. A matrix $\underline{U} \in \mathsf{U}(2^N)$, however, does not generally correspond to an entity $\boldsymbol{U}$ in the *even* subalgebra $(\mathcal{G}_3^+)^{\otimes N}/C$, so that $\boldsymbol{U}\boldsymbol{\Psi} \notin (\mathcal{G}_3^+)^{\otimes N}/D$ for a general spinor $\boldsymbol{\Psi} \in (\mathcal{G}_3^+)^{\otimes N}/D$. Nevertheless, letting $\boldsymbol{E}_- \equiv \prod_q \boldsymbol{E}_-^q$ be the idempotent "opposite" to $\boldsymbol{E}_+$, and noting that this satisfies $\boldsymbol{E}_+\boldsymbol{E}_- = 0$, the product of $\boldsymbol{U}\boldsymbol{\Psi}$ with $\boldsymbol{E}_+$ may be written as

$$(4.1) \qquad \boldsymbol{U}\boldsymbol{\Psi}\boldsymbol{E}_+ \;=\; (\boldsymbol{U}\boldsymbol{\Psi}\boldsymbol{E}_+ + \hat{\boldsymbol{U}}\boldsymbol{\Psi}\boldsymbol{E}_-)\boldsymbol{E}_+ \;=\; 2\langle \boldsymbol{U}\boldsymbol{\Psi}\boldsymbol{E}_+\rangle_+\,\boldsymbol{E}_+\,,$$

where the "hat" on $\hat{\boldsymbol{U}}$ denotes its image under inversion in the origin (so that $\hat{\boldsymbol{E}}_+ = \boldsymbol{E}_-$), and hence $\langle\_\rangle_+$ is a projection onto the even subalgebra. Because $(\mathcal{G}_3^+)^{\otimes N}/C$ and $\mathsf{U}(2^N)$ are both (2^{2N})-dimensional, nothing is lost in this projection! Thus we can drop the right-factor of $\boldsymbol{E}_+$ as usual, and define the action of $\boldsymbol{U}$ on $\boldsymbol{\Psi} \in (\mathcal{G}_3^+)^{\otimes N}/D$ as

$$(4.2) \qquad \boldsymbol{U} \circ \boldsymbol{\Psi} \;\equiv\; 2\langle \boldsymbol{U}\boldsymbol{\Psi}\boldsymbol{E}_+\rangle_+\,.$$

More generally, the usual action of the Pauli matrices on spinors corresponds to the following action of the basis vectors on the reduced even subalgebra [**DLG93**]:

$$(4.3) \qquad \boldsymbol{\sigma}_\mu \circ \boldsymbol{\Psi} \;\equiv\; \boldsymbol{\sigma}_\mu \boldsymbol{\Psi} \boldsymbol{\sigma}_z\,, \qquad \iota \circ \boldsymbol{\Psi} \;\equiv\; \iota \boldsymbol{\Psi} \boldsymbol{\sigma}_z$$

The simplest logic gate is the NOT of a single qubit, which operates on the computational basis as follows:

$$(4.4) \qquad \underline{\boldsymbol{N}}|0\rangle \;=\; |1\rangle \;\leftrightarrow\; -\iota\boldsymbol{\sigma}_y\,, \qquad \underline{\boldsymbol{N}}|1\rangle \;=\; |0\rangle \;\leftrightarrow\; 1$$

Thus it might appear reasonable to represent the NOT by $\boldsymbol{N} \equiv \iota\boldsymbol{\sigma}_y \in \mathsf{SU}(2)$, but when $\iota\boldsymbol{\sigma}_y$ is applied a superposition $(1 - \iota\boldsymbol{\sigma}_y)/\sqrt{2} \leftrightarrow (|0\rangle + |1\rangle)/\sqrt{2}$, we get $(1 + \iota\boldsymbol{\sigma}_y)/\sqrt{2} \leftrightarrow (|0\rangle - |1\rangle)/\sqrt{2}$ instead of $(|0\rangle + |1\rangle)/\sqrt{2}$ again. For a single qubit this difference is just an overall rotation by π about $\boldsymbol{\sigma}_z$, but a second qubit can be affected by this phase difference between the first qubit's states. Therefore the correct representation of the NOT gate in SU(2) is actually $\mathbf{N} \equiv \pm \iota \boldsymbol{\sigma}_x$, which preserves this superposition up to an irrelevant overall phase shift: $\iota \boldsymbol{\sigma}_x(|0\rangle + |1\rangle)/\sqrt{2} \leftrightarrow$

$$
\begin{aligned}
(\iota \boldsymbol{\sigma}_x) \circ (1 - \iota \boldsymbol{\sigma}_y)/\sqrt{2} &= \iota \circ \boldsymbol{\sigma}_x \circ (1 - \iota \boldsymbol{\sigma}_y)/\sqrt{2} \\
&= \iota \circ (\boldsymbol{\sigma}_x(1 - \iota \boldsymbol{\sigma}_y)\boldsymbol{\sigma}_z/\sqrt{2}) = \iota \circ (-\iota \boldsymbol{\sigma}_y + 1)/\sqrt{2} \\
&= (-\iota \boldsymbol{\sigma}_y + 1)\iota \boldsymbol{\sigma}_z/\sqrt{2} \leftrightarrow -\iota(|0\rangle + |1\rangle)/\sqrt{2}
\end{aligned}
\tag{4.5}
$$

More interesting logical operations on the qubits must be able to transform the state of one *conditional* on that of another. The usual way in which this is done is via the c-NOT or *controlled-NOT* gate. As a matrix in SU(4), this is represented in the computational basis by

$$
\underline{\mathbf{N}}^{2|1} \equiv \sqrt{\iota} \begin{bmatrix} 1 & 0 & 0 & 0 \\ 0 & 1 & 0 & 0 \\ 0 & 0 & 0 & 1 \\ 0 & 0 & 1 & 0 \end{bmatrix},
\tag{4.6}
$$

which makes it clear that this operation NOT's the second qubit whenever the first is 1. The corresponding operator in geometric algebra is

$$
\mathbf{N}^{2|1} \equiv (1 + \iota \boldsymbol{\sigma}_z^1)/\sqrt{2} \left(\mathbf{E}_+^1 + \mathbf{E}_-^1 \iota \boldsymbol{\sigma}_x^2 \right).
\tag{4.7}
$$

This may also be written in exponential form as

$$
e^{\iota \pi \boldsymbol{\sigma}_z^1/4} e^{\iota \pi \mathbf{E}_-^1 \boldsymbol{\sigma}_x^2/2} = e^{-\iota \pi (\mathbf{E}_-^1 (1 - \boldsymbol{\sigma}_x^2)/2 - 1/4)}.
\tag{4.8}
$$

Physical implementations of this operation by e.g. NMR typically expand the exponential into a product of relatively simple commuting factors which can be performed sequentially [**HSTC00**].

Note that since $(1 - \boldsymbol{\sigma}_x^2)/2$ is also an idempotent, $\mathbf{N}^{2|1}$ differs from $\mathbf{N}^{1|2}$ by a swap of the x and z axes for both qubits. This self-inverse operation, called the *Hadamard transform* $\mathbf{H}$, is simply a rotation by π about the $(\boldsymbol{\sigma}_x + \boldsymbol{\sigma}_z)/\sqrt{2}$ axis. Sandwiching $\mathbf{N}^{2|1}$ by Hadamards $\mathbf{H}^2 = \iota(\boldsymbol{\sigma}_x^2 + \boldsymbol{\sigma}_z^2)/\sqrt{2}$ to just the second qubit gives

$$
\mathbf{H}^2 \mathbf{N}^{2|1} \mathbf{H}^2 = e^{-\iota \pi (\mathbf{E}_-^1 \mathbf{H}^2 (1 - \boldsymbol{\sigma}_x^2) \mathbf{H}^2/2 - 1/4)} = e^{-\iota \pi (\mathbf{E}_-^1 \mathbf{E}_-^2 - 1/4)},
\tag{4.9}
$$

so the c-NOT can also be viewed as a rotated phase shift of the state $|11\rangle$ by π. The Hadamard gate has the important feature of transforming basis states into superpositions thereof; indeed, as an element of the even subalgebra, it actually represents the spinor of a uniform superposition directly:

$$
\begin{aligned}
\underline{\mathbf{H}}|0\rangle &= \iota(|0\rangle + |1\rangle)/\sqrt{2} \leftrightarrow (\iota \boldsymbol{\sigma}_z + \iota \boldsymbol{\sigma}_x)/\sqrt{2}\, D \\
\underline{\mathbf{H}}|1\rangle &= \iota(|0\rangle - |1\rangle)/\sqrt{2} \leftrightarrow (\iota \boldsymbol{\sigma}_z - \iota \boldsymbol{\sigma}_x)/\sqrt{2}\, D
\end{aligned}
\tag{4.10}
$$

Thus, by using the relations (4.3), we can show that applying a Hadamard to one of two qubits in the state $|11\rangle$ followed by a c-NOT gate to the other yields the entangled singlet state: $\underline{\mathbf{N}}^{2|1} \underline{\mathbf{H}}^1 |11\rangle \leftrightarrow$

$$
\begin{aligned}
&\tfrac{1}{2} \left((1 + \iota) \mathbf{E}_+^1 + (1 - \iota) \mathbf{E}_-^1 \iota \boldsymbol{\sigma}_x^2 \right) \circ \left((\iota \boldsymbol{\sigma}_z^1 - \iota \boldsymbol{\sigma}_x^1)(-\iota \boldsymbol{\sigma}_y^2) \right) \\
&= \tfrac{1}{2} \left(\left((1+\iota) \circ (\iota \boldsymbol{\sigma}_z^1)\right)(-\iota \boldsymbol{\sigma}_y^2) + \left((1-\iota) \circ (-\iota \boldsymbol{\sigma}_x^1)\right)\left((\iota \boldsymbol{\sigma}_x^2) \circ (-\iota \boldsymbol{\sigma}_y^2)\right) \right) \\
&= \tfrac{1}{2}(1 - \iota \boldsymbol{\sigma}_z^1)\iota \boldsymbol{\sigma}_y^2 - \tfrac{1}{2}(\iota \boldsymbol{\sigma}_x^1 - \iota \boldsymbol{\sigma}_y^1)\iota \boldsymbol{\sigma}_z^2 \xleftrightarrow{D} \tfrac{\iota}{2}(\boldsymbol{\sigma}_y^2 + \boldsymbol{\sigma}_x^2 - \boldsymbol{\sigma}_y^1 - \boldsymbol{\sigma}_x^1)
\end{aligned}
\tag{4.11}
$$

$\leftrightarrow \sqrt{-\imath}(|10\rangle - |01\rangle)/\sqrt{2} \equiv \sqrt{-\imath}|\psi_-\rangle$. "Quantum" gates like $\boldsymbol{H}$ are not, of course, found in conventional boolean logic, and are an essential component of all quantum algorithms that are more efficient than their classical counterparts [**EJ98, CEMM98**]. Indeed, the c-NOT gate together with general single qubit rotations are known to generate $\mathsf{SU}(2^N)$, and hence are *universal* for quantum logic [**BBC**[+]**95**].

It turns out that unitary transformations are not the most general sort of operation that can be applied to a quantum system. Most such *quantum operations*, however, produce a statistical outcome, and the ensemble of possible outcomes must be described by a density operator. The previous definition (Eq. (2.20)) of the density operator of an ensemble of identical and noninteracting qubits may be extended to an ensemble of multi-qubit systems as follows:

$$(4.12) \qquad \underline{\boldsymbol{\rho}} \equiv \overline{|\psi\rangle\langle\psi|} \;\;\leftrightarrow\;\; \boldsymbol{\rho} \equiv \overline{(\boldsymbol{\Psi}\boldsymbol{D})\boldsymbol{E}_+(\boldsymbol{\Psi}\boldsymbol{D})^\sim} \;=\; \overline{\boldsymbol{\Psi}\boldsymbol{E}_+\tilde{\boldsymbol{\Psi}}}\boldsymbol{C}$$

Suppressing the correlator $\boldsymbol{C}$ as usual, $\boldsymbol{\rho}$ may also be expressed in diagonal form as

$$(4.13) \qquad \boldsymbol{\rho} \;=\; \boldsymbol{R}\left(\sum_{k=0}^{2^N} \rho_k \boldsymbol{E}_{(k)}\right)\boldsymbol{R}^\dagger \;=\; \sum_{k=0}^{2^N} \rho_k \boldsymbol{r}_k \boldsymbol{r}_k^\dagger \;.$$

where $\boldsymbol{R} \in \mathcal{G}_3^{\otimes N}/\boldsymbol{C}$ corresponds to a unitary matrix $\underline{\boldsymbol{R}} \in \mathsf{U}(2^N)$ (in the usual $\boldsymbol{\sigma}_z$ coordinate system), and $0 \leq \rho_k \leq 1$ are the eigenvalues of $\underline{\boldsymbol{\rho}}$. The idempotents $\boldsymbol{E}_{(k)}$ are given by $\prod_q \boldsymbol{E}_{\epsilon_k^q}^q \leftrightarrow |\chi_k^1 \cdots \chi_k^N\rangle\langle\chi_k^1 \cdots \chi_k^N|$, where $\epsilon_k^q \equiv 1 - 2\chi_k^q$ with χ_k^q equal to the q-th bit in the binary expansion of $k \in \{0, \ldots, 2^N - 1\}$. It follows that $|\rho_k\rangle \leftrightarrow \boldsymbol{r}_k \equiv \boldsymbol{R}\boldsymbol{E}_{(k)}$ for $\rho_k > 0$ are the spinors of the (unique, if $\rho_k \neq \rho_\ell \;\forall k \neq l$) minimal ensemble that realizes $\boldsymbol{\rho}$, which therefore describes a pure state if and only if it has rank 1 as an operator.

Note that by Eq. (4.2), the density operator transforms under unitary operations as

$$(4.14) \qquad \boldsymbol{\rho} \;\mapsto\; \overline{(\boldsymbol{U} \circ \boldsymbol{\Psi})\boldsymbol{E}_+(\boldsymbol{U} \circ \boldsymbol{\Psi})^\sim} \;=\; \boldsymbol{U}\,\overline{\boldsymbol{\Psi}\boldsymbol{E}_+\tilde{\boldsymbol{\Psi}}}\,\boldsymbol{U}^\dagger \;=\; \boldsymbol{U}\boldsymbol{\rho}\boldsymbol{U}^\dagger \;.$$

Similarly, the ensemble-average expectation value of any observable $\boldsymbol{O} = \boldsymbol{O}^\dagger \in \mathcal{G}_3^{\otimes N}/\boldsymbol{C}$ is

$$(4.15) \qquad \overline{\langle\psi|\underline{\boldsymbol{O}}|\psi\rangle} \;\leftrightarrow\; 2^N \overline{\langle \boldsymbol{E}_+\tilde{\boldsymbol{\Psi}}\boldsymbol{O}\boldsymbol{\Psi}\boldsymbol{E}_+\rangle}_0 \;=\; 2^N \langle \boldsymbol{O}\,\overline{\boldsymbol{\Psi}\boldsymbol{E}_+\tilde{\boldsymbol{\Psi}}}\rangle_0 \;\equiv\; 2^N \langle \boldsymbol{O}\boldsymbol{\rho}\rangle_0 \;,$$

just as shown in Eq. (2.17) for single qubit ensembles. In contrast to the case of a single qubit, however, the geometric interpretation of these observables is not straightforward. While one can certainly express $\boldsymbol{\rho}$ as a finite ensemble average $\sum_k p_k \boldsymbol{\Psi}_k \boldsymbol{E}_+ \tilde{\boldsymbol{\Psi}}_k$ (where the $p_k > 0$ are probabilities with $\sum_k p_k = 1$), this decomposition is highly nonunique. The minimal ensemble obtained by diagonalization, on the other hand, will generally include entangled spinors $\boldsymbol{r}_k$, for which the expectation value $\langle \boldsymbol{O}\,\boldsymbol{r}_k \boldsymbol{r}_k^\dagger \rangle_0$ cannot be expressed as a product of inner products of the factors of $\boldsymbol{O} = \boldsymbol{O}^1 \cdots \boldsymbol{O}^N$ with the polarization vectors of the individual qubits (indeed, $\boldsymbol{O}$ itself need not be factorizable!).

The best one can do is to expand $\boldsymbol{\rho}$ in the *product operator* basis consisting of all 2^{2N} products of the basis vectors $\boldsymbol{\sigma}_\mu^q$, i.e.

$$(4.16) \qquad \boldsymbol{\rho} \;=\; \sum_{\mu^1, \ldots, \mu^N \in \{0,x,y,z\}} \rho_{\mu^1 \cdots \mu^N}\, \boldsymbol{\sigma}_{\mu^1}^1 \cdots \boldsymbol{\sigma}_{\mu^N}^N \;,$$

where $\rho_{\mu^1 \cdots \mu^N} \in \mathbb{R}$ and $\boldsymbol{\sigma}_0^q \equiv 1$ for notational convenience. The utility of this basis is most simply demonstrated via a concrete example, namely NMR spectroscopy. Here one is given a liquid sample consisting of identical molecules whose nuclear spins are chemically distinguishable, and hence constitutes an ensemble of multi-qubit systems (see [**CPH98, HCST00, HSTC00**] and references therein). The energy of interaction between the spins and an external magnetic field along z is given by an observable called the Zeeman Hamiltonian, $\boldsymbol{Z} \equiv (\omega^1 \boldsymbol{\sigma}_z^1 + \cdots + \omega^N \boldsymbol{\sigma}_z^N)/2$, where ω^q is the energy difference between the $|0\rangle$ and $|1\rangle$ states of the q-th spin in the field. In thermal equilibrium at room temperatures, the polarization of the spins relative to the strongest available fields is typically $\alpha \sim 10^{-6}$, and the density operator of the ensemble is essentially $\boldsymbol{\rho}_{eq} = 2^{-N}(1 + \alpha(\boldsymbol{\sigma}_z^1 + \cdots + \boldsymbol{\sigma}_z^N))$. Via a suitable pulse of radio-frequency radiation, this may be rotated to $\boldsymbol{\rho}'_{eq} \equiv 2^{-N}(1 + \alpha(\boldsymbol{\sigma}_x^1 + \cdots + \boldsymbol{\sigma}_x^N))$, which evolves under the interaction with the field as

$$(4.17) \quad e^{-\iota \boldsymbol{Z} t} \boldsymbol{\rho}'_{eq} e^{\iota \boldsymbol{Z} t} = 2^{-N} \left(1 + \alpha(\cos(\omega^1 t)\boldsymbol{\sigma}_x^1 - \sin(\omega^1 t)\boldsymbol{\sigma}_y^1 + \cdots \right.$$
$$\left. \cdots + \cos(\omega^N t)\boldsymbol{\sigma}_x^N - \sin(\omega^N t)\boldsymbol{\sigma}_y^N) \right) .$$

Thus on measuring the total magnetization M_x along the x axis, $\boldsymbol{O} \equiv \gamma(\boldsymbol{\sigma}_x^1 + \cdots + \boldsymbol{\sigma}_x^N)$ (where γ is the nuclear gyromagnetic ratio), one obtains the sum of the projections of the rotating magnetization vectors of the spins along the x-axis, i.e.

$$(4.18) \quad M_x(t) = \alpha\gamma(\cos(\omega^1 t) + \cdots + \cos(\omega^N t)) ,$$

whose Fourier transform reveals the contribution from each spin. The way in which the factors of product operators transform like vectors under rotations accounts in large part for the computational utility of the product operator basis. Of course, unless it is a natural part of the problem at hand (as in NMR), one is better off not chosing a basis at all!

A *stochastic quantum operation* is a linear transformation of the density operator that may be written in operator sum form as [**Kra83**]

$$(4.19) \quad \boldsymbol{\rho} \mapsto \Omega(\boldsymbol{\rho}) \equiv \sum_k \boldsymbol{Q}_k \boldsymbol{\rho} \boldsymbol{Q}_k^\dagger ,$$

where the *Kraus operators* $\boldsymbol{Q}_k \in \mathcal{G}_3^{\otimes N}/\boldsymbol{C}$ satisfy $\sum_k \boldsymbol{Q}_k^\dagger \boldsymbol{Q}_k = 1$. The term "stochastic" here[3] refers to the fact that such an operation preserves the scalar part of $\boldsymbol{\rho}$, since

$$(4.20) \quad \langle \Omega(\boldsymbol{\rho}) \rangle_0 = \sum_k \left\langle \boldsymbol{Q}_k \boldsymbol{\rho} \boldsymbol{Q}_k^\dagger \right\rangle_0 = \left\langle \boldsymbol{\rho} \sum_k \boldsymbol{Q}_k^\dagger \boldsymbol{Q}_k \right\rangle_0 = 2^{-N} .$$

It is also easily seen that such quantum operations are *positive*, in that they preserve the positive-definiteness of $\boldsymbol{\rho}$; in fact, these operations have a yet stronger property known as *complete positivity*, meaning that if the qubits to which Ω applies are embedded in a larger system, then applying Ω to just those qubits preserves the positive-definiteness of the larger system's density operator. That this is a nontrivial extension of positivity is shown by the two-qubit "partial transpose" operator Ω_T^1, which carries $\boldsymbol{\sigma}_y^1 \mapsto -\boldsymbol{\sigma}_y^1$ but leaves all the other operator factors unchanged; this is clearly positive on density operators not involving the second qubit, but acts on the density operator of the singlet state (Eq. (4.11)) as

$$(4.21) \quad \begin{aligned} \boldsymbol{\psi}_- &\equiv \tfrac{1}{4}(\boldsymbol{\sigma}_x^1 + \boldsymbol{\sigma}_y^1 - \boldsymbol{\sigma}_x^2 - \boldsymbol{\sigma}_y^2) \boldsymbol{E}_+^1 \boldsymbol{E}_+^2 (\boldsymbol{\sigma}_x^1 + \boldsymbol{\sigma}_y^1 - \boldsymbol{\sigma}_x^2 - \boldsymbol{\sigma}_y^2) \\ &= \tfrac{1}{4}(1 - \boldsymbol{\sigma}_x^1 \boldsymbol{\sigma}_x^2 - \boldsymbol{\sigma}_y^1 \boldsymbol{\sigma}_y^2 - \boldsymbol{\sigma}_z^1 \boldsymbol{\sigma}_z^2) \\ &\mapsto \Omega_T^1(\boldsymbol{\psi}_-) = \tfrac{1}{4}(1 - \boldsymbol{\sigma}_x^1 \boldsymbol{\sigma}_x^2 + \boldsymbol{\sigma}_y^1 \boldsymbol{\sigma}_y^2 - \boldsymbol{\sigma}_z^1 \boldsymbol{\sigma}_z^2) , \end{aligned}$$

---

[3] We prefer to avoid the more common but clumsy and matrix-bound term "trace-preserving".

which has eigenvalues $[1/2, 1/2, 1/2, -1/2]$.

A quantum operation Ω is called *unital* if it preserves the identity itself, i.e. $\Omega(1) = 1$, or equivalently, $\sum_k \boldsymbol{Q}_k \boldsymbol{Q}_k^\dagger = 1$. Perhaps the most important example of a stochastic unital operation is found in the *contraction*[4] by a single qubit $q \in \{1, \ldots, N\}$, which may be written in operator sum form as [**SCH98**]:

(4.22) $$2\langle \boldsymbol{\rho} \rangle^q \equiv \boldsymbol{E}_+^q \boldsymbol{\rho} \boldsymbol{E}_+^q + \boldsymbol{E}_-^q \boldsymbol{\rho} \boldsymbol{E}_-^q + \boldsymbol{\sigma}_\mathsf{x}^q (\boldsymbol{E}_+^q \boldsymbol{\rho} \boldsymbol{E}_+^q + \boldsymbol{E}_-^q \boldsymbol{\rho} \boldsymbol{E}_-^q) \boldsymbol{\sigma}_\mathsf{x}^q$$

This may also be expressed by dropping all terms in the product operator expansion of $\boldsymbol{\rho}$ depending on q, and multiplying the remaining terms by a factor of 2. Note that, while $\langle \_ \rangle^q$ is stochastic and unital, this factor means that the contraction itself is neither. The factor is nevertheless required if the result is to be interpreted as a density operator for the remaining qubits, since the contraction by the second qubit of the above singlet state is $\langle \boldsymbol{\psi}_- \rangle^2 = 1/4$ (not $1/2$).

This example also illustrates an important way in which general quantum operations are realized in practice, despite the fact that the universe as a whole evolves unitarily. As shown previously, the superposition state with spinor $\boldsymbol{\Psi}^1 = (1 - \iota \boldsymbol{\sigma}_\mathsf{y}^1)/\sqrt{2}$ is converted into the singlet state with density operator $\boldsymbol{\psi}_-$ by letting it interact with a second qubit so as to effect the c-NOT operation $\boldsymbol{N}^{2|1}$. The contraction then corresponds to "discarding" the second qubit (i.e. ensuring that it does not further interact with the first and hence can be ignored), which yields the density operator $1/2$ of the totally mixed state for the first qubit. Since the basis states $\boldsymbol{E}_\pm^1$ are unaffected by $\boldsymbol{N}^{2|1}$, the net quantum operation on the first qubit corresponds to what is known in quantum communications theory as the *phase damping channel*

(4.23) $$\boldsymbol{\rho} \mapsto (1-p)\boldsymbol{\rho} + p\,\boldsymbol{E}_+ \boldsymbol{\rho} \boldsymbol{E}_+ + p\,\boldsymbol{E}_- \boldsymbol{\rho} \boldsymbol{E}_-$$

with damping parameter $p = 1$. Phase damping is also known as T_2 relaxation in NMR, and as *decoherence* in quantum information processing; it is widely believed to be the dominant mechanism by which classical statistical mechanics arises from the underlying unitary dynamics [**GJK**[+]**96**].

To illustrate the utility of geometric algebra in the study of general quantum operations, an eigenvalue characterization of stochastic, unital, one-bit quantum operations Ω will now be derived. This characterization was originally given by Fujiwara & Algoet [**FA99**], although the derivation here parallels that more recently obtained using matrix methods King & Ruskai [**KR00**]. This derivation will regard the Kraus operators $\boldsymbol{Q}_k \in \mathcal{G}_3$ as "complex quaternions" $\boldsymbol{A}_k + \iota \boldsymbol{B}_k$ with $\boldsymbol{A}_k, \boldsymbol{B}_k \in \mathcal{G}_3^+$, and consider the action of an arbitrary operation Ω on the scalar and vector parts of $\boldsymbol{\rho} \equiv (1 + \boldsymbol{r})/2$ separately.

First, the action on 1 is

(4.24) $$\begin{aligned}\Omega(1) &= \sum_k (\boldsymbol{A}_k + \iota \boldsymbol{B}_k)(\boldsymbol{A}_k + \iota \boldsymbol{B}_k)^\dagger \\ &= \sum_k \left(\boldsymbol{A}_k \tilde{\boldsymbol{A}}_k + \boldsymbol{B}_k \tilde{\boldsymbol{B}}_k\right) + \iota \sum_k \left(\boldsymbol{B}_k \tilde{\boldsymbol{A}}_k - \boldsymbol{A}_k \tilde{\boldsymbol{B}}_k\right)\,.\end{aligned}$$

The first summation is symmetric with respect to spatial reversion and inversion, i.e. scalar, while the second (excluding the ι) is reversion antisymmetric but inversion symmetric, i.e. a bivector. Writing $\boldsymbol{A} \equiv \alpha + \iota \boldsymbol{a}$ and $\boldsymbol{B} \equiv \beta + \iota \boldsymbol{b}$, so that

---

[4]Otherwise known as the "partial trace".

$A_k \tilde{A}_k + B_k \tilde{B}_k = \alpha_k^2 + \|a_k\|^2 + \beta_k^2 + \|b_k\|^2$, we may further expand the second summation as follows:

$$
\begin{aligned}
&\sum_k \left[(\beta_k + \iota b_k)(\alpha_k - \iota a_k) - (\alpha_k + \iota a_k)(\beta_k - \iota b_k) \right] \\
&= \sum_k \left[(\alpha_k \beta_k + \iota(\alpha_k b_k - \beta_k a_k) + b_k a_k) - \right. \\
&\qquad\qquad \left. (\alpha_k \beta_k + \iota(\beta_k a_k - \alpha_k b_k) + a_k b_k) \right] \\
&= \sum_k \left[2\iota(\alpha_k b_k - \beta_k a_k) - 2 a_k \wedge b_k \right]
\end{aligned}
\tag{4.25}
$$

Thus Ω is unital if and only if

$$
\sum_k (\alpha_k b_k - \beta_k a_k) = \sum_k a_k \times b_k ,
\tag{4.26}
$$

(where "$\times$" denotes the cross product) and

$$
\sum_k \left(\alpha_k^2 + \|a_k\|^2 + \beta_k^2 + \|b_k\|^2 \right) = 1 .
\tag{4.27}
$$

Similarly, the action on $r \in \mathfrak{R}_3$ is

$$
\begin{aligned}
\Omega(r) &= \sum_k (A_k + \iota B_k) r (\tilde{A}_k - \iota \tilde{B}_k) \\
&= \sum_k \left(A_k r \tilde{A}_k + B_k r \tilde{B}_k \right) + \iota \sum_k \left(B_k r \tilde{A}_k - A_k r \tilde{B}_k \right) .
\end{aligned}
\tag{4.28}
$$

The first summation is over different dilation/rotations of r; the second summation (excluding the ι) is reversion and inversion antisymmetric, i.e. a trivector, and may be further expanded as above:

$$
\begin{aligned}
&\sum_k \left[(\beta_k + \iota b_k) r (\alpha_k - \iota a_k) - (\alpha_k + \iota a_k) r (\beta_k - \iota b_k) \right] \\
&= \sum_k \left[(\alpha_k \beta_k r + \iota(\alpha_k b_k r - \beta_k r a_k) + b_k r a_k) - \right. \\
&\qquad\qquad \left. (\alpha_k \beta_k r + \iota(\beta_k a_k r - \alpha_k r b_k) + a_k r b_k) \right] \\
&= \sum_k \left[2\iota \alpha_k b_k \cdot r - 2\iota \beta_k a_k \cdot r + 2 b_k \wedge r \wedge a_k \right]
\end{aligned}
\tag{4.29}
$$

Multiplying through by $-\iota/2$ converts this to

$$
\sum_k \left[\alpha_k b_k \cdot r - \beta_k a_k \cdot r - (\iota(a_k \wedge b_k)) \cdot r \right] ,
\tag{4.30}
$$

which vanishes if and only if

$$
\sum_k (\alpha_k b_k - \beta_k a_k) \cdot r = \sum_k (b_k \times a_k) \cdot r .
\tag{4.31}
$$

If Ω is stochastic, this must be true for all r, which is equivalent to

$$
\sum_k (\alpha_k b_k - \beta_k a_k) = \sum_k b_k \times a_k
\tag{4.32}
$$

A comparison with Eq. (4.26) shows further that Ω is both unital and stochastic if and only if $\sum_k \alpha_k b_k = \sum_k \beta_k a_k$ and $\sum_k a_k \times b_k = 0$.

If we regard a stochastic Ω as an affine transformation, i.e.

$$
\Omega(\tfrac{1}{2}(1+r)) = \tfrac{1}{2}(1 + t + \sum_\mu (\sigma_\mu \cdot r) s_\mu)
\tag{4.33}
$$

($s_\mu, t \in \mathfrak{R}_3$), we see from the derivation leading up to (4.26) that

$$
\begin{aligned}
t &= 2 \sum_k (\alpha_k b_k - \beta_k a_k - a_k \times b_k) \\
&= 4 \sum_k (\alpha_k b_k - \beta_k a_k) = 4 \sum_k b_k \times a_k \quad \text{(by (4.32))} .
\end{aligned}
\tag{4.34}
$$

Similarly, the above vectors $s_\mu = \langle \Omega(\sigma_\mu) \rangle_1$ are

$$
s_\mu = \sum_k \left(A_k \sigma_\mu \tilde{A}_k + B_k \sigma_\mu \tilde{B}_k \right) \quad (\mu \in \{\mathsf{x}, \mathsf{y}, \mathsf{z}\}) ,
\tag{4.35}
$$

i.e. a sum of independent dilation/rotations of each basis vector.

It follows that stochastic and unital quantum operations Ω may be characterized by finding conditions for the linear map $\boldsymbol{r} \mapsto \langle \Omega(\boldsymbol{r}) \rangle_1 = \Omega(\boldsymbol{r})$ to be written as a sum of dilation/rotations of $\boldsymbol{r}$. To this end, we expand $\boldsymbol{A}_k \boldsymbol{r} \tilde{\boldsymbol{A}}_k$ as

$$(4.36) \quad \begin{aligned} (\alpha_k + \iota \boldsymbol{a}_k)\, \boldsymbol{r}\, (\alpha_k - \iota \boldsymbol{a}_k) &= \alpha_k^2 \boldsymbol{r} + \alpha_k \iota (\boldsymbol{a}_k \boldsymbol{r} - \boldsymbol{r} \boldsymbol{a}_k) + \boldsymbol{a}_k \boldsymbol{r} \boldsymbol{a}_k \\ &= \alpha_k^2 \boldsymbol{r} + 2\alpha_k \boldsymbol{r} \times \boldsymbol{a}_k + 2(\boldsymbol{r} \cdot \boldsymbol{a}_k)\boldsymbol{a}_k - \|\boldsymbol{a}_k\|^2 \boldsymbol{r} \,, \end{aligned}$$

with a similar expansion for $\boldsymbol{B}_k \boldsymbol{r} \tilde{\boldsymbol{B}}_k$. Thus on assuming that Ω is diagonal, i.e. $\boldsymbol{s}_\mu \equiv \lambda_\mu \boldsymbol{\sigma}_\mu = \Omega(\boldsymbol{\sigma}_\mu)$, we get

$$(4.37) \quad \lambda_\mu \boldsymbol{\sigma}_\mu = \boldsymbol{\sigma}_\mu \sum_k (\alpha_k^2 + \beta_k^2 - \|\boldsymbol{a}_k\|^2 - \|\boldsymbol{b}_k\|^2) + 2\,\boldsymbol{\sigma}_\mu \times \sum_k (\alpha_k \boldsymbol{a}_k + \beta_k \boldsymbol{b}_k) \\ + 2 \sum_k \big((\boldsymbol{\sigma}_\mu \cdot \boldsymbol{a}_k)\boldsymbol{a}_k + (\boldsymbol{\sigma}_\mu \cdot \boldsymbol{b}_k)\boldsymbol{b}_k\big) \,.$$

Dotting both sides by $\boldsymbol{\sigma}_\mu$ now yields

$$(4.38) \quad \begin{aligned} \lambda_\mu &= \sum_k \big(\alpha_k^2 + \beta_k^2 - \|\boldsymbol{a}_k\|^2 - \|\boldsymbol{b}_k\|^2 + 2\,(\boldsymbol{\sigma}_\mu \cdot \boldsymbol{a}_k)^2 + 2\,(\boldsymbol{\sigma}_\mu \cdot \boldsymbol{b}_k)^2\big) \\ &= 1 - 2 \sum_k \big(\|\boldsymbol{a}_k\|^2 + \|\boldsymbol{b}_k\|^2 - (\boldsymbol{\sigma}_\mu \cdot \boldsymbol{a}_k)^2 - (\boldsymbol{\sigma}_\mu \cdot \boldsymbol{b}_k)^2\big) \\ &= 1 + 2 \sum_k \big((\boldsymbol{\sigma}_\mu \wedge \boldsymbol{a}_k)^2 + (\boldsymbol{\sigma}_\mu \wedge \boldsymbol{b}_k)^2\big) \,, \end{aligned}$$

so we have simple expressions for the eigenvalues. Now consider the vector obtained from the first line of this equation, i.e. $\sum_{\mu \in \{x,y,z\}} \lambda_\mu \boldsymbol{\sigma}_\mu$

$$(4.39) \quad \begin{aligned} &= \sum_\mu \boldsymbol{\sigma}_\mu \sum_k \big(\alpha_k^2 + \beta_k^2 - \|\boldsymbol{a}_k\|^2 - \|\boldsymbol{b}_k\|^2 + 2\,(\boldsymbol{\sigma}_\mu \cdot \boldsymbol{a}_k)^2 + 2\,(\boldsymbol{\sigma}_\mu \cdot \boldsymbol{b}_k)^2\big) \\ &= \boldsymbol{p}_0 \sum_k \big(\alpha_k^2 + \beta_k^2\big) + \sum_\mu \boldsymbol{p}_\mu \sum_k \big((\boldsymbol{\sigma}_\mu \cdot \boldsymbol{a}_k)^2 + (\boldsymbol{\sigma}_\mu \cdot \boldsymbol{b}_k)^2\big) \,, \end{aligned}$$

where

$$(4.40) \quad \begin{aligned} \boldsymbol{p}_0 &\equiv \boldsymbol{\sigma}_x + \boldsymbol{\sigma}_y + \boldsymbol{\sigma}_z \,, & \boldsymbol{p}_x &\equiv \boldsymbol{\sigma}_x - \boldsymbol{\sigma}_y - \boldsymbol{\sigma}_z \,, \\ \boldsymbol{p}_y &\equiv -\boldsymbol{\sigma}_x + \boldsymbol{\sigma}_y - \boldsymbol{\sigma}_z \,, & \boldsymbol{p}_z &\equiv -\boldsymbol{\sigma}_x - \boldsymbol{\sigma}_y + \boldsymbol{\sigma}_z \,. \end{aligned}$$

Since the coefficients of the $\boldsymbol{p}$'s are nonnegative and sum to 1 by (4.27), this shows that the vector $\sum_\mu \lambda_\mu \boldsymbol{\sigma}_\mu$ lies within the tetrahedron $\langle \boldsymbol{p}_0, \boldsymbol{p}_x, \boldsymbol{p}_y, \boldsymbol{p}_z \rangle$, which is the condition on the eigenvalues found by Fujiwara and Algoet [**FA99**] as well as by King and Ruskai [**KR00**].

It is also known that an arbitrary linear map Ω has an operator sum representation if and only if it is completely positive [**Sch96**], so the above can also be viewed as a characterization of complete positivity for stochastic and unital maps of a single qubit's density operator. Finally, it is worth stressing once again that, because of the isomorphisms which exist between the Pauli algebra and the even subalgebra of the Dirac algebra, every step of the above derivation carries with it a natural interpretation in space-time, and is in fact even easier to carry out when the full power of the Dirac algebra is used.

In conclusion, it is hoped that the forgoing has given the reader a taste of the new insights which geometric algebra can provide into quantum information processing — and an appetite for more!

References

[Alt86] S. L. Altmann, *Rotations, Quaternions and Double Groups*, Oxford Univ. Press, 1986.
[Alt89] S. L. Altmann, *Hamilton, Rodrigues, and the quaternion scandal*, Math. Mag. **62** (1989), 291–308.
[Bay96] W. E. Baylis (ed.), *Clifford (Geometric) Algebras, with Applications in Physics, Mathematics, and Engineering*, Birkhauser, Boston MA, 1996.

[BBC+95] A. Barenco, C. H. Bennett, R. Cleve, D. P. DiVincenzo, N. Margolus, P. Shor, T. Sleator, J. A. Smolin, and H. Weinfurter, *Elementary gates for quantum computation*, Phys. Rev. A **52** (1995), 3457–3467.

[BBDH93] W. E. Baylis, J. Bonenfant, J. Derbyshire, and J. Huschilt, *Light polarization: A geometric algebra approach*, Am. J. Phys. **61** (1993), 534–545.

[BD00] C. H. Bennett and D. P. DiVincenzo, *Quantum information and computation*, Nature **404** (2000), 247–255.

[Blo46] F. Bloch, *Nuclear induction*, Phys. Rev. **70** (1946), 460–474.

[Bro99] M. Brooks (ed.), *Quantum Computing and Communications*, Springer-Verlag, London, U.K., 1999.

[CEMM98] R. Cleve, A. Ekert, C. Macchiavello, and M. Mosca, *Quantum algorithms revisited*, Proc. R. Soc. A **454** (1998), 339–354.

[Cli78] W. K. Clifford, *Applications of Grassmann's extensive algebra*, Am. J. Math. **1** (1878), 350–358; see also *Mathematical Papers by W. K. Clifford*, (R. Tucker, ed.), Chelsea Publ. Co., Bronx NY, 1968.

[CPH98] D. G. Cory, M. D. Price, and T. F. Havel, *Nuclear magnetic resonance spectroscopy: An experimentally accessible paradigm for quantum computing*, Physica D **120** (1998), 82–101.

[DHSvA93] C. J. L. Doran, D. Hestenes, F. Sommen and N. Van Acker, *Lie groups as spin groups*, J. Math. Phys. **34** (1993), 3642–3669.

[DLG93] C. J. L. Doran, A. N. Lasenby, and S. F. Gull, *States and operators in the spacetime algebra*, Found. Phys. **23** (1993), 1239–1264.

[DLG+96] C. J. L. Doran, A. N. Lasenby, S. F. Gull, S. S. Somaroo, and A. D. Challinor, *Spacetime algebra and electron physics*, in *Advances in Imaging and Electron Physics* (P. Hawkes, ed.), Academic Press, Englewood Cliffs, NJ, 1996, pp. 271–386.

[EJ96] A. Ekert and R. Jozsa, *Quantum computation and Shor's factorizing algorithm*, Rev. Mod. Phys. **68** (1996), 733–753.

[EJ98] A. Ekert and R. Jozsa, *Quantum algorithms: Entanglement enhanced information processing*, Phil. Trans. R. Soc. Lond. A **356** (1998), 1769–1782.

[EK95] A. Ekert and P. L. Knight, *Entangled quantum systems and the Schmidt decomposition*, Am. J. Phys. **63** (1995), 415–423.

[FA99] A. Fujiwara and P. Algoet, *One-to-one parametrization of quantum channels*, Phys. Rev. A **59** (1999), 3290–3295.

[FH81] F. A. M. Frescura and B. J. Hiley, *Geometric interpretation of the Pauli spinor*, Am. J. Phys. **49** (1981), 152–157.

[GJK+96] D. Giulini, E. Joos, C. Kiefer, J. Kupsch, I.-O. Stamatescu, and H. D. Zeh, *Decoherence and the Appearance of a Classical World in Quantum Theory*, Springer-Verlag, Berlin, FRG, 1996.

[HCST00] T. F. Havel, D. G. Cory, S. S. Somaroo, and C.-H. Tseng, *Geometric algebra methods in quantum information processing by NMR spectroscopy*, in *Geometric Algebra with Applications in Science and Engineering* (E. Bayro Corrochano and G. Sobczyk, eds.), Birkhauser, Boston, USA, 2001.

[HSTC00] T. F. Havel, S. S. Somaroo, C.-H. Tseng, and D. G. Cory, *Principles and demonstrations of quantum information processing by NMR spectroscopy*, Applicable Algebra in Engineering, Communications and Computing (T. Beth and M. Grassl, special edition eds.) **10** (2000), 339–374 (see also `quant-ph/9812086`).

[Hav01] T. F. Havel, *Qubit logic, algebra and geometry*, in *Automated Deduction in Geometry* (J. Richter-Gebert and D. Wang, eds.), Lect. Notes Artif. Intel., vol. **2061**, Springer-Verlag, Berlin/Heidelberg, FRG, 2001.

[Hes66] D. Hestenes, *Space-Time Algebra*, Gordon and Breach, New York, NY, 1966.

[Hes99] D. Hestenes, *New Foundations for Classical Mechanics (2nd ed.)*, Kluwer Academic Pub., Amsterdam, NL, 1999.

[HS84] D. Hestenes and G. Sobczyk, *Clifford Algebra to Geometric Calculus*, D. Reidel Pub. Co., Dordrecht, NL, 1984.

[Jan89] B. Jancewicz, *Multivectors and Clifford Algebra in Electrodynamics*, World Scientific Pub. Co., 1989.

[KR00] C. King and M. B. Ruskai, *Minimal entropy of states emerging from noisy quantum channels*, IEEE Trans. Info. Th. **47** (2001), 192–209 (see also `quant/ph-9911079`).

[Kra83] K. Kraus, *States, Effects and Operations*, Springer-Verlag, Berlin, FRG, 1983.

[Llo95] S. Lloyd, *Quantum-mechanical computers*, Sci. Am. **273** (1995), 140–145.

[Per93] A. Peres, *Quantum Theory: Concepts and Methods*, Kluwer Academic Pub., Amsterdam, NL, 1993.

[Rie58] M. Riesz, *Clifford Numbers and Spinors*, Inst. for Fluid Dynamics & Appl. Math., Univ. of Maryland, 1958; reprinted by Kluwer (eds. E. F. Bolinder & P. Lounesto), Dordrecht, NL, 1993.

[Sak94] J. J. Sakurai, *Modern Quantum Mechanics* (revised ed.), Addison-Wesley Pub. Co., Reading, MA, 1994.

[Sch96] B. Schumacher, *Sending entanglement through noisy quantum channels*, Phys. Rev. A **54** (1996), 2614–2628.

[SCH98] S. S. Somaroo, D. G. Cory, and T. F. Havel, *Expressing the operations of quantum computing in multiparticle geometric algebra*, Phys. Lett. A **240** (1998), 1–7.

[SLD99] S. Somaroo, A. Lasenby, and C. Doran, *Geometric algebra and the causal approach to multiparticle quantum mechanics*, J. Math. Phys. **40** (1999), 3327–3340.

[Ste98] A. M. Steane, *Quantum computing*, Rep. Prog. Theor. Phys. **61** (1998), 117–173.

[Tol34] R. C. Tolman, *Relativity, Thermodynamics and Cosmology*, 1934; reprinted by Dover Publications, Inc., New York, NY, 1980.

[Tol38] R. C. Tolman, *The Principles of Statistical Mechanics*, 1938; reprinted by Dover Publications, Inc., New York, NY, 1980.

[vdW85] B. L. van der Waerden, *A History of Algebra*, Springer-Verlag, Berlin, FRG, 1985.

[Vla01] A. Y. Vlasov, *Clifford algebras and universal sets of quantum gates*, Phys. Rev. A **63** (1999), 054302.

[WC98] C. P. Williams and S. H. Clearwater, *Ultimate Zero and One: Computing at the Quantum Frontier*, Copernicus Books, Springer-Verlag, New York, NY, 1999.

T. F. Havel, BCMP, Harvard Medical School, Boston, MA, USA
Current address: NED, Massachusetts Institute of Technology, Cambridge, MA, USA
E-mail address, Timothy F. Havel: `tfhavel@mit.edu`

C. Doran, MRAO, Cavendish Laboratory, Madingley Road, Cambridge, CB3 0HE, United Kingdom
E-mail address, Chris J. L. Doran: `C.Doran@mrao.cam.ac.uk`

Quantum Computing and the Jones Polynomial

Louis H. Kauffman

ABSTRACT. This paper is an exploration of relationships between the Jones polynomial and quantum computing. We discuss the structure of the Jones polynomial in relation to representations of the Temperley Lieb algebra, and give an example of a unitary representation of the braid group. We discuss the evaluation of the polynomial as a generalized quantum amplitude and show how the braiding part of the evaluation can be construed as a quantum computation when the braiding representation is unitary. The question of an efficient quantum algorithm for computing the whole polynomial remains open.

CONTENTS

1. Introduction
2. Dirac Brackets
3. Braiding, Projectors and the Temperley Lieb Algebra
4. The Bracket Polynomial
5. Knot Amplitudes
6. Quantum Computing
7. Summary
References

1. Introduction

This paper is an exploration of issues interrelating the Jones polynomial [10] and quantum computing. In section 2 of the paper we review the formalism of Dirac brackets and some of the quantum physics associated with this formalism. The section ends with a brief description of the concept of quantum computer that we shall use in this paper. In section 3 we discuss the Jones and Temperley Lieb algebras and how they can be used to produce representations of the Artin

---

2000 *Mathematics Subject Classification.* Primary 81P68; Secondary 81-01.

Key words and phrases. Quantum computing, quantum topology, Jones polynomial.

Research on this paper was partially supported by National Science Foundation Grant DMS 9802859, and by the Defense Advanced Research Projects Agency (DARPA) and Air Force Materiel Command USAF under agreement number F30602-01-0522.

Braid group. While most of these representations are not unitary, we show how to construct non-trivial unitary representations of the three-strand braid group by considering the structure of two projectors. It turns out that two elementary projectors naturally generate a Temperley Lieb algebra. This provides a way to make certain unitary representations and to motivate the construction of both the Alexander and the Jones polynomial. In regard to the Alexander polynomial, we end this section with a representation of the Artin Braid Group, constructed using projectors, that is equivalent to the classical Burau representation. In section 4 we construct the bracket polynomial model for the Jones polynomial and relate its structure to the representations discussed in the previous section. Section 5 shows how to reformulate the bracket state sum in terms of discrete quantum amplitudes. This sets the stage for our proposal, explained in section 6, for regarding knot invariants as quantum computers. This proposal needs unitary braiding (a special condition) and the results of the computer are probabilistic. Nevertheless, I believe that this model deserves consideration. The dialogue between topology and quantum computing is just beginning.

2. Dirac Brackets

We begin with a discussion of Dirac's notation, $<b|a>$, [**4**]. In this notation $<a|$ and $|b>$ are covectors and vectors respectively. $<b|a>$ is the evaluation of $|a>$ by $<b|$, hence it is a scalar, and in ordinary quantum mechanics it is a complex number. One can think of this as the amplitude for the state to begin in "a" and end in "b". That is, there is a process that can mediate a transition from state a to state b. Except for the fact that amplitudes are complex valued, they obey the usual laws of probability. This means that if the process can be factored into a set of all possible intermediate states $c_1, c_2, ..., c_n$, then the amplitude for $a \longrightarrow b$ is the sum of the amplitudes for $a \longrightarrow c_i \longrightarrow b$. Meanwhile, the amplitude for $a \longrightarrow c_i \longrightarrow b$ is the product of the amplitudes of the two subconfigurations $a \longrightarrow c_i$ and $c_i \longrightarrow b$. Formally we have

$$<b|a> = \Sigma_i <b|c_i><c_i|a>$$

where the summation is over all the intermediate states $i = 1, ..., n$.
In general, the amplitude for mutually disjoint processes is the sum of the amplitudes of the individual processes. The amplitude for a configuration of disjoint processes is the product of their individual amplitudes.

Dirac's division of the amplitudes into bras $<b|$ and kets $|a>$ is done mathematically by taking a vector space V (a Hilbert space, but it can be finite dimensional) for the kets: $|a>$ belongs to V. The dual space V^* is the home of the bras. Thus $<b|$ belongs to V^* so that $<b|$ is a linear mapping $<b| : V \longrightarrow C$ where C denotes the complex numbers. We restore symmetry to the definition by realizing that an element of a vector space V can be regarded as a mapping from the complex numbers to V. Given $|a>: C \longrightarrow V$, the corresponding element of V is the image of 1 (in C) under this mapping. In other words, $|a>(1)$ is a member of V. Now we have $|a>: C \longrightarrow V$ and $<b| : V \longrightarrow C$. The composition $<b| \circ |a> = <b|a>: C \longrightarrow C$ is regarded as an element of C by taking the specific value $<b|a>(1)$. The complex numbers are regarded as the "vacuum", and the entire amplitude $<b|a>$ is a "vacuum to vacuum" amplitude for a process that includes the creation of the state a, its transition to b, and the annihilation of b to the vacuum once more.

Dirac notation has a life of its own. Let
$$P = |y><x|.$$
Let
$$<x||y> = <x|y>.$$
Then
$$PP = |y><x||y><x| = |y><x|y><x| = <x|y>P.$$
Up to a scalar multiple, P is a projection operator. That is, if we let
$$Q = P/<x|y>,$$
then
$$QQ = PP/<x|y><x|y> = <x|y>P/<x|y><x|y> = P/<x|y> = Q.$$
Thus $QQ = Q$. In this language, the completeness of intermediate states becomes the statement that a certain sum of projections is equal to the identity: Suppose that $\Sigma_i |c_i><c_i| = 1$ (summing over i) with $<c_i|c_i> = 1$ for each i. Then

$$<b|a> = <b||a> = <b|\Sigma_i|c_i><c_i||a> = \Sigma_i <b||c_i><c_i||a>$$

$$<b|a> = \Sigma_i <b|c_i><c_i|a>$$

Iterating this principle of expansion over a complete set of states leads to the most primitive form of the Feynman integral [5]. Imagine that the initial and final states a and b are points on the vertical lines $x = 0$ and $x = n+1$ respectively in the $x-y$ plane, and that $(c(k)_{i(k)}, k)$ is a given point on the line $x = k$ for $0 < i(k) < m$. Suppose that the sum of projectors for each intermediate state is complete. That is, we assume that following sum is equal to one, for each k from 1 to $n-1$:

$$|c(k)_1><c(k)_1| + ... + |c(k)_m><c(k)_m| = 1.$$

Applying the completeness iteratively, we obtain the following expression for the amplitude $<b|a>$;

$$<b|a> = \Sigma\Sigma\Sigma...\Sigma <b|c(1)_{i(1)}><c(1)_{i(1)}|c(2)_{i(2)}> ... <c(n)_{i(n)}|a>$$

where the sum is taken over all $i(k)$ ranging between 1 and m, and k ranging between 1 and n. Each term in this sum can be construed as a combinatorial path from a to b in the two dimensional space of the $x-y$ plane. Thus the amplitude for going from a to b is seen as a summation of contributions from all the "paths" connecting a to b. See Figure 1.

Feynman used this description to produce his famous path integral expression for amplitudes in quantum mechanics. His path integral takes the form

$$\int dP exp(iS)$$

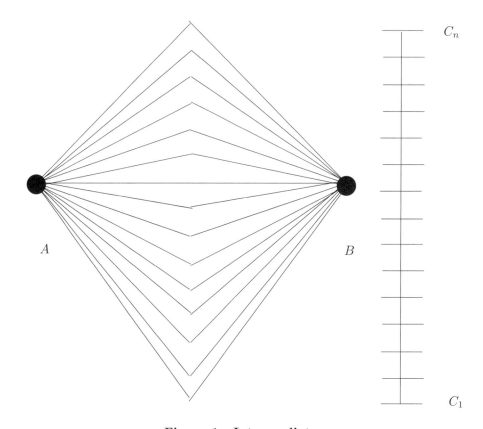

Figure 1 - Intermediates

where i is the square root of minus one, the integral is taken over all paths from point a to point b, and S is the action for a particle to travel from a to b along a given path. For the quantum mechanics associated with a classical (Newtonian) particle the action S is given by the integral along the given path from a to b of the difference $T - V$ where T is the classical kinetic energy and V is the classical potential energy of the particle.

2.1. What is a Quantum Computer? We are now in a position to explain the definition of quantum computer that will be used in this paper. Let H be a given finite dimensional vector space over the complex numbers C. Let $\{W_0, W_1, ..., W_n\}$ be an orthonormal basis for H so that with $|i> := |W_i>$ denoting W_i and $<i|$ denoting the conjugate transpose of $|i>$, we have

$$<i|j> = \delta_{ij}$$

where δ_{ij} denotes the Kronecker delta (equal to one when its indices are equal to one another, and equal to zero otherwise). Given a vector v in H let $|v|^2 := <v|v>$. Note that $<i|v$ is the i-th coordinate of v.

An *observation of* v returns one of the coordinates $|i>$ of v with probability $|<i|v>|^2$. This model of observation is a simple instance of the situation with a quantum mechanical system that is in a mixed state until it is observed. The result of observation is to put the system into one of the basis states.

When the dimension of the space H is two ($n = 1$), a vector in the space is called a *qubit*. A qubit represents one quantum of binary information. On observation, one obtains either the ket $|0>$ or the ket $|1>$. This constitutes the binary distinction that is inherent in a qubit. Note however that the information obtained is probabilistic. If the qubit is

$$\psi = \alpha|0> + \beta |1>,$$

then the ket $|0>$ is observed with probability $|\alpha|^2$, and the ket $|1>$ is observed with probability $|\beta|^2$. In speaking of an idealized quantum computer, we do not specify the nature of measurement process beyond these probability postulates.

In the case of general dimension n of the space H, we will call the vectors in H *qunits*. It is quite common to use spaces H that are tensor products of two-dimensional spaces (so that all computations are expressed in terms of qubits) but this is not necessary in principle. One can start with a given space, and later work out factorizations into qubit transformations.

A *quantum computation* consists in the application of a unitary transformation U to an initial qunit $\psi = a_0|0> +...+ a_n|n>$ with $|\psi|^2 = 1$, plus an observation of $U\psi$. An observation of $U\psi$ returns the ket $|i>$ with probability $|U\psi|^2$. In particular, if we start the computer in the state $|i>$, then the probability that it will return the state $|j>$ is $|<j|U|i>|^2$.

It is the necessity for writing a given computation in terms of unitary transformations, and the probabilistic nature of the result that characterizes quantum computation. Such computation could be carried out by an idealized quantum mechanical system. It is hoped that such systems can be physically realized.

3. Braiding, Projectors and the Temperley Lieb Algebra

The Jones polynomial is one of the great mathematical breakthroughs of the twentieth century, and like many such breakthroughs it appears basically simple in retrospect. I will tell two stories in this section. The first story is a capsule summary of how Jones discovered the polynomial by way of an apparently strange algebraic structure that first appeared in his research on von Neumann algebras, and then was pointed out to be an algebra known to experts in the Potts model in statistical mechanics. The second story shows that the essential algebra for the needed representation of the braid group is present in the algebra generated by any two simple projectors (see below for the definitions of these terms) and that it is graphically illustrated by the Dirac bra-ket notation for these operators.

Jones was studying the inclusion of one von Neumann algebra N in another one M. In this context there is a projection $e_1 : M \longrightarrow N$ so that the restriction of e to N is the identity mapping, and so that $e_1^2 = e_1$. In his context the algebra M could be extended to include this projector to an algebra $M_1 = M \bigcup \{e_1\}$. Then we have

$$N \subset M \subset M_1$$

and the construction can be continued inductively to produce
$$N \subset M \subset M_1 \subset M_2 \subset M_3 \subset ...$$
and an algebra of projectors
$$e_1, e_2, e_3, ...$$
such that
$$e_i^2 = e_i, i = 1, 2, 3, ...$$
$$e_i e_{i\pm 1} e_i = \kappa e_i, i = 2, 3, ...$$
$$e_i e_j = e_j e_i, |i - j| > 1.$$

We will call an algebra that can be expressed with generators and relations as above a *Jones algebra*. J_∞ will denote a Jones algebra on infinitely many generators as above. J_n will denote the Jones algebra generated by an identity element 1 and generators $e_1, ..., e_{n-1}$.

It was pointed out that the relations
$$e_i e_{i\pm 1} e_i = \kappa e_i, i = 2, 3, ...$$
$$e_i e_j = e_j e_i, |i - j| > 1$$
look suspiciously like the basic braiding relations in the Artin Braid group which read
$$\sigma_i \sigma_{i\pm 1} \sigma_i = \sigma_{i\pm 1} \sigma_i \sigma_{i\pm 1}, i = 2, 3, ...$$
$$\sigma_i \sigma_j = \sigma_j \sigma_i, |i - j| > 1.$$

This pattern led Jones to first construct a representation of the Artin Braid Group to his algebra, and then to discover an invariant of knots and links that is related to this representation. Figure 2 illustrates the generators of the braid group. The second and third Reidemeister moves shown in Figure 6 illustrate the braiding relations except for commutativity of distant generators.

The representation that Jones discovered is a linear one in the form of
$$\rho : B_\infty \longrightarrow J_\infty$$
where
$$\rho(\sigma_i) = \alpha 1 + \beta e_i$$
for appropriate constants α and β. We will elaborate on this representation shortly. Here J_∞ denotes the algebra generated by the e_i for $i = 1, 2, 3, ...$ It seems an amazing coincidence that a representation algebra for the Artin Braid group would appear in a context that seems so far away from this structure. The complex source of Jones' algebra makes this connection seem quite mysterious, and the fact that this same algebra appears in statistical mechanics also seems mysterious. What is the source of this apparent connection of the Artin Braid group with algebras and structures coming from quantum physics?

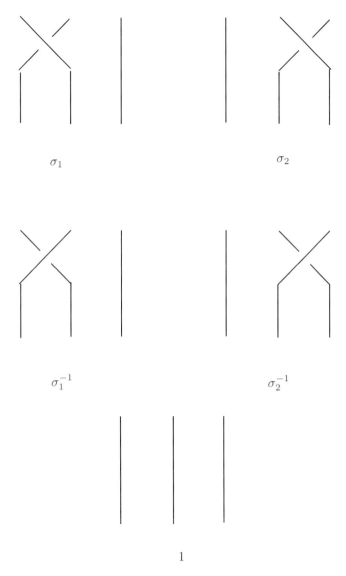

Figure 2 - Braid Group Generators

Remark. In the discussion to follow, we will use the bra and ket notations of Dirac and we will write $< v|$ for v^t, using the notation v^t for the transpose of a vector v. It is to be understood that in the case of a complex vector space, this is the conjugate transpose, but that in the generalizations that we use (over more general rings) we will simply take the formal transpose without conjugation. Later we will construct real-valued representations of the Temperley-Lieb algebra, and there transpose will be the same as conjugate transpose.

For the purpose of this discussion it will be useful to define a *projector* to be a linear map $P : V \longrightarrow V$ where V is a vector space or a module over a ring k, and

P^2 is a non-zero multiple of P. Shall call a projector *simple* if, in a basis, it takes the form $P = vv^t$ where v is a column vector and v^t is its transpose. Then v^tv is the dot product of v with itself and hence a scalar. Therefore

$$P^2 = PP = vv^tvv^t = v[v^tv]v^t$$

$$= [v^tv]vv^t = [v^tv]P.$$

Because of the ubiquity of projectors in quantum physics, the physicist P.A.M. Dirac devised a beautiful notation for this situation. Dirac would write $|v>$ for v and $<v|$ for v^t. He would write

$$<v||w> = <v|w> = v^tw$$

for the dot product of two vectors in a given basis.

Then one can write P in Dirac notation by the formula

$$P = |v><v|$$

and we have

$$P^2 = PP = |v><v||v><v| = |v><v|v><v|$$

$$= <v|v>|v><v| = <v|v>P.$$

Now *consider the algebra generated by two simple projectors* $P = |v><v|$ and $Q = |w><w|$. We have

$$P^2 = <v|v>P,$$

$$Q^2 = <w|w>Q$$

and

$$PQP = |v><v||w><w||v><v|$$

$$= |v><v|w><w|v><v|$$

$$= <v|w><w|v>|v><v|$$

$$= <v|w><w|v>P$$

while

$$QPQ = |w><w||v><v||w><w|$$

$$= |w><w|v><v|w><w|$$

$$= <w|v><v|w>|w><w|$$

$$= <w|v><v|w>Q$$

$$= <v|w><w|v>Q.$$

Thus, with $\lambda = <v|w><w|v>$ we have that

$$PQP = \lambda P$$

$$QPQ = \lambda Q.$$

We can define $e = P/<v|v>$ and $f = Q/<w|w>$ and find
$$e^2 = e$$
$$f^2 = f$$
$$efe = \kappa e$$
$$fef = \kappa f$$
where $\kappa = \lambda/(<v|v><w|w>)$. In this way we see that *any two simple projectors generate a Jones algebra of type J_2*. In this sense the appearance of such algebras is quite natural. The relationship with braiding remains as remarkable as ever.

In order to see how these representations work, it is useful to discuss the combinatorics of these algebras a bit further. The *Temperley Lieb algebra TL_n* [11] is an algebra over a commutative ring k with generators $\{1, U_1, U_2, ..., U_{n-1}\}$ and relations

$$U_i^2 = \delta U_i,$$

$$U_i U_{i\pm 1} U_i = U_i,$$

$$U_i U_j = U_j U_i, |i - j| > 1,$$

where δ is a chosen element of the ring k. These equations give the multiplicative structure of the algebra. The algebra is a free module over the ring k with basis the equivalence classes of these products modulo the given relations.

We will make the ground ring specific in the examples to follow. It is clear that the concepts of Temperley Lieb algebra and Jones algebra are interchangeable. Given a Jones algebra J_∞, with $e_i e_{i\pm 1} e_i = \kappa e_i$, let $\delta = 1/\sqrt{\kappa}$ (assuming that this square root exists in the ground ring k. Then let $U_i = \delta e_i$ and we find that $U_i^2 = \delta U_i$ with

$$U_i U_{i\pm 1} U_i = (1/\sqrt{\kappa})^3 e_i e_{i\pm 1} e_i$$
$$= (1/\sqrt{\kappa})^3 \kappa e_i = (1/\sqrt{\kappa}) e_i - U_i,$$

converting the Jones algebra to a Temperley Lieb algebra.

It is useful to see the bare bones of the algebra of two projectors. For this purpose, lets write
$$P = ><$$
and
$$Q =][.$$
Then
$$PP = ><>< = <>><= <> P$$
and
$$QQ =][][= []Q$$

while
$$PQP =><][><=<][>P$$
$$QPQ =][><][=[><]Q =<][>Q.$$

To see how the representation of the braid group is constructed, lets assume that the scalars $<]$ and $[>$ are both equal to 1 and that $\delta =<>= []$. Then P and Q form a two-generator Temperley Lieb algebra TL_3. We will illustrate how to represent the three strand Artin braid group B_3 to TL_2.

It is useful to use the iconic symbol $><$ for a projector and to choose another iconic symbol $\asymp$ for the identity operator in the algebra. With these choices we have

$$\asymp\asymp \;=\; \asymp$$

$$\asymp >< \;=\; >< \asymp \;=\; ><$$

$$\asymp][\;=\;][\asymp \;=\;][$$

We define the representation $\rho: B_3 \longrightarrow TL_3$ on the generators $\sigma_1 = \sigma$ and $\sigma_2 = \tau$ of the three strand braid group, whose relations are $\sigma\tau\sigma = \tau\sigma\tau$ plus the invertibility of the generators. We define

$$\rho(\sigma) = A1 + BP = A\asymp + B><$$

$$\rho(\sigma^{-1}) = B1 + AP = B\asymp + A><$$

and

$$\rho(\tau) = A1 + BQ = A\asymp + B][$$

$$\rho(\tau^{-1}) = B1 + AQ = B\asymp + A][.$$

where A and B are commuting indeterminates.

With these definitions, we have

$$\rho(\sigma) = A\asymp + B><$$

$$\rho(\sigma^{-1}) = B\asymp + A><.$$

Thus

$$\asymp = (A\asymp + B><)(B\asymp + A><)$$

$$= AB\asymp\asymp + A^2\asymp >< +B^2 ><\asymp + AB ><><$$

$$= AB\asymp + A^2 >< +B^2 >< +AB\delta ><$$

$$\asymp = AB\asymp + (A^2 + B^2 + AB\delta) ><$$

Consequently, we will have $1 = \rho(\sigma)\rho(\sigma^{-1})$ if we take $B = A^{-1}$ and $\delta = -A^2 - A^{-2}$. We shall take these values from now on so that

$$\rho(\sigma) = A\asymp + A^{-1} ><= A1 + A^{-1}P$$

and
$$\rho(\tau) = A \asymp + A^{-1}] [= A1 + A^{-1}Q.$$

With these specializations of A and B, it is easy to verify that ρ is a representation of the Artin Braid Group. Note that $P^2 = \delta P$, $Q^2 = \delta Q$ and $PQP = P$.

$$\rho(\sigma)\rho(\tau)\rho(\sigma) = (A + A^{-1}P)(A + A^{-1}Q)(A + A^{-1}P)$$
$$= (A^2 + Q + P + A^{-2}PQ)(A + A^{-1}P)$$
$$= A^3 + AQ + AP + A^{-1}PQ + AP + A^{-1}QP + A^{-1}P^2 + A^{-3}PQP$$
$$= A^3 + AQ + AP + A^{-1}PQ + AP + A^{-1}QP + A^{-1}\delta P + A^{-3}P$$
$$= A^3 + (2A + +A^{-1}(-A^2 - A^{-2}) + A^{-3})P + AQ + A^{-1}(PQ + QP)$$
$$= A^3 + AP + AQ + A^{-1}(PQ + QP)$$
$$\rho(\sigma)\rho(\tau)\rho(\sigma) = A^3 + A(P + Q) + A^{-1}(PQ + QP)$$

Since this last expression is symmetric in P and Q, we conclude that
$$\rho(\sigma)\rho(\tau)\rho(\sigma) = \rho(\tau)\rho(\sigma)\rho(\tau).$$
Hence ρ is a representation of the Artin Braid Group.

This argument generalizes to yield a corresponding representation of the Artin Braid Group B_n to the Temperley Lieb algebra TL_n for each $n = 2, 3, \ldots$. We will discuss the structure of these representations below. In the next section we show how the Jones polynomial can be constructed by a state summation model. This model can be also be viewed as a generalization of the above representation of the Temperley Lieb algebra.

The very close relationship between elementary quantum mechanics and topology is very well illustrated by the structure and representations of the Temperley Lieb algebra.

Figure 3 illustrates a diagrammatic interpretation of the Temperley Lieb algebra. In this interpretation, the multiplicative generators of the module are collections of strands connecting n top points and n bottom points. Top points can be connected either to top or to bottom points. Bottom points can be connected to either bottom or to top points. All connections are made in the plane with no overlapping lines and no lines going above the top row of points or below the bottom row of points. Multiplication is accomplished by connecting the bottom row of one configuration with the top row of another. In Figure 3 we have illustrated the types of special configurations that correspond to the U_i, and we have shown that δ is interpreted as a closed loop.

One way to make a matrix representation of the Temperley Lieb algebra (and a corresponding representation of the braid group) is to use the matrix M defined as follows

$$M = \begin{bmatrix} 0 & iA \\ -iA^{-1} & 0 \end{bmatrix}.$$

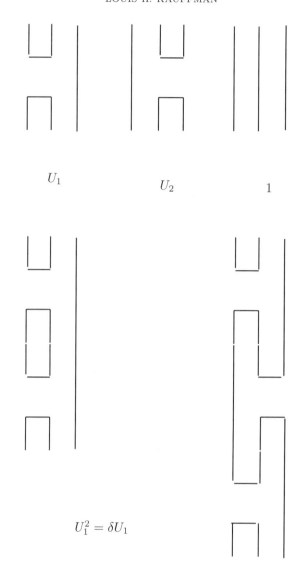

Figure 3 - Diagrammatic Temperley Lieb Algebra

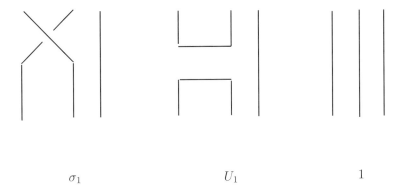

$$\rho(\sigma_1) = AU_1 + A^{-1}\mathbf{1}$$

Figure 4 - Braid Group Representation

Note that $M^2 = 1$ where 1 denotes the (2×2) identity matrix. We will use M with either upper or lower indices so that $M^{ab} = M_{ab}$. M will represent both the cup and the cap in the Temperley Lieb diagrams, with M_{ab} representing the cap and M^{ab} representing the cup. If U denotes a cup over a cap, then

$$U^{ab}_{cd} = M^{ab} M_{cd}.$$

Note that

$$(U^2)^{ab}_{cd} = \Sigma_{ij} U^{ab}_{ij} U^{ij}_{cd} =$$

$$= \Sigma_{ij} M^{ab} M_{ij} M^{ij} M_{cd} = [\Sigma_{ij} M_{ij} M^{ij}] M^{ab} M_{cd}$$

$$= [\Sigma_{ij} M_{ij} M^{ij}] U^{ab}_{cd}.$$

Note that

$$\Sigma_{ij} M_{ij} M^{ij} = \Sigma_{ij}(M_{ij})^2 = -A^2 - A^{-2}.$$

Thus, letting $\delta = -A^2 - A^{-2}$, we have

$$U^2 = \delta U.$$

Then we take U_i as a tensor product of identity matrices corresponding to the vertical lines in the diagram for this element and one factor of U for the placement

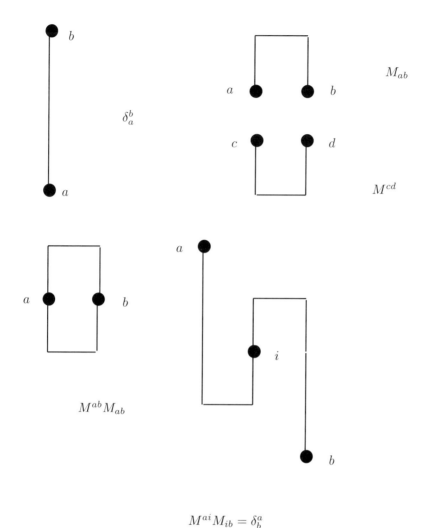

Figure 5 - Abstract Tensors

of the cup-cap at the locations i and $i+1$. To see how this works to give the relation $U_i U_{i\pm 1} U_i = U_i$, we verify that $U_1 U_2 U_1 = U_1$ in TL_3. In the calculation to follow we will use the Einstein summation convention. Repeated upper and lower indices are summed across the index set $\{1, 2\}$.

$$U_1 = U \otimes 1$$

and

$$U_2 = 1 \otimes U$$

so that

$$(U_1)_{def}^{abc} = M^{ab} M_{de} \delta_f^c$$

and
$$(U_2)_{def}^{abc} = \delta_d^a M^{bc} M_{ef}.$$

Therefore
$$(U_1 U_2 U_1)_{def}^{abc} = (U_1)_{ijk}^{abc}(U_2)_{rst}^{ijk}(U_1)_{def}^{rst}$$

$$= (M^{ab}M_{ij}\delta_k^c)(\delta_r^i M^{jk}M_{st})(M^{rs}M_{de}\delta_f^t)$$

$$= M^{ab}(M_{rj}M^{jc})(M_{sf}M^{rs})M_{de} = M^{ab}(\delta_r^c)(\delta_f^r)M_{de} =$$

$$= M^{ab}M_{de}\delta_f^c = (U_1)_{def}^{abc}$$

Thus
$$U_1 U_2 U_1 = U_1.$$

This representation of the Temperley Lieb algebra is useful for knot theory and it is conjectured to be a faithful representation. One may also conjecture that the corresponding braid group representation is faithful.

Remark. The reader should note that the diagrammatic interpretation of the Temperley Lieb algebra gives a clear way to follow the index details of the calculation we have just performed. In the diagrams an index that is not on a free end is summed over just as in the Einstein summation convention. An index at the end of a line is a free index and does not receive summation. See Figure 5 for an illustration of the index algebra in relation to these diagrams. We will generalize the diagrammatic algebra in section 5.

3.1. Two Projectors and a Unitary Representation of the Three Strand Braid Group.
The Temperley Lieb representation of the braid group that we have described is not a unitary representation except when $A^2 = -1$, a value that is not of interest in the knot theory. In order to find elementary unitary representations of the braid group, one has to go deeper.

It is useful to think of the Temperley Lieb algebra as generated by projections $e_i = U_i/\delta$ so that $e_i^2 = e_i$ and $e_i e_{i+1} e_i = \tau e_i$ where $\tau = \delta^{-2}$ and e_i and e_j commute for $|i - j| > 1$.

With this in mind, consider elementary projectors $e = |A><A|$ and $f = |B><B|$. We assume that $<A|A> = <B|B> = 1$ so that $e^2 = e$ and $f^2 = f$. Now note that
$$efe = |A><A|B><B|A><A| = <A|B><B|A> e = \tau e$$

Thus
$$efe = \tau e$$
where $\tau = <A|B><B|A>$.

This algebra of two projectors is the simplest instance of a representation of the Temperley Lieb algebra. In particular, this means that a representation of the three-strand braid group is naturally associated with the algebra of two projectors, a simple toy model of quantum physics!

Quite specifically if we let $<A| = (a,b)$ and $|A> = (a,b)^t$ the transpose of this row vector, then

$$e = |A><A| = \begin{bmatrix} a^2 & ab \\ ab & b^2 \end{bmatrix}$$

is a standard projector matrix when $a^2 + b^2 = 1$. To obtain a specific representation, let

$$e_1 = \begin{bmatrix} 1 & 0 \\ 0 & 0 \end{bmatrix}$$

and

$$e_2 = \begin{bmatrix} a^2 & ab \\ ab & b^2 \end{bmatrix}.$$

It is easy to check that

$$e_1 e_2 e_1 = a^2 e_1$$

and that

$$e_2 e_1 e_2 = a^2 e_2.$$

Note also that

$$e_1 e_2 = \begin{bmatrix} a^2 & ab \\ 0 & 0 \end{bmatrix}$$

and

$$e_2 e_1 = \begin{bmatrix} a^2 & 0 \\ ab & 0 \end{bmatrix}.$$

We define

$$U_i = \delta e_i$$

for $i = 1, 2$ with $a^2 = \delta^{-2}$. Then we have, for $i = 1, 2$

$$U_i^2 = \delta U_i$$

$$U_1 U_2 U_1 = U_1$$

$$U_2 U_1 U_2 = U_2$$

and

$$trace(U_1) = trace(U_2) = \delta$$

while
$$trace(U_1U_2) = trace(U_2U_1) = 1.$$
We will use these results on the traces of these matrices in Section 6.

Now we return to the matrix parameters: Since $a^2 + b^2 = 1$ this means that $\delta^{-2} + b^2 = 1$ whence
$$b^2 = 1 - \delta^{-2}.$$
Therefore b is real when δ^2 is greater than or equal to 1.

We are interested in the case where $\delta = -A^2 - A^{-2}$ and *A is a unit complex number*. Under these circumstances the braid group representation
$$\rho(\sigma_i) = AU_i + A^{-1}1$$
will be unitary whenever U_i is a real symmetric matrix. Thus we will obtain a unitary representation of the three-strand braid group B_3 when $\delta^2 \geq 1$. Specifically, let $A = e^{i\theta}$. Then $\delta = -2cos(2\theta)$, so the condition $\delta^2 \geq 1$ is equivalent to $cos^2(2\theta) \geq 1/4$. Thus we get the specific range of angles $|\theta| \leq \pi/6$ and $|\theta - \pi| \leq \pi/6$ that gives unitary representations of the three-strand braid group.

3.2. Pairs of Projectors and the Alexander Polynomial.

Just for the record we note a more general braid group representation that is available via our remarks about the structure of two projectors. Let $\{W_1, W_2, ..., W_{n-1}, W_n\}$ be the standard basis of *column* vectors for a module of dimension n over $k = C[A, A^{-1}]$ where C denotes the complex numbers and W_k is an n-tuple whose entries are zero in all places except the k-th place where the entry is one. We shall refer to linear combinations of the W_k as *vectors* over k. Given any vector v over k, let $|v>$ denote v as a column vector, and let $<v| = v^t$ denote its transpose (just the transpose, as in our previous remarks), the corresponding row vector. Then $P(v) = |v><v|$ is a matrix such that $P^2 = <v|v> P$, and $<v|v> = v^tv$ is equal to the sum of the squares of the entries of v.

For $k = 1, 2, ..., n-1$ and $i^2 = -1$, let
$$v_k = iAW_k - iA^{-1}W_{k+1}$$
and
$$U_k = |v_k><v_k|.$$
Then, with $\delta = -A^2 - A^{-2}$,
$$U_k^2 = \delta U_k$$
$$U_k U_{k\pm1} U_k = U_k$$
$$U_k U_l = U_l U_k = 0, |k - l| > 1.$$

Thus these matrices give a special representation of the Temperley Lieb algebras TL_n for each n. Since the loop value is as given above, we can make corresponding representations of the Artin Braid Groups B_n by the formulas

$$\rho(\sigma_k) = AI_n + A^{-1}U_k,$$

$$\rho(\sigma_k^{-1}) = A^{-1}I_n + AU_k$$

where I_n denotes the $n \times n$ identity matrix. It is not hard to verify that this representation of B_n is equivalent to the classical Burau representation (See [**15**]) of the braid group. This shows that there is a pathway from the algebra of projectors to the Alexander polynomial! We will treat this theme in a separate paper.

4. The Bracket Polynomial

In this section we shall discuss the structure of the bracket state model for the Jones polynomial [**11**]. In this way, we will explicitly construct the Jones polynomial by using a state summation that is closely related to the braid group representation described in the last section.

Before discussing the bracket polynomial we recall the basic theorem of Reidemeister [**23**] about knot and link diagrams. Reidemeister proved that the three local moves on diagrams illustrated in Figure 6 capture combinatorially the notion of ambient isotopy of links and knots in three-dimensional space. That is, if two diagrams represent knots or links that are isotopic in three-dimensional space, then the one diagram can be obtained from the other by a sequence of Reidemeister moves. It is understood that a Reidemeister move is a local change on the diagram and that it is locally just as indicated by the picture of the move. That is, a type one move adds or eliminates a loop in the underlying 4-regular graph of the knot diagram. A type two move operates on a two sided region and a type three move operates on a three sided region. It is also understood that one can simplify a diagram by a homeomorphism of the plane. This could be called the type zero move, but it is always available. The equivalence relation generated by the type two and type three moves is called *regular isotopy*. The bracket polynomial is a regular isotopy invariant that can be normalized to produce an invariant of all three Reidemeister moves.

The *bracket polynomial*, $<K> = <K>(A)$, assigns to each unoriented link diagram K a Laurent polynomial in the variable A such that

1. If K and K' are regularly isotopic links, then $<K> = <K'>$.
2. If $K \ O$ denotes the disjoint union of K with an extra unknotted and unlinked component O, then

$$<K \ O> = \delta <K>$$

where

$$\delta = -A^2 - A^{-2}.$$

3. $<K>$ satisfies the following formula where in Figure 7 the small diagrams represent parts of larger diagrams that are identical except at the site indicated in the bracket. In the text formula we have used the notations $S_A K$

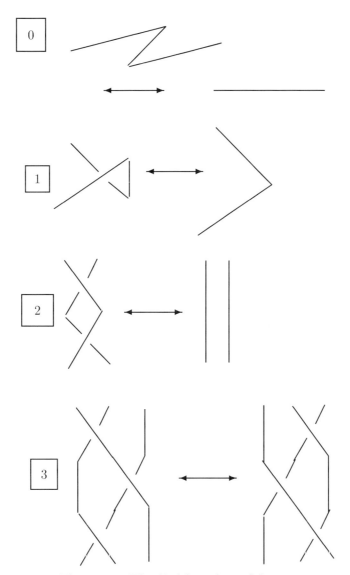

Figure 6 - The Reidemeister Moves

and $S_B K$ to indicate the two smoothings of a single crossing in the diagram K. That is, $K, S_A K$ and $S_B K$ differ at the site of one crossing in the diagram K. The convention for these smoothings is indicated in Figure 7.

$$<K> = A<S_A K> + A^{-1}<S_B K>$$

This formula for expanding the bracket polynomial can be indicated symbolically in the same fashion that we used in the previous section to indicate the representation of the Artin Braid Group to the Temperley Lieb algebra. We will denote a crossing in the link diagram by the letter chi, χ . The letter itself denotes a crossing where *the curved line in the letter chi is crossing over the straight segment in the letter*. The barred letter denotes the switch of this crossing where *the curved line*

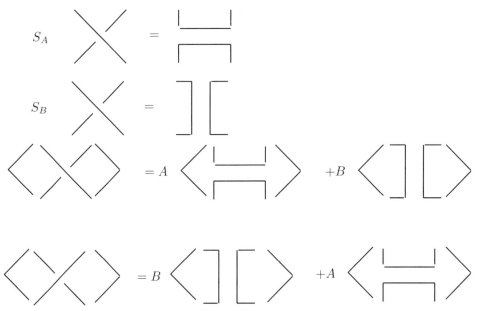

Figure 7 - Bracket Expansion

in the letter chi is undercrossing the straight segment in the letter. In the state model a crossing in a diagram for the knot or link is expanded into two possible states by either smoothing (reconnecting) the crossing horizontally, $\asymp$, or vertically $\rangle\langle$. Coefficients in this expansion correspond exactly to our representation of the braid group so that any closed loop (without crossings) in the plane has value $\delta = -A^2 - A^{-2}$ and the crossings expand according to the formulas

$$\chi = A \asymp + A^{-1} \rangle\langle$$

$$\overline{\chi} = A^{-1} \asymp + A \rangle\langle .$$

The verification that the bracket is invariant under the second Reidemeister move is then identical to our proof in the previous section that

$$\chi\overline{\chi} = \asymp.$$

Knowing that the bracket is invariant under the second Reidemeister move allows us to verify directly that it is invariant under the third Reidemeister move. This is illustrated in Figure 8. In this Figure we show the two equivalent configurations in the third Reidemeister move vertically on the left, with arrows point to the right of each configuration to an expansion via the bracket at one crossing. The expansions give the same bracket calculation due to invariance under the second Reidemeister move. Since the bracket is invariant under the second and third Reidemeister moves, *property 1. is a direct consequence of properties 2. and 3.*. The second two properties define the bracket on arbitrary link diagrams.

In fact we could have begun with the following more general definition: Let K be any unoriented link diagram. Define a *state* of K to be a choice of smoothings for all the crossings of K. There are 2^N states of a diagram with N crossings. A *smoothing* of a crossing is a local replacement of that crossing with two arcs that

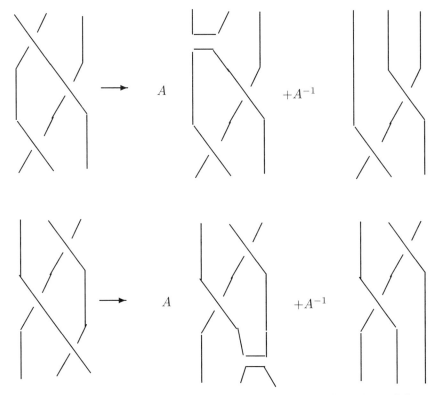

Figure 8 - Invariance of Bracket under Third Reidemeister Move

do not cross one another, as shown below. There are two choices for smoothing a given crossing. In illustrating a state it is convenient to label the smoothing with A or B to indicate the crossing from which it was smoothed. The A or B is called a *vertex weight* of the state.

Label each state with *vertex weights* A or B as illustrated in Figure 9. Here A and B are commuting polynomial variables. Define two evaluations related to the state: The first evaluation is the product of the vertex weights, denoted

$$[K|S].$$

The second evaluation is the number of loops (Jordan curves) in the state S, denoted

$$||S||.$$

Define the *state summation*, $[K]$, by the formula

$$[K] = \sum_S [K|S]\delta^{||S||-1}.$$

It follows from this definition, that $[K]$ satisfies the formulas

$$[\chi] = A[\asymp] + B[><]$$

$$[O \ K] = \delta[K],$$

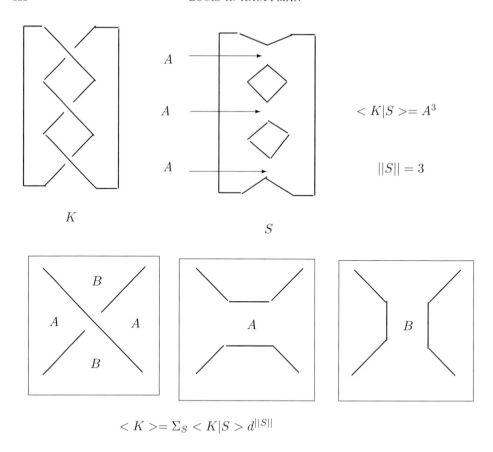

Figure 9 - Bracket States

and
$$[O] = 1.$$

The demand that $[K]$ be invariant under the *second* Reidemeister move leads to the conditions $B = A^{-1}$ and $\delta = -A^2 + A^{-2}$. This specialization is easily seen to be invariant under the third Reidemeister move. Calling this specialization the *topological* bracket, and denoting it (as above) by $< K >$ one finds the following behavior under the first Reidemeister move

$$< \gamma > = -A^3 < \smile >$$

and

$$< \bar{\gamma} > = -A^{-3} < \smile >$$

where γ denotes a curl of positive type as indicated in Figure 10, and $\bar{\gamma}$ indicates a curl of negative type as also seen in this Figure.

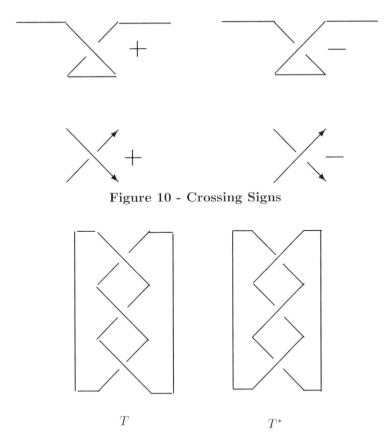

Figure 10 - Crossing Signs

Figure 11 - Trefoil and Mirror Image

The topological bracket is invariant under regular isotopy and can be normalized to an invariant of ambient isotopy by the definition

$$f_K(A) = (-A^3)^{-w(K)} <K>(A)$$

where $w(K)$ is the sum of the crossing signs of the oriented link K. $w(K)$ is called the writhe of K. The convention for crossing signs is shown in Figure 10.

By a change of variables one obtains the original Jones polynomial, $V_K(t)$ [10] from the normalized bracket:

$$V_K(t) = f_K(t^{-1/4}).$$

The bracket model for the Jones polynomial is quite useful both theoretically and in terms of practical computations. One of the neatest applications is to simply compute $f_K(A)$ for the trefoil knot T and determine that $f_K(A)$ is not equal to $f_K(A^{-1})$. This shows that the trefoil is not ambient isotopic to its mirror image (See Figure 11), a fact that is quite tricky to prove by classical methods.

Remark. The relationship of the Temperley Lieb algebra with the bracket polynomial comes through the basic bracket identity. This identity, interpreted in the context of the diagrammatic Temperley Lieb algebra becomes a representation ρ of the Artin braid group B_n on n strands to the Temperley Lieb algebra TL_n defined by the formulas

$$\rho(\sigma_i) = AU_i + A^{-1}1$$

$$\rho(\sigma_i^{-1}) = A^{-1}U_i + A1.$$

Here $sigma_i$ denotes the braid generator that twists strands i and $i+1$. For this representation of the Temperley Lieb algebra, the loop value δ is $-A^2 - A^{-2}$ and the ring k is $Z[A, A^{-1}]$, the ring of Laurent polynomials in A with integer coefficients.

Remark. There are hints of quantum mechanical interpretations in the combinatorics of this state sum model for the Jones polynomial. The expansion formula for the bracket polynomial

$$<K> = A<S_AK> + A^{-1}<S_BK>$$

suggests that the diagram K should be thought of as a superposition of the diagrams S_AK and S_BK. That is, we can think of a knot diagram with respect to a given crossing as the superposition of the diagrams obtained by smoothing that crossing. Then, with respect to all the crossings, one can think of the diagram as a superposition of the states obtained by smoothing each crossing in one of its two possible ways. This is a superposition view of the bracket state sum as a whole.

$$<K> = \sum_S <K|S> \delta^{||S||-1}$$

In this sense the bracket polynomial evaluation is directly analogous to an amplitude in quantum mechanics. We shall make this analogy more precise in the sections to follow. However, the topological information is contained in this amplitude as whole, and not in any specific state evaluation. Thus the topological model ignores the standard measurement situation in quantum mechanics where one gets at best information about one state at a time when a measurement is taken. This means that a quantum computational model of the bracket polynomial will be essentially probabilistic, only giving partial information at each measurement.

As a result of this discussion, it is natural to ask to what extent one can extract partial topological information from an incomplete summation over the states of the bracket polynomial. It is not clear at this stage what this answer is to this question. It may require a new exploration of the properties of the state sum and its corresponding polynomial.

5. Knot Amplitudes

At the end of the first section we said: the connection of quantum mechanics with topology is an amplification of Dirac notation. In this section we begin the process of amplification!

Consider first a circle in a spacetime plane with time represented vertically and space horizontally. The circle represents a vacuum to vacuum process that includes

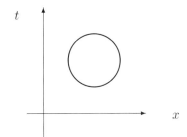

Figure 12 - Circle in Spacetime

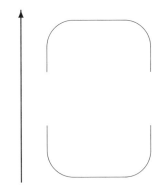

Figure 13 - Creation and Annihilation

the creation of two "particles", and their subsequent annihilation. See Figures 12 and 13.

In accord with our previous description, we could divide the circle into these two parts (creation(a) and annihilation (b)) and consider the amplitude $<b|a>$. Since the diagram for the creation of the two particles ends in two separate points, it is natural to take a vector space of the form $V \otimes V$ as the target for the bra and as the domain of the ket.

We imagine at least one particle property being catalogued by each dimension of V. For example, a basis of V could enumerate the spins of the created particles. If $\{e_a\}$ is a basis for V then $\{e_u \otimes e_v\}$ forms a basis for $V \otimes V$. The elements of this new basis constitute all possible combinations of the particle properties. Since such combinations are multiplicative, the tensor product is the appropriate construction.

In this language the creation ket is a map cup,
$$cup = |a>: C \longrightarrow V \otimes V,$$
and the annihilation bra is a mapping cap,
$$cap = <b| : V \otimes V \longrightarrow C.$$

The first hint of topology comes when we realize that it is possible to draw a much more complicated simple closed curve in the plane that is nevertheless decomposed with respect to the vertical direction into many cups and caps. In fact, any simple (no self-intersections) differentiable curve can be rigidly rotated until it is in general position with respect to the vertical. It will then be seen to be decomposed into these minima and maxima. Our prescriptions for amplitudes suggest that we regard any such curve as an amplitude via its description as a mapping from C to C.

Each simple closed curve gives rise to an amplitude, but any simple closed curve in the plane is isotopic to a circle, by the Jordan Curve Theorem. If these are topological amplitudes, then they should all be equal to the original amplitude for the circle. Thus the question: What condition on creation and annihilation will insure topological amplitudes? The answer derives from the fact that all isotopies of the simple closed curves are generated by the cancellation of adjacent maxima and minima as illustrated below.

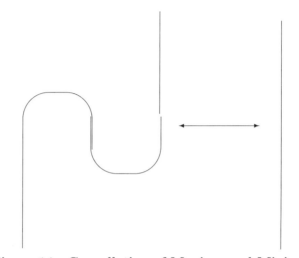

Figure 14 - Cancellation of Maxima and Minima

In composing mappings it is necessary to use the identifications $(V \otimes V) \otimes V = V \otimes (V \otimes V)$ and $V \otimes k = k \otimes V = V$. Thus in the illustration above, the composition on the left is given by

$$V = V \otimes k - 1 \otimes cup \to V \otimes (V \otimes V)$$

$$= (V \otimes V) \otimes V - cap \otimes 1 \to k \otimes V = V.$$

This composition must equal the identity map on V (denoted 1 here) for the amplitudes to have a proper image of the topological cancellation. This condition is said very simply by taking a matrix representation for the corresponding operators.

Specifically, let $\{e_1, e_2, ..., e_n\}$ be a basis for V. Let $e_{ab} = e_a \otimes e_b$ denote the elements of the tensor basis for $V \otimes V$. Then there are matrices M_{ab} and M^{ab} such that

$$cup(1) = \Sigma M^{ab} e_{ab}$$

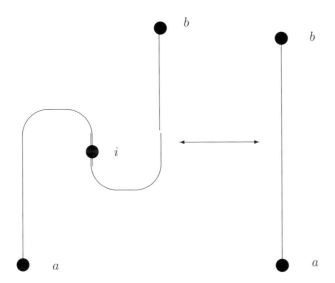

Figure 15 - Algebraic Cancellation of Maxima and Minima

with the summation taken over all values of a and b from 1 to n. Similarly, cap is described by

$$cap(e_{ab}) = M_{ab}.$$

Thus the amplitude for the circle is

$$cap[cup(1)] = cap \Sigma M^{ab} e_{ab} = \Sigma M^{ab} M_{ab}.$$

In general, the value of the amplitude on a simple closed curve is obtained by translating it into an "abstract tensor expression" in the M_{ab} and M^{ab}, and then summing over these products for all cases of repeated indices.

Returning to the topological conditions we see that they are just that the matrices (M_{ab}) and (M^{ab}) are inverses in the sense that $\Sigma M_{ai} M^{ib} = \delta_a^b$ and $\Sigma M^{ai} M_{ib} = \delta_b^a$ where δ_a^b denotes the (identity matrix) Kronecker delta that is equal to one when its two indices are equal to one another and zero otherwise.

In Figure 15, we show the diagrammatic representative of the equation $\Sigma M_{ai} M^{ib} = \delta_a^b$.

In the simplest case cup and cap are represented by 2×2 matrices. The topological condition implies that these matrices are inverses of each other. Thus the problem of the existence of topological amplitudes is very easily solved for simple closed curves in the plane.

Now we go to knots and links. Any knot or link can be represented by a picture that is configured with respect to a vertical direction in the plane. The picture will decompose into minima (creations) maxima (annihilations) and crossings of the two types shown below. (Here I consider knots and links that are unoriented. They do not have an intrinsic preferred direction of travel.) See Figure 16. In Figure 16 we have indicated the crossings as mappings of $V \otimes V$ to itself, called R and

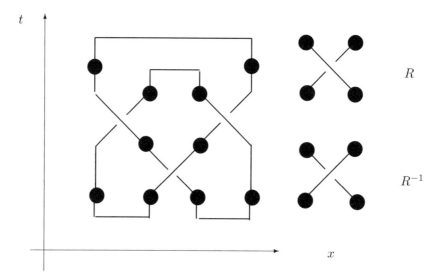

Figure 16 - Morse Knot Decomposition

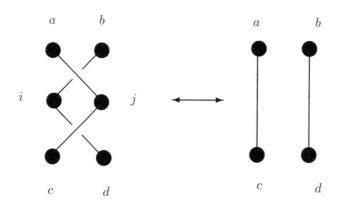

Figure 17 - Braiding Cancellation

R^{-1} respectively. These mappings represent the transitions corresponding to these elementary configurations.

That R and R^{-1} really must be inverses follows from the isotopy shown in Figure 17 (This is the second Reidemeister move.)

We now have the vocabulary of *cup*, *cap*, R and R^{-1}. Any knot or link can be written as a composition of these fragments, and consequently a choice of such mappings determines an amplitude for knots and links. In order for such an amplitude to be topological we want it to be invariant under the list of local moves on the diagrams shown in Figure 18. These moves are an augmented list of the Reidemeister moves, adjusted to take care of the fact that the diagrams are arranged with respect to a given direction in the plane. The equivalence relation generated by

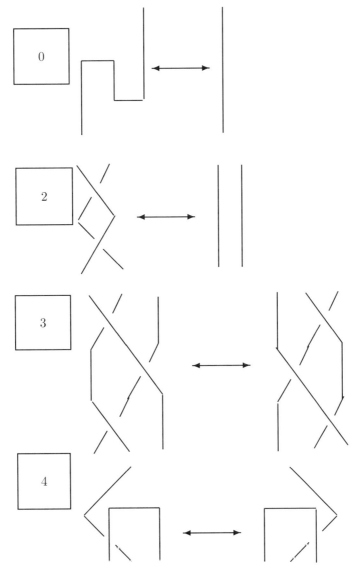

Figure 18- Moves for Regular Isotopy of Morse Diagrams

these moves is called regular isotopy. It is one move short of the relation known as ambient isotopy. The missing move is the first Reidemeister move shown in Figure 6.

In the first Reidemeister move, a curl in the diagram is created or destroyed. Ambient isotopy (generated by all the Reidemeister moves) corresponds to the full topology of knots and links embedded in three dimensional space. Two link diagrams are ambient isotopic via the Reidemeister moves if and only if there is a continuous family of embeddings in three dimensions leading from one link to the other. The moves give us a combinatorial reformulation of the spatial topology of knots and links.

By ignoring the first Reidemeister move, we allow the possibility that these diagrams can model framed links, that is links with a normal vector field or, equivalently, embeddings of curves that are thickened into bands. It turns out to be fruitful to study invariants of regular isotopy. In fact, one can usually normalize an invariant of regular isotopy to obtain an invariant of ambient isotopy. We shall see an example of this phenomenon with the bracket polynomial in a few paragraphs.

As the reader can see, we have already discussed the algebraic meaning of moves 0. and 2. The other moves translate into very interesting algebra. Move 3., when translated into algebra, is the famous Yang-Baxter equation. The Yang-Baxter equation occurred for the first time in problems related to exactly solved models in statistical mechanics (See [19].). All the moves taken together are directly related to the axioms for a quasi-triangular Hopf algebra (aka quantum group). We shall not go into this connection here.

There is an intimate connection between knot invariants and the structure of generalized amplitudes, as we have described them in terms of vector space mappings associated with link diagrams. This strategy for the construction of invariants is directly motivated by the concept of an amplitude in quantum mechanics. It turns out that the invariants that can actually be produced by this means (that is by assigning finite dimensional matrices to the caps, cups and crossings) are incredibly rich. They encompass, at present, all of the known invariants of polynomial type (Alexander polynomial, Jones polynomial and their generalizations.).

It is now possible to indicate the construction of the Jones polynomial via the bracket polynomial as an amplitude, by specifying its matrices. The cups and the caps are defined by $(M_{ab}) = (M^{ab}) = M$ where M is the 2×2 matrix (with $ii = -1$).

$$M = \begin{bmatrix} 0 & iA \\ -iA^{-1} & 0 \end{bmatrix}$$

Note that $MM = I$ where I is the identity matrix. Note also that the amplitude for the circle is

$$\Sigma M_{ab} M^{ab} = \Sigma M_{ab} M_{ab} = \Sigma M_{ab}^2$$

$$= (iA)^2 + (-iA^{-1})^2 = -A^2 - A^{-2}.$$

The matrix R is then defined by the equation

$$R_{cd}^{ab} = A M^{ab} M_{cd} + A^{-1} \delta_c^a \delta_d^b,$$

Since, diagrammatically, we identify R with a (right handed) crossing, this equation can be written diagrammatically as the generating identity for the bracket polynomial:

$$\chi = A \asymp + A^{-1} ><$$

Taken together with the loop value of $-A^2 - A^{-2}$ that is a consequence of this matrix choice, these equations can be regarded as a recursive algorithm for computing the amplitude. This algorithm is the bracket state model for the (unnormalized) Jones polynomial [11]. We have discussed this model in the previous sections.

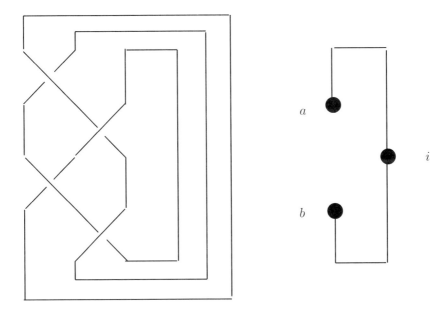

$$\eta_a^b = M_{ai} M^{bi}$$

Figure 19 - Pairing Maxima and Minima for Braid Closures

The upshot of these remarks is that the bracket state summation can be reformulated as a matrix model as described in this section. Thus the values of the bracket polynomial can be regarded as generalized quantum amplitudes. Note also that the model that we have described in this section can be seen as a generalization of the representation of the Temperley Lieb algebra from Section 3 with basic matrix M as above. In fact, in the case where the knot or link is a closure of a braid, we can say even more. Suppose that $K = \bar{b}$ where b is an n-strand braid in B_n. Then the cups and caps can be paired off as shown in Figure 19 so that

$$Z(K) = \delta <K> = Trace(\eta^{\otimes n} \rho(b)).$$

Here

$$\eta_b^a = \Sigma_i M_{bi} M^{ai}$$

so that $\eta = MM^t$ where M^t denotes the transpose of the matrix M.
and $\rho : B_n \longrightarrow TL_n$ is the matrix representation of the Temperley Lieb algebra specified in Section 3. The key to the workings of this representation of the bracket calculation for braids is the fact that for any pure connection element Q (obtained

from a product of the $U_i's$) in the Temperley-Lieb algebra TL_n, the evaluation
$$Trace(\eta^{\otimes n} Q)$$
is equal to $\delta^{\lambda(Q)}$ where $\lambda(Q)$ is the number of loops in the (braid) closure of the diagram for Q.

In general, if we have a linear function $TR : TL_n \longrightarrow Z[\delta]$ such that $TR(Q) = \delta^{\lambda(Q)}$ for elements Q as above, then $TR(\rho(b)) = \delta <\bar{b}>$ for any n-strand braid b.

6. Quantum Computing

In this paper I have concentrated on giving a picture of the general framework of the Jones polynomial and how it is related to a very general, in fact categorical, view of quantum mechanics. Many algorithms in quantum topology are configured without regard to unitary evolution of the amplitude since the constraint has been topological invariance rather than conformation to physical reality. This gives rise to a host of problems of attempting to reformulate topological amplitudes as quantum computations. A particular case in point is the bracket model for the Jones polynomial. It would be of great interest to see a reformulation of this algorithm that would make it a quantum computation in the strict sense of quantum computing. One way to think about this is to view the bracket model as a vacuum-vacuum amplitude as we have done in the last section of this paper. Then it can be configured as a composition of operators (cups, caps and braiding). If the braiding is unitary. Then at least this part can be viewed as a quantum computation.

To see how this can be formulated consider the vacuum-vacuum computation of a link amplitude as we have described it in section 4. In Figure 20 we have indicated an amplitude where the temporal decomposition consists first in a composition of cups (creations), then braiding and then caps (annihilations). Thus we can write the amplitude in the form
$$Z_K = <CUP|M|CAP>$$
where $<CUP|$ denotes the composition of cups, M is the composition of elementary braiding matrices and $|CAP>$ is the composition of caps. We then regard $<CUP|$ as the preparation of this state and $|CAP>$ as the detection of this state. In order to view Z_K as a quantum computation, we need that M be a unitary operator. This will be the case if the R-matrices (the solutions to the Yang-Baxter equation used in the model for this amplitude) are unitary. In this case, each R-matrix can be viewed as a quantum gate (or possibly a composition of quantum gates) and the vacuum-vacuum diagram for the knot is interpreted as a quantum computer. This quantum computer will probabalistically compute the values of the states in the state sum for Z_K. In order to do so, we would need to specify those observations and preparations that correspond to the cups and the caps in the diagram. A more modest proposal is to regard the braiding sector of the diagram as a quantum computer. That braiding sector will represent a unitary evolution, and one can ask more generally what can be computed by using such a gate derived from a braid.

It should be noted that because the quantum computer gives probabilistic data, it cannot compute the knot invariants exactly. In fact, the situation is more serious than that. If we assume that the parameters in the knot invariant are complex, then

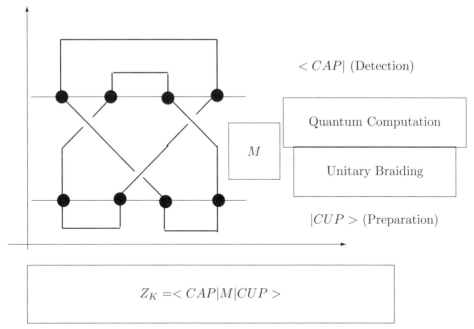

Figure 20 - $Z_K = <CUP|M|CAP>$ - **A Knot Quantum Computer**

the computer will only find (probabilistically) the absolute squares of the various complex parameters. Important phase information will be lost, and it is not obvious that topologically invariant information about the knot can be extracted.

6.1. A Unitary Representation of the Three Strand Braid Group and the Corresponding Quantum Computer. Many questions are raised by the formulation of a quantum computer associated with a given Morse link diagram. First of all, unitary solutions to the Yang-Baxter equation (or unitary representations of the Artin braid group) that also give link invariants are not so easy to come by. We gave a small example of a unitary representation of the three-strand braid group in the first section of this paper. Thus we are prepared to look at some aspects of the computation of a knot invariant as a quantum computation. In fact, we can use this representation to compute the Jones polynomial for closures of 3-braids, and therefore this representation provides a test case for the corresponding quantum computation. We now analyze this case by first making explicit how the bracket polynomial is computed from this representation.

First recall that the representation depends on two matrices U_1 and U_2 with

$$U_1 = \begin{bmatrix} \delta & 0 \\ 0 & 0 \end{bmatrix}$$

and

$$U_2 = \begin{bmatrix} \delta^{-1} & \sqrt{1-\delta^{-2}} \\ \sqrt{1-\delta^{-2}} & \delta - \delta^{-1} \end{bmatrix}.$$

The representation is given on the two braid generators by

$$\rho(\sigma_1) = AI + A^{-1}U_1$$

$$\rho(\sigma_2) = AI + A^{-1}U_2$$

for any A with $\delta = -A^2 - A^{-2}$, and with $A = e^{i\theta}$, then $\delta = -2cos(2\theta)$. We get the specific range of angles $|\theta| \leq \pi/6$ and $|\theta - \pi| \leq \pi/6$ that give unitary representations of the three-strand braid group.

Note that $tr(U_1) = tr(U_2) = \delta$ while $tr(U_1 U_2) = tr(U_2 U_1) = 1$. If b is any braid, let $I(b)$ denote the sum of the exponents in the braid word that expresses b. For b a three-strand braid, it follows that

$$\rho(b) = A^{I(b)} I + \tau(b)$$

where I is the 2×2 identity matrix and $\tau(b)$ is a sum of products in the Temperley Lieb algebra involving U_1 and U_2. Since the Temperley Lieb algebra in this dimension is generated by $I, U_1, U_2, U_1 U_2$ and $U_2 U_1$, it follows that

$$<\bar{b}> = A^{I(b)} \delta^2 + tr(\tau(b))$$

where $\bar{b}$ denotes the standard braid closure of b, and the sharp brackets denote the bracket polynomial as described in previous sections. From this we see at once that

$$<\bar{b}> = tr(\rho(b)) + A^{I(b)}(\delta^2 - 2).$$

It follows from this calculation that the question of computing the bracket polynomial for the closure of the three-strand braid b is mathematically equivalent to the problem of computing the trace of the matrix $\rho(b)$. To what extent can our quantum computer determine the trace of this matrix?

The matrix in question is a product of unitary matrices, the quantum gates that we have associated with the braids σ_1 and σ_2. The entries of the matrix $\rho(b)$ are the results of preparation and detection for the two dimensional basis of qubits for our machine:

$$<i|\rho(b)|j>.$$

Given that the computer is prepared in $|j>$, the probability of observing it in state $|i>$ is equal to $|<i|\rho(b)|j>|^2$. Thus we can, by running the quantum computation repeatedly, estimate the absolute squares of the entries of the matrix $\rho(b)$. This will not yield the complex phase information that is needed for either the trace of the matrix or the absolute value of that trace. Thus we conclude that our quantum computer can compute information relating to the braiding process, but that it cannot approximate the full value of the bracket polynomial.

Note that our quantum computer does indeed have the capability to detect three strand braiding, since for a braid b the matrix $\rho(b)$ can have non-trivial off-diagaonal elements. The absolute squares of these elements are approximated by successive runs of the quantum computer. In this quantum computer, braiding corresponds to entangled quantum states and is delectable by that token. The bracket polynomial itself depends upon subtler phase relationships and is not detectable by this quantum computer.

6.2. Comments. These results are less than satisfying since there does not seem to be a way to calculate the entire knot polynomial even by probabilisitic approximations. It is not clear what the practical value of such a computation will be for understanding a given link invariant. Nevertheless, it is to be expected that a close relationship between quantum link invariants and quantum computing will be fruitful for both fields.

There are other ideas in the topology that deserve comparison with the quantum states. For example, topological entanglement in the sense of linking and braiding is intuitively related to the entanglement of quantum states. In our general model using a unitary representation of the braid group, topological entanglement entails quantum entanglement. The quantum topological states associated with the bracket polynomial would certainly figure strongly in a quantum computing model of this algorithm. The specific model that we have given uses only one qubit and so does not produce entanglement. There are other representations of the Artin Braid group that do produce quantum entanglements corresponding to topological braiding. These phenomena will be the subject of a subsequent paper [20].

We mention one further possibility. In the paper [21] by Lidar and Biham the authors show how to simulate special cases of the Ising model on a quantum computer. Their method is more combinatorial and less algebraic than the approach sketched in this section using braiding. It is possible that a generalization of their approach will work for the state sum of the bracket polynomial. This is a topic for further research.

6.3. And Quantum Field Theory. Finally, it is important to remark that there is an interpretation of the Jones polynomial in terms of quantum field theory. Witten [26] writes down a functional integral for link invariants in a 3-manifold M:

$$Z(M,K) = \int dA exp[(ik/4\pi)S(M,A)] tr(P exp(\int_K A)).$$

Here M denotes a 3-manifold without boundary and A is a gauge field (gauge connection) defined on M. The gauge field is a one-form on M with values in a representation of a Lie algebra. $S(M,A)$ is the integral over M of the trace of the Chern-Simons three-form $CS = AdA + (2/3)AAA$. (The product is the wedge product of differential forms.)

With the standard representation of the Lie algebra of $SU(2)$ as 2×2 complex matrices, one can see that the formalism of $Z(S,K)$ (S^3 denotes the three-dimensional sphere.) yields the Jones polynomial with the basic properties as we have discussed. See Witten's paper or [26] or [15],[16].

The question is: How does the quantum field theory approach to the Jones polynomial relate to quantum computing? One way to discuss this question is to reformulate (topological) quantum field theories as state summations, as we did for the Jones polynomial, and then proceed in a fashion analogous to our amplitudes discussion above. It is more challenging to try to imagine reformulating quantum computing at the level of quantum field theory. If this were accomplished, the subject of quantum computing and the Jones polynomial might well take a new road.

7. Summary

In relating quantum computing with knot polynomials the key themes are unitarity and measurement. Much is now surely unforeseen. For a good survey of quantum computing we recommend [1] and [22], and for another view of topological issues see [6] and [7]. See [25] for an excellent treatment of measurement theory in quantum mechanics and a usage of the Dirac formalism that is in resonance with the concerns of this paper.

References

[1] D. Aharonov. Quantum computation. quant-phys/9812037 15Dec 1998.
[2] M.F. Atiyah. The Geometry and Physics of Knots. Cambridge University Press (1990).
[3] R.J. Baxter. Exactly Solved Models in Statistical Mechanics. Acad. Press (1982).
[4] P.A.M. Dirac. Principles of Quantum Mechanics. Oxford University Press (1958).
[5] R. Feynman and A.R. Hibbs. Quantum Mechanics and Path Integrals. McGraw Hill (1965).
[6] M. Freedman. Topological Views on Computational Complexity. Documenta Mathematica - Extra Volume ICM (1998) , pp. 453-464.
[7] M. Freedman, M. Larsen and Z. Wang. A modular functor which is universal for quantum computation. arXiv:quant-ph/0001108v2 1 Feb 2000.
[8] V.F.R.Jones.A polynomial invariant for links via von Neumann algebras. Bull.Amer.Math.Soc. 129 (1985) 103-112.
[9] D.S.Freed and R.E.Gompf. Computer Calculation of Witten's 3-Manifold Invariant. Commun. Math. Phys. 141 (1991), pp. 79-117.
[10] V.F.R.Jones. A new knot polynomial and von Neumann algebras. Notices of AMS 33 (1986) 219-225.
[11] L.H.Kauffman. State Models and the Jones Polynomial. Topology 26 (1987) 395-407.
[12] L.H. Kauffman. On Knots. Annals of Mathematics Studies Number 115, Princeton University Press (1987).
[13] L.H.Kauffman. New invariants in the theory of knots. Amer. Math. Monthly Vol.95,No.3,March 1988. pp 195-242.
[14] L.H.Kauffman. Statistical mechanics and the Jones polynomial. AMS Contemp. Math. Series (1989), Vol. 78. pp. 263-297.
[15] L.H. Kauffman. Knots and Physics , World Scientific Pub. (1991 and 1993).
[16] L. H. Kauffman. Functional integration and the theory of knots. J. Math. Phys. Vol. 36, No.5 (19950, pp. 2402-2429.
[17] L. H. Kauffman. Witten's Integral and the Kontsevich Integral. In Particles Fields and Gravitation. edited by Jakub Rembielinski, AIP Proceedings No. 453, American Inst. of Physics Pub. (1998), pp. 368 - 381.
[18] L.H.Kauffman (Editor). Knots and Applications, World Scientific Pub. Co. (1995).
[19] L.H.Kauffman. Knots and Statistical Mechanics, Proceedings of Symposia in Applied Mathematics - The Interface of Knots and Physics (edited by L. Kauffman), Vol.51 (1996), pp. 1-87.
[20] L.H.Kauffman and S. Lomonaco. Topological Entanglement and Quantum Entanglement. (in preparation)
[21] D. Lidar and O. Biham. Simulating ising spin glasses on a quantum computer. quant-ph/9611038v6 23 Sept. 1997.
[22] S. Lomonaco. *A rosetta stone for quantum mechanics with an introduction to quantum computiation.* quant-ph/0007045 (2000).
[23] K. Reidemeister. *Knotentheorie.* Chelsea, New York (1948), Julius Springer (1932).

[24] N.Yu.Reshetikhin and V.G. Turaev. Invariants of 3-manifolds via link polynomials and quantum groups. Invent. Math. Vol. 103 (1991), pp. 547-597.
[25] J. Schwinger. *Quantum Mechanics: Symbolism of Atomic Measurement.* Springer-Verlag, 2001.
[26] E. Witten. Quantum field theory and the Jones polynomial. Commun.Math.Phys. 121 , 351-399 (1989).

(Louis H. Kauffman) DEPARTMENT OF MATHEMATICS, STATISTICS, AND COMPUTER SCIENCE, UNIVERSITY OF ILLINOIS AT CHICAGO, 851 SOUTH MORGAN STREET, CHICAGO, IL 60607-7045
  E-mail address, Louis H. Kauffman: kauffman@uic.edu
  URL: http://math.uic.edu/~kauffman

Quantum Hidden Subgroup Algorithms: A Mathematical Perspective

Samuel J. Lomonaco, Jr. and Louis H. Kauffman

ABSTRACT. The ultimate objective of this paper is to create a stepping stone to the development of new quantum algorithms. The strategy chosen is to begin by focusing on the class of abelian quantum hidden subgroup algorithms, i.e., the class of abelian algorithms of the Shor/Simon genre. Our strategy is to make this class of algorithms as mathematically transparent as possible. By the phrase "mathematically transparent" we mean to expose, to bring to the surface, and to make explicit the concealed mathematical structures that are inherently and fundamentally a part of such algorithms. In so doing, we create symbolic abelian quantum hidden subgroup algorithms that are analogous to the those symbolic algorithms found within such software packages as Axiom, Cayley, Maple, Mathematica, and Magma.

As a spin-off of this effort, we create three different generalizations of Shor's quantum factoring algorithm to free abelian groups of finite rank. We refer to these algorithms as wandering (or vintage $\mathbb{Z}_Q$) Shor algorithms. They are essentially quantum algorithms on free abelian groups A of finite rank n which, with each iteration, first select a random cyclic direct summand $\mathbb{Z}$ of the group A and then apply one iteration of the standard Shor algorithm to produce a random character of the "approximating" finite group $\widetilde{A} = \mathbb{Z}_Q$, called the group probe. These characters are then in turn used to find either the order P of a maximal cyclic subgroup $\mathbb{Z}_P$ of the hidden quotient group H_φ, or the entire hidden quotient group H_φ. An integral part of these wandering quantum algorithms is the selection of a very special random transversal $\iota_\mu : \widetilde{A} \longrightarrow A$, which we refer to as a Shor transversal. The algorithmic time complexity of the first of these wandering Shor algorithms is found to be $O\left(n^2 (\lg Q)^3 (\lg \lg Q)^{n+1}\right)$.

1991 *Mathematics Subject Classification.* Primary 81-01, 81P68.

Key words and phrases. Shor's algorithm, hidden subgroup algorithms, quantum computation, quantum algorithms.

This effort partially supported by the Defense Advanced Research Projects Agency (DARPA) and Air Force Research Laboratory, Air Force Materiel Command, USAF, under agreement number F30602-01-2-0522, the National Institute for Standards and Technology (NIST), and by L-O-O-P Fund Grant WADC2000. The U.S. Government is authorized to reproduce and distribute reprints for Government purposes notwithstanding any copyright annotations thereon. The views and conclusions contained herein are those of the authors and should not be interpreted as necessarily representing the official policies or endorsements, either expressed or implied, of the Defense Advanced Research Projects Agency, the Air Force Research Laboratory, or the U.S. Government. (Copyright 2002 by authors. Reproduction of this article, in its entirety, is permitted for non-commercial purposes.)

©2002 by authors. Reproduction of this article, in its entirety, is permitted for non-commercial purposes.

Contents

Part 1. Preamble

1. Introduction
2. An example of Shor's quantum factoring algorithm
3. Definition of the hidden subgroup problem (HSP) and hidden subgroup algorithms (HSAs)

Part 2. Algebraic Preliminaries

4. The Character Group
5. Fourier analysis on a finite abelian group
6. Implementation issues: Group algebras as Hilbert spaces

Part 3. $QRand_\varphi()$: The Progenitor of All QHSAs

7. Implementing $Prob_\varphi(\chi)$ with quantum subroutine $\text{QRAND}_\varphi()$

Part 4. Vintage Simon Algorithms

8. Properties of the probability distribution $Prob_\varphi(\chi)$ when φ has a hidden subgroup
9. A Markov process $\mathcal{M}_\varphi$ induced by $Prob_\varphi$
10. Vintage Simon quantum hidden subgroup algorithms (QHSAs)

Part 5. Vintage Shor Algorithms

11. Vintage Shor quantum hidden subgroup algorithms(QHSAs)
12. Direct summand structure
13. Vintage Shor QHSAs with group probe $\widetilde{A} = \mathbb{Z}_Q$.
14. Finding Shor transversals for vintage $\mathbb{Z}_Q$ Shor algorithms
15. Maximal Shor transversals
16. Identifying characters of cyclic groups with points on the unit circle $\mathbb{S}^1$ in the complex plane $\mathbb{C}$.
17. Group norms
18. Vintage $\mathbb{Z}_Q$ Shor QHSAs (Cont.)
19. When are characters of $\widetilde{A} = \mathbb{Z}_Q$ close to some character of a maximal cyclic subgroup $\mathbb{Z}_P$ of H_φ?
20. Summary of Vintage $\mathbb{Z}_Q$ Shor QHSAs
21. A cursory analysis of complexity
22. Two alternative vintage $\mathbb{Z}_Q$ Shor algorithms

Part 6. Epilogue

23. Conclusion
24. Acknowledgement
25. Appendix A. Continued fractions
26. Appendix B. Probability Distributions on Integers

References

Part 1. Preamble

1. Introduction

The ultimate objective of this paper is to create a stepping stone to the development of new quantum algorithms. The strategy chosen is to begin by focusing on the class of abelian quantum hidden subgroup algorithms (QHSAs), i.e., the class of abelian algorithms of the Shor/Simon genre. Our strategy is to make this class of algorithms as mathematically transparent as possible. By the phrase "mathematically transparent," we mean to expose, to bring to the surface, and to make explicit the concealed mathematical structures that are inherently and fundamentally a part of such algorithms. In so doing, we create a class of symbolic abelian QHSAs that are analogous to those symbolic algorithms found within such software packages as Axiom, Cayley, Magma, Maple, and Mathematica.

During this mathematical analysis, the differences between the Simon and Shor quantum algorithms become dramatically apparent. This is in spite of the fact that these two share a common ancestor, namely, the quantum random group character generator QRAND, described herein. While the Simon algorithm is a QHSA on finite abelian groups which produces random characters of the hidden quotient group, the Shor algorithm is a QHSA on free abelian finite rank groups which produces random characters of a group which "approximate" the hidden quotient group. It is misleading, and a frequent cause of much confusion in the open literature, to call them both essentially the same QHSA.

Surprisingly, these two very different algorithms touch an amazing array of different mathematical disciplines, from the obvious to the not-so-obvious, requiring the integration of many diverse fields of mathematics. Shor's quantum factoring algorithm, for example, depends heavily on the interplay of two metrics on the unit circle $\mathbb{S}^1$, namely the arclength metric $\text{ARC}_{2\pi}$ and the chordal metric $\text{CHORD}_{2\pi}$. This observation greatly simplifies the analysis of the Shor factoring algorithm, while at the same time revealing more of the structure concealed within the algorithm.

As a spin-off of this effort, we create three different generalizations of Shor's quantum factoring algorithm to free abelian groups of finite rank, found in sections 20 and 22. We refer to these algorithms as wandering (or vintage $\mathbb{Z}_Q$) Shor algorithms. They are essentially QHSAs on free abelian finite rank n groups A which, with each iteration, first select a random cyclic direct summand $\mathbb{Z}$ of the group A and then apply one iteration of the standard Shor algorithm to produce a random character of the "approximating" finite group $\widetilde{A}$, called a group probe. These algorithms find either the order P of a maximal cyclic subgroup $\mathbb{Z}_P$ of the hidden quotient group H_φ, or the entire hidden quotient group H_φ. An integral part of these wandering algorithms is the selection of a very special random transversal $\iota_\mu : \widetilde{A} \longrightarrow A$, which we refer to as a Shor transversal. The algorithmic time complexity of the first of these wandering (or vintage $\mathbb{Z}_Q$) algorithms is found in theorem 11 of section 21 to be $O\left(n^2 \left(\lg Q\right)^3 \left(\lg \lg Q\right)^{n+1}\right)$, where n denotes the fixed finite rank of the free abelian group A. Theorem 11 is based on the assumptions also found in section 21. This asymptotic bound is by no means the tightest possible.

Throughout this paper, it is assumed that the reader is familiar with the class of quantum hidden subgroup algorithms. For an introductions to this subject, please refer, for example, to any one of the references [8], [25], [26], [29], [33], [36], [43], [44]. This paper focuses, in particular, on the abelian hidden subgroup problem (HSP), with eye toward future work by the authors on the non-abelian HSP. There is a great deal of literature on the abelian HSP, for example, [5], [14], [25], [26], [27], [29], [35], [36], [43], [44], [45]. For literature on the non-abelian hidden subgroup problem, see for example, [16], [24], [14], [26], [33], [36], [38], [39], [41], [48].

2. An example of Shor's quantum factoring algorithm

As an example of what we would like to make mathematically transparent, consider the following instance of Peter Shor's quantum factoring algorithm. A great part of this paper is devoted to exposing and bringing to the surface the many concealed mathematical structures that are inherently and fundamentally part of this example.

Perhaps you see them? Perhaps you find them to be self evident? If you do, then you need read no more of this paper, although you are most certainly welcome to read on. If, on the other hand, the following example leaves you with a restless, uneasy feeling of not fully understanding what is really going on (i.e., of not fully understanding what concealed mathematical structures are lurking underneath these calculations), then you are invited to read the remainder of this paper.

Peter Shor's quantum factoring algorithm reduces the task of factoring a positive integer N to first finding a random integer a relatively prime to N, and then next to determining the period P of the following function

$$\begin{aligned} \mathbb{Z} &\xrightarrow{\varphi} \mathbb{Z} \bmod N \\ x &\longmapsto a^x \bmod N \;, \end{aligned}$$

where $\mathbb{Z}$ denotes the additive group of integers, and where $\mathbb{Z} \bmod N$ denotes the integers $\bmod N$ under multiplication[1].

Since $\mathbb{Z}$ is an infinite group, Shor chooses to work instead with the finite additive cyclic group $\mathbb{Z}_Q$ of order $Q = 2^m$, where $N^2 \leq Q < 2N^2$, and with the

---

[1] A random integer a with $\gcd(a, N) = 1$ is found by selecting a random integer, and then applying the Euclidean algorithm to determine whether or not it is relatively prime to N. If not, then the gcd is a non-trivial factor of N, and there is no need to proceed futher. However, this possibility is highly unlikely if N is large.

"approximating" map

$$\begin{array}{ccc} \mathbb{Z}_Q & \xrightarrow{\widetilde{\varphi}} & \mathbb{Z} \bmod N \\ x & \longmapsto & a^x \bmod N, \quad 0 \le x < Q \end{array}$$

Shor begins by constructing a quantum system with two quantum registers

$$|\text{LEFT\_REGISTER}\rangle \, |\text{RIGHT\_REGISTER}\rangle \, ,$$

the left intended to hold the arguments x of $\widetilde{\varphi}$, the right to hold the corresponding values of $\widetilde{\varphi}$. This quantum system has been constructed with a unitary transformation

$$U_{\widetilde{\varphi}} : |x\rangle \, |1\rangle \longmapsto |x\rangle \, |\widetilde{\varphi}(x)\rangle$$

implementing the "approximating" map $\widetilde{\varphi}$.

As an example, let us use Shor's algorithm to factor the enormous ☺ integer $N = 21$, assuming that $a = 2$ has been randomly chosen. Thus, $Q = 2^9 = 512$.

Unknown to Peter Shor, the period is $P = 6$, and hence, $Q = 6 \cdot 85 + 2$.

Shor proceeds by executing the following steps:

$\boxed{\text{STEP 0}}$ Initialize

$$|\psi_0\rangle = |0\rangle \, |1\rangle$$

$\boxed{\text{STEP 1}}$ Apply the Fourier transform

$$\mathcal{F} : |u\rangle \longmapsto \frac{1}{\sqrt{512}} \sum_{x=0}^{511} \omega^{ux} |x\rangle$$

to the left register, where $\omega = \exp(2\pi i/512)$ is a primitive 512-th root of unity, to obtain

$$|\psi_1\rangle = \frac{1}{\sqrt{512}} \sum_{x=0}^{511} |x\rangle \, |1\rangle$$

$\boxed{\text{STEP 2}}$ Apply the unitary transformation

$$U_{\widetilde{\varphi}} : |x\rangle \, |1\rangle \longmapsto |x\rangle \, |2^x \bmod 21\rangle$$

to obtain

$$|\psi_2\rangle = \frac{1}{\sqrt{512}} \sum_{x=0}^{511} |x\rangle \, |2^x \bmod 21\rangle$$

STEP 3 Once again apply the Fourier transform

$$\mathcal{F}: |x\rangle \longmapsto \frac{1}{\sqrt{512}} \sum_{y=0}^{511} \omega^{xy} |y\rangle$$

to the left register to obtain

$$|\psi_3\rangle = \frac{1}{512} \sum_{x=0}^{511} \sum_{y=0}^{511} \omega^{xy} |y\rangle |2^x \bmod 21\rangle = \frac{1}{512} \sum_{y=0}^{511} |y\rangle \left(\sum_{x=0}^{511} \omega^{xy} |2^x \bmod 21\rangle \right)$$

$$= \frac{1}{512} \sum_{y=0}^{511} |y\rangle |\Upsilon(y)\rangle$$

where

$$|\Upsilon(y)\rangle = \sum_{x=0}^{511} \omega^{xy} |2^x \bmod 21\rangle$$

STEP 4 Measure the left register. Then with Probability

$$Prob_{\widetilde{\varphi}}(y) = \frac{\langle \Upsilon(y) | \Upsilon(y) \rangle}{(512)^2}$$

the state will "collapse" to $|y\rangle$ with the value measured being the integer y, where $0 \leq y < Q$.

Let us digress for a moment to find a more usable expression for the probability distribution $Prob_{\widetilde{\varphi}}(y)$.

$$|\Upsilon(y)\rangle = \sum_{x=0}^{511} \omega^{xy} |2^x \bmod 21\rangle$$

$$= \sum_{x_1=0}^{85-1} \sum_{x_0=0}^{6-1} \omega^{(6x_1+x_0)y} |2^{6x_1+x_0} \bmod 21\rangle + \sum_{x_0=0}^{2-1} \omega^{(6 \cdot 85 + x_0)y} |2^{6 \cdot 85 + x_0} \bmod 21\rangle$$

But the order of $a = 2$ modulo 21 is $P = 6$, i.e., $P = 6$ is the smallest positive integer such that $2^6 = 1 \bmod 21$. Hence, the above expression becomes

$$|\Upsilon(y)\rangle = \left(\sum_{x_1=0}^{84} \omega^{6x_1 y} \right) \sum_{x_0=0}^{5} \omega^{x_0 y} |2^{x_0} \bmod 21\rangle + \omega^{6 \cdot 85 y} \sum_{x_0=0}^{1} \omega^{x_0 y} |2^{x_0} \bmod 21\rangle$$

$$= \left(\sum_{x_1=0}^{85} \omega^{6x_1 y} \right) \sum_{x_0=0}^{1} \omega^{x_0 y} |2^{x_0} \bmod 21\rangle + \left(\sum_{x_1=0}^{84} \omega^{6x_1 y} \right) \sum_{x_0=2}^{5} \omega^{x_0 y} |2^{x_0} \bmod 21\rangle$$

Since the kets $\{\ |2^{x_0} \bmod 21\rangle \ |\ 0 \leq x_0 < 6\ \}$ are all distinct, we have

$$\langle \Upsilon(y) \mid \Upsilon(y) \rangle = 2 \left| \sum_{x_1=0}^{85} \omega^{6x_1 y} \right|^2 + 4 \left| \sum_{x_1=0}^{84} \omega^{6x_1 y} \right|^2.$$

After a little algebraic manipulation, we finally have the following expression for $Prob_{\widetilde{\varphi}}(y)$:

$$Prob_{\varphi}(y) = \frac{\langle \Upsilon(y) \mid \Upsilon(y) \rangle}{(512)^2} = \begin{cases} \dfrac{\sin^2\left(\frac{\pi y}{128}\right) + 2\sin^2\left(\frac{\pi y}{256}\right)}{(131072)\sin^2\left(\frac{3\pi y}{256}\right)} & \text{if } y \neq 0 \text{ or } 256 \\ \dfrac{10923}{65536} & \text{if } y = 0 \text{ or } 256 \end{cases}$$

A plot of $Prob_{\widetilde{\varphi}}(y)$ is shown in Figure 1.

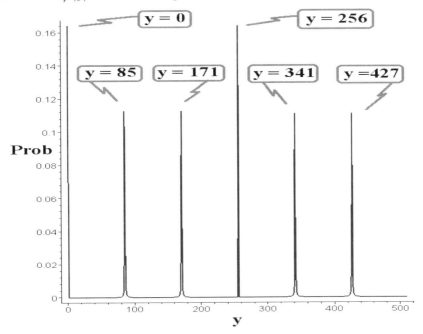

Figure 1. A plot of $\mathbf{Prob}_{\widetilde{\varphi}}(\mathbf{y})$.

The peaks in the above plot of $Prob_{\widetilde{\varphi}}(y)$ occur at the integers

$$y = 0,\ 85,\ 171,\ 256,\ 341,\ 427.$$

The probability that at least one of these six integers will occur is quite high. It is actually 0.78^+. Indeed, the probability distribution has been intentionally engineered to make the probability of these particular integers as high as possible. And there is a good reason for doing so.

The above six integers are those for which the corresponding rational y/Q is "closest" to a rational of the form d/P. By "closest" we mean that

$$\left| \frac{y}{Q} - \frac{d}{P} \right| < \frac{1}{2Q} < \frac{1}{2P^2}.$$

In particular,
$$\frac{0}{512}, \frac{85}{512}, \frac{171}{512}, \frac{256}{512}, \frac{341}{512}, \frac{427}{512}$$
are rationals respectively "closest" to the rationals
$$\frac{0}{6}, \frac{1}{6}, \frac{2}{6}, \frac{3}{6}, \frac{4}{6}, \frac{5}{6}.$$
So by theorem 12 of Appendix A, the six rational numbers $0/6$, $1/6$, ..., $5/6$ are convergents of the continued fraction expansions of $0/512$, $85/512$, ..., $427/512$, respectively. Hence, each of the six rationals $0/6$, $1/6$, ..., $5/6$ can be found with the recursion given in Appendix A.

But ..., we are not searching for rationals of the form d/P. Instead, we seek only the denominator $P = 6$.

Unfortunately, the denominator $P = 6$ can only be gotten from the continued fraction recursion when the numerator and denominator of d/P are relatively prime. Given that the algorithm has selected one of the random integers 0, 85, ..., 427, the probability that the corresponding rational d/P has relatively prime numerator and denominator is $\phi(6)/6 = 1/3$, where $\phi(-)$ denotes the Euler totient function. So the probability of finding $P = 6$ is actually not 0.78^+, but is instead 0.23^-.

From Peter Shor's perspective, the expression for the probability distribution is not known, since the period P is not known. All that Peter sees is a random integer y produced by the probability distribution $Prob_{\widetilde{\varphi}}$. However, he does know an approximate lower bound for the probability that the random y produced by $Prob_{\widetilde{\varphi}}$ is a "closest" one, namely the approximate lower bound $4/\pi^2 = 0.41^-$. Also, because[2]
$$\liminf \frac{\phi(N) \ln \ln N}{N} = e^{-\gamma},$$
where $\gamma = 0.5772\cdots$ denotes Euler's constant, he knows that
$$\frac{\phi(P)}{P} = \Omega\left(\frac{1}{\lg \lg N}\right).$$
Hence, if he repeats the algorithm $O(\lg \lg N)$ times[3], he will obtain one of the desired integers y with probability bounded below by approximately $4/\pi^2$.

However, once he has in his possession a candidate P' for the actual period $P = 6$, the only way he can be sure he has the correct period P is to test P' by computing $2^{P'} \bmod 21$. If the result is 1, he is certain he has found the correct period P. This last part of the computation is done by the repeated squaring algorithm[4].

---

[2]Please refer to reference [**21**, Theorem 328, Section 18.4].

[3]For even tighter asymptotic bounds, please refer to [**9**] and [**37**].

[4]By the repeated squaring algorithm, we mean the algorithm which computes $a^{P'} \bmod N$ via the expression
$$a^{P'} = \prod_j \left(a^{2^j}\right)^{P'_j},$$
where $P' = \sum_j P'_j 2^j$ is the radix 2 expansion of P'.

3. Definition of the hidden subgroup problem (HSP) and hidden subgroup algorithms (HSAs)

We now proceed by defining what is meant by a hidden subgroup problem (HSP) and a corresponding hidden subgroup algorithm. For other perspectives on HSPs, please refer to [**29**], [**27**], [**35**].

DEFINITION 1. *A map $\varphi : A \longrightarrow S$ from a group A into a set S is said to have* **hidden subgroup structure** *if there exists a subgroup K_φ of A, called a* **hidden subgroup**, *and an injection $\iota_\varphi : A/K_\varphi \longrightarrow S$, called a* **hidden injection**, *such that the diagram*

$$\begin{array}{ccc} A & \xrightarrow{\varphi} & S \\ {\scriptstyle \nu} \searrow & & \nearrow {\scriptstyle \iota_\varphi} \\ & A/K_\varphi & \end{array}$$

is commutative, where A/K_φ denotes the collection of right cosets of K_φ in A, and where $\nu : A \longrightarrow A/K_\varphi$ is the natural map of A onto A/K_φ. We refer to the group A as the **ambient group** *and to the set S as the* **target set**. *If K_φ is a normal subgroup of A, then $H_\varphi = A/K_\varphi$ is a group, called the* **hidden quotient group**, *and $\nu : A \longrightarrow A/K_\varphi$ is an epimorphism, called the* **hidden epimorphism**.

The hidden subgroup problem can be expressed as follows:

PROBLEM 1 (**Hidden Subgroup Problem (HSP)**). *Given a map with hidden subgroup structure*

$$\varphi : A \longrightarrow S \ ,$$

determine a hidden subgroup K_φ of A. An algorithm solving this problem is called a **hidden subgroup algorithm (HSA)**.

The corresponding quantum form of this HSP is stated as follows:

PROBLEM 2 (**Hidden Subgroup Problem: Quantum Version**). *Let*

$$\varphi : A \longrightarrow S$$

be a map with hidden subgroup structure. Construct a quantum implementation of the map φ as follows:

Let $\mathcal{H}_A$ and $\mathcal{H}_S$ be Hilbert spaces defined respectively by the orthonormal bases

$$\{\ |a\rangle \mid a \in A\ \} \quad \text{and} \quad \{\ |s\rangle \mid s \in S\ \} \ ,$$

and let $s_0 = \varphi(0)$, where 0 denotes the identity of the ambient group A. Finally, let U_φ be the unitary transformation

$$\begin{array}{rcl} U_\varphi : \mathcal{H}_A \otimes \mathcal{H}_S & \longrightarrow & \mathcal{H}_A \otimes \mathcal{H}_S \\ |a\rangle |s_0\rangle & \longmapsto & |a\rangle |\varphi(a)\rangle \end{array},$$

Determine the hidden subgroup K_φ with bounded probability of error by making as few queries as possible of the blackbox U_φ. A quantum algorithm solving this problem is called a **quantum hidden subgroup algorithm (QHSA)**.

In this paper, we focus on the **abelian hidden subgroup problem (AHSP)**, i.e., the HSP with the ambient group A assumed to be a finitely generated abelian group, and where the image of the hidden morphism φ is a **finite subset** of S. (We will also on occasion assume that the entire set S is finite.)

In this paper we focus on the following two classes of abelian hidden subgroup problems:[5]

- **Vintage Simon AHSP.** The ambient group A is finite and abelian.
- **Vintage Shor AHSP.** The ambient group A is free abelian of finite rank.

$\boxed{\text{NOTATION CONVENTION}}$ For notational simplicity, *throughout this paper we will use additive notation for both the ambient group A and the hidden subgroup K_φ, and multiplicative notation for the hidden quotient group $H_\varphi = A/K_\varphi$.*[6]

Part 2. Algebraic Preliminaries

4. The Character Group

Let G be an abelian group. Then the **character group** (or, **dual group**) $\widehat{G}$ of G is defined as the group of all morphisms of G into the group $\mathbb{S}^1$, i.e.,

$$\widehat{G} = Hom\left(G, \mathbb{S}^1\right)$$

where $\mathbb{S}^1$ denotes the group of orientation preserving symmetries of the standard circle, and where multiplication on $\widehat{G}$ is defined as:

$$(f_1 f_2)(g) = f_1(g) f_2(g) \quad \text{for all } f_1, f_2 \in \widehat{G}$$

The elements of $\widehat{G}$ are called **characters**.[7]

REMARK 1. *The group $\mathbb{S}^1$ can be identified with*
1) *The multiplicative group $U(1) = \{\ e^{2\pi i x}\ |\ x \in \mathbb{R}\ \}$, i.e., with multiplication defined by $e^{2\pi i \alpha} \cdot e^{2\pi i \beta} = e^{2\pi i (\alpha + \beta)}$*

2) *The additive group $2\pi \mathbb{R}/2\pi \mathbb{Z}$, i.e., the reals modulo 2π under addition, i.e., with addition defined as*

$$2\pi\alpha + 2\pi\beta \mod 2\pi = 2\pi (\alpha + \beta \mod 1)$$

---

[5]For the general abelian HSP, please refer to [**8**] and [**29**].
[6]This follows the notational convention found in [**43**].
[7]More generally, for non-abelian groups, a character is defined as the trace of a representation of the group.

REMARK 2. *Please note that the 1-sphere $\mathbb{S}^1$ can be thought of as a $\mathbb{Z}$-module under the action*

$$(n, 2\pi\alpha) \longmapsto 2\pi(n\alpha \bmod 1)$$

THEOREM 1. *Every finite abelian group G is isomorphic to the direct product of cyclic groups, i.e.,*

$$G \cong \mathbb{Z}_{m_1} \times \mathbb{Z}_{m_2} \times \ldots \times \mathbb{Z}_{m_\ell},$$

where $\mathbb{Z}_{m_j}$ denotes the cyclic group of order m_j.

THEOREM 2. *Let G be a finite abelian group. If $G = G_1 \times G_2$, then $\widehat{G} = \widehat{G}_1 \times \widehat{G}_2$.*

THEOREM 3. $\widehat{\mathbb{Z}}_m \cong \mathbb{Z}_m$

COROLLARY 1. *If G is a finite abelian group, then $G \cong \widehat{G}$.*

REMARK 3. *The isomorphism $G \cong \widehat{G}$ can be expressed more explicitly as follows:*

Let $G \cong \mathbb{Z}_{m_1} \times \mathbb{Z}_{m_2} \times \ldots \times \mathbb{Z}_{m_\ell}$, and let $g_1, g_2, \ldots, g_\ell$ denote generators of $\mathbb{Z}_{m_1}, \mathbb{Z}_{m_2}, \ldots, \mathbb{Z}_{m_\ell}$ respectively. Moreover, let $\omega_1, \omega_2, \ldots, \omega_\ell$ be m_1-th, m_2-th, $\ldots$, m_ℓ-th primitive roots of unity, respectively. Then the character $\widetilde{\chi}_j$ of $\mathbb{Z}_{m_j}$ defined by

$$\widetilde{\chi}_j(g_j) = \omega_j$$

generates $\widehat{\mathbb{Z}}_{m_j}$ as a cyclic group, i.e., the powers $(\widetilde{\chi}_j)^k$ generate $\widehat{\mathbb{Z}}_{m_j}$. Moreover, the characters χ_j of G defined by

$$\chi_j = \left(\prod_{i=0}^{j-1} \widetilde{\chi}_i^0\right) \widetilde{\chi}_j \left(\prod_{i=j+1}^{\ell} \widetilde{\chi}_i^0\right)$$

generate $\widehat{G}$. It follows that an isomorphism $G \cong \widehat{G}$ is given by

$$g_j \longleftrightarrow \chi_j$$

NOTATION CONVENTION *In general, we will not need to represent the isomorphism $G \cong \widehat{G}$ as explicitly as stated above. We will use the following convention. Let $\{g_1, g_2, \ldots, g_\ell\}$ and $\{\chi_1, \chi_2, \ldots, \chi_\ell,\}$ denote respectively the set of elements of G and $\widehat{G}$ indexed in such a way that*

$$g_j \longleftrightarrow \chi_j$$

is the chosen isomorphism of G and $\widehat{G}$. We will at times use the notation

$$\begin{cases} g \longleftrightarrow \chi_g \\ \chi \longleftrightarrow g_\chi \end{cases}$$

5. Fourier analysis on a finite abelian group

As in the previous section, let G be a **finite abelian group**[8] and let $\widehat{G}$ denote is character group. Let g and χ denote respectively elements of the groups G and $\widehat{G}$.

Let $\mathbb{C}G$ and $\mathbb{C}\widehat{G}$ denote the corresponding group algebras of G and $\widehat{G}$ over the complex numbers $\mathbb{C}$. Hence, $\mathbb{C}G$ consists of all maps $f : G \longrightarrow \mathbb{C}$. Addition '+', multiplication '•', and scalar multiplication are defined as:

$$\begin{cases} (f_1 + f_2)(g) & = & f_1(g) + f_2(g) & \forall g \in G \\ \\ (f_1 \bullet f_2)(g) & = & \displaystyle\sum_{h \in G} f_1(h) f_2(h^{-1}g) & \forall g \in G \qquad \textbf{(Convolution)} \\ \\ (\lambda f)(g) & = & \lambda f(g) & \forall \lambda \in \mathbb{C} \text{ and } \forall g \in G \end{cases}$$

Caveat. *Please note that the symbol g has at least three different meanings:*

♦ INTERPRETATION 1. *The symbol g denotes an element of the group G*

♦ INTERPRETATION 2. *The symbol g denotes a pointwise map*

$$g : G \longrightarrow \mathbb{C}$$

defined by

$$g(g') = \begin{cases} 1 & \text{if } g = g' \\ 0 & \text{otherwise} \end{cases}$$

Thus,

$$f = \sum_{g \in G} f(g) g \text{ denotes } g \longmapsto f(g)$$

Hence, $g \in \mathbb{C}G$. Since G is isomorphic as a group to the set of pointwise maps $\{g : G \longrightarrow \mathbb{C} \mid g \in G\}$ under convolution, we can and do identify the group elements of G with the pointwise maps $g \in \mathbb{C}G$. Thus, INTERPRETATIONS 1 *and* 2 *lead to no ambiguity at the algebraic level.*

♦ INTERPRETATION 3. *The symbol g denotes a character of $\widehat{G}$ defined by*

$$g(\chi) = \chi(g)$$

Thus, with this interpretation, $g \in \widehat{\widehat{G}} \subset \mathbb{C}\widehat{\widehat{G}}$. This third interpretation can, in some instances, lead to some unnecessary confusion. When this intended interpretation is possibly not clear from context, we will resort to the notation

$$g^\bullet$$

---

[8]If G is infinite, then ring multiplication '•' is not always well defined. So $\mathbb{C}G$ is not a ring, but a $\mathbb{Z}$-module, with a group of operators. One way of of making $\mathbb{C}G$ into a ring, is to restrict the maps on G, e.g., to maps with compact support, to maps with L^2 norm, etc.

for INTERPRETATION 3 of the symbol g. Thus, for example,
$$f^{\bullet} = \sum_{g \in G} f(g) g^{\bullet} \text{ denotes the map } \chi \longmapsto \sum_{g \in G} f(g) \chi(g)$$

In like manner, *the symbol χ has at least three different meanings:*
- INTERPRETATION $\widehat{1}$. The symbol χ denotes an element of the group $\widehat{G}$
- INTERPRETATION $\widehat{2}$. The symbol χ denotes a pointwise map
$$\chi : \widehat{G} \longrightarrow \mathbb{C}$$
defined by
$$\chi(\chi') = \begin{cases} 1 & \text{if } \chi = \chi' \\ 0 & \text{otherwise} \end{cases}$$

Thus,
$$\widehat{f} = \sum_{\chi \in \widehat{G}} \widehat{f}(\chi) \chi \text{ denotes } \chi \longmapsto \widehat{f}(\chi)$$

Hence, $\chi \in \mathbb{C}\widehat{G}$. Since $\widehat{G}$ *is isomorphic as a group to the set of pointwise maps* $\left\{ \chi : \widehat{G} \longrightarrow \mathbb{C} \mid \chi \in \widehat{G} \right\}$ *under convolution, we can and do identify the group elements of $\widehat{G}$ with the pointwise maps $\chi \in \mathbb{C}\widehat{G}$. Thus,* INTERPRETATIONS $\widehat{1}$ *and* $\widehat{2}$ *lead to no ambiguity at the algebraic level.*

- INTERPRETATION $\widehat{3}$. The symbol χ denotes a character map of G onto $\mathbb{C}$ defined by
$$g \longmapsto \chi(g)$$
Thus, with this interpretation, $\chi \in \mathbb{C}G$. This third interpretation can, in some instances, also lead to some unnecessary confusion. When this intended interpretation is possibly not clear from context, we will resort to the notation
$$\chi^{\bullet}$$
for INTERPRETATION $\widehat{3}$ *of the symbol χ. Thus, for example,*
$$\widehat{f}^{\bullet} = \sum_{\chi \in \widehat{G}} \widehat{f}(\chi) \chi^{\bullet} \text{ denotes the map } g \longmapsto \sum_{\chi \in \widehat{G}} \widehat{f}(\chi) \chi(g)$$

We define complex inner products on the group algebras $\mathbb{C}G$ and $\mathbb{C}\widehat{G}$ as follows:
$$\begin{cases} (f_1, f_2) &= \frac{1}{|G|} \sum_{g \in G} f_1(g) \overline{f_2(g)} \quad \forall f_1, f_2 \in \mathbb{C}G \\ \left(\widehat{f_1}, \widehat{f_2}\right) &= \frac{1}{|G|} \sum_{\chi \in \widehat{G}} \widehat{f_1}(\chi) \overline{\widehat{f_2}(\chi)} \quad \forall \widehat{f_1}, \widehat{f_2} \in \mathbb{C}\widehat{G} \end{cases}$$

where $\overline{f_2(g)}$ and $\overline{\widehat{f_2}(\chi)}$ denote respectively the complex conjugates of $f_2(g)$ and $\widehat{f_2}(\chi)$.

The corresponding norms are defined as

$$\begin{cases} \|f\| = \sqrt{(f,f)} & \forall f \in \mathbb{C}G \\ \|\widehat{f}\| = \sqrt{(\widehat{f},\widehat{f})} & \forall \widehat{f} \in \mathbb{C}\widehat{G} \end{cases}$$

As an immediate consequence of the above definitions, we have:

$$(g_1, g_2) = \begin{cases} 1 & \text{if } g_1 = g_2 \\ 0 & \text{otherwise} \end{cases} \quad \text{and} \quad (\chi_1, \chi_2) = \begin{cases} 1 & \text{if } \chi_1 = \chi_2 \\ 0 & \text{otherwise} \end{cases}$$

It also follows from the standard character identities that

$$(g_1^\bullet, g_2^\bullet) = \begin{cases} 1 & \text{if } g_1^\bullet = g_2^\bullet \\ 0 & \text{otherwise} \end{cases} \quad \text{and} \quad (\chi_1^\bullet, \chi_2^\bullet) = \begin{cases} 1 & \text{if } \chi_1^\bullet = \chi_2^\bullet \\ 0 & \text{otherwise} \end{cases}$$

We are now in a position to define the Fourier transform on a finite abelian group G.

DEFINITION 2. *The **Fourier transform** $\mathcal{F}$ for a finite abelian group G is defined as*

$$\mathbb{C}G \xrightarrow{\mathcal{F}} \mathbb{C}\widehat{G}$$

$$f \longmapsto \widehat{f} = \frac{1}{\sqrt{|G|}} \sum_{g \in G} f(g)\overline{g^\bullet} = \frac{1}{\sqrt{|G|}} \sum_{\chi \in \widehat{G}} \left(\sum_{g \in G} f(g)\overline{\chi}(g) \right) \chi$$

Hence,

$$\widehat{f}(\chi) = \sqrt{|G|}\,(f, \chi^\bullet) = \frac{1}{\sqrt{|G|}} \sum_{g \in G} f(g)\overline{\chi(g)}$$

PROPOSITION 1.

$$f = \frac{1}{\sqrt{|G|}} \sum_{\chi \in \widehat{G}} \widehat{f}(\chi)\chi^\bullet$$

PROOF.

$$\frac{1}{\sqrt{|G|}} \sum_{\chi \in \widehat{G}} \widehat{f}(\chi)\chi(g_0) = \frac{1}{\sqrt{|G|}} \sum_{\chi \in \widehat{G}} \frac{1}{\sqrt{|G|}} \sum_{g \in G} f(g)\overline{\chi(g)}\chi(g_0)$$

$$= \frac{1}{|G|} \sum_{g \in G} f(g) \sum_{\chi \in \widehat{G}} \overline{\chi(g)}\chi(g_0) = f(g_0)$$

$\square$

We define the inverse Fourier transform as follows:

DEFINITION 3. *The inverse Fourier transform $\mathcal{F}^{-1}$ is defined as*

$$\begin{array}{ccc} \mathbb{C}\widehat{G} & \stackrel{\mathcal{F}^{-1}}{\longrightarrow} & \mathbb{C}G \\ \widehat{f} & \longmapsto & f = \frac{1}{\sqrt{|G|}} \sum_{\chi \in \widehat{G}} \widehat{f}(\chi) \chi^{\bullet} \end{array}$$

Hence,

$$f(g) = \sqrt{|G|}\left(\widehat{f}, \overline{g^{\bullet}}\right) = \frac{1}{\sqrt{|G|}} \sum_{\chi \in \widehat{G}} \widehat{f}(\chi)\chi(g)$$

THEOREM 4 (Plancherel identity).

$$\|f\| = \left\|\widehat{f}\right\|$$

PROOF.

$$\|f\|^2 = (f,f) = \frac{1}{|G|}\sum_{g \in G} |f(g)|^2$$

$$= \frac{1}{|G|}\sum_{g \in G} \frac{1}{\sqrt{|G|}} \sum_{\chi \in \widehat{G}} \widehat{f}(\chi)\chi(g) \overline{\left(\frac{1}{\sqrt{|G|}} \sum_{\chi' \in \widehat{G}} \widehat{f}(\chi')\chi'(g)\right)}$$

$$= \frac{1}{|G|^2}\sum_{g \in G}\sum_{\chi \in \widehat{G}}\sum_{\chi' \in \widehat{G}} \widehat{f}(\chi)\overline{\widehat{f}(\chi')}\chi(g)\overline{\chi'}(g)$$

$$= \frac{1}{|G|^2}\sum_{\chi \in \widehat{G}}\sum_{\chi' \in \widehat{G}} \widehat{f}(\chi)\overline{\widehat{f}(\chi')}\left(\sum_{g \in G}\chi(g)\overline{\chi'}(g)\right)$$

$$= \frac{1}{|G|}\sum_{\chi \in \widehat{G}} \left|\widehat{f}(\chi)\right|^2 = \left\|\widehat{f}\right\|^2$$

□

6. Implementation issues: Group algebras as Hilbert spaces

For implementation purposes, we will need to view group algebras also as Hilbert spaces.[9]

In particular, $\mathbb{C}G$ and $\mathbb{C}\widehat{G}$ can be respectively viewed as the Hilbert spaces $\mathcal{H}_G$ and $\mathcal{H}_{\widehat{G}}$ defined by the respective orthonormal bases

$$\{|g\rangle \mid g \in G\} \text{ and } \left\{|\chi\rangle \mid \chi \in \widehat{G}\right\}.$$

---

[9]Category theorists will recognize this as a forgetful functor.

In this context, the **Fourier transform** $\mathcal{F}$ becomes

$$\mathcal{H}_G \xrightarrow{\mathcal{F}} \mathcal{H}_{\widehat{G}}$$

$$|f\rangle = \sum_{g \in G} f(g) |g\rangle \longmapsto \left|\widehat{f}\right\rangle = \frac{1}{\sqrt{|G|}} \sum_{\chi \in \widehat{G}} \left(\sum_{g \in G} f(g) \overline{\chi}(g) \right) |\chi\rangle$$

and the **inverse Fourier transform** $\mathcal{F}^{-1}$ becomes

$$\mathcal{H}_{\widehat{G}} \xrightarrow{\mathcal{F}^{-1}} \mathcal{H}_G$$

$$\left|\widehat{f}\right\rangle \longmapsto |f\rangle = \frac{1}{\sqrt{|G|}} \sum_{g \in G} \left(\sum_{\chi \in \widehat{G}} \widehat{f}(\chi) \chi(g) \right) |g\rangle$$

One important and useful identification is to use the Hilbert space isomorphism

$$\begin{array}{ccc} \mathcal{H}_G & \longleftrightarrow & \mathcal{H}_{\widehat{G}} \\ |g\rangle & \longleftrightarrow & |\chi_g\rangle \\ |g_\chi\rangle & \longleftrightarrow & |\chi\rangle \end{array}$$

to identify the two Hilbert spaces $\mathcal{H}_G$ and $\mathcal{H}_{\widehat{G}}$. As a result, the Fourier transform $\mathcal{F}$ and it's inverse $\mathcal{F}^{-1}$ can both be viewed as transforms taking the Hilbert space $\mathcal{H}_G$ to itself, i.e.,

$$\mathcal{H}_G \underset{\mathcal{F}^{-1}}{\overset{\mathcal{F}}{\rightleftarrows}} \mathcal{H}_G$$

REMARK 4. *This last identification is crucial for the implementation of hidden subgroup algorithms.*

Part 3. QRand$_\varphi$(): The Progenitor of All QHSAs

7. Implementing $Prob_\varphi(\chi)$ with quantum subroutine QRand$_\varphi$()

Let
$$\varphi : A \longrightarrow S$$
be a map from a finite abelian group A into a finite set S.

We use additive notation for the group A; and let $s_0 = \varphi(0)$ denote the image of the identity 0 of A under the map φ.

Let $\mathcal{H}_A$, $\mathcal{H}_{\widehat{A}}$, and $\mathcal{H}_S$ denote the Hilbert spaces respectively defined by the orthonormal bases

$$\left\{ |a\rangle \mid a \in A \right\}, \left\{ |\chi\rangle \mid \chi \in \widehat{A} \right\}, \text{ and } \left\{ |s\rangle \mid s \in S \right\}.$$

We assume that we are given a quantum system which implements the unitary transformation U_φ defined by

$$\mathcal{H}_A \otimes \mathcal{H}_S \xrightarrow{U_\varphi} \mathcal{H}_A \otimes \mathcal{H}_S$$

$$|a\rangle |s_0\rangle \longmapsto |a\rangle |\varphi(a)\rangle$$

We will use the above implementation to construct a quantum subroutine $\mathrm{QRAND}_\varphi()$ which produces a probability distribution

$$Prob_\varphi : \widehat{A} \longrightarrow [0,1]$$

on the character group $\widehat{A}$ of the group A.

Before doing so, we will, as explained in the previous section, make use of various identifications, such as respectively identifying the Fourier and inverse Fourier transforms $\mathcal{F}_A$ and $\mathcal{F}_A^{-1}$ on the group A

$$\mathbb{C}A = \mathcal{H}_A \underset{\mathcal{F}_A^{-1}}{\overset{\mathcal{F}_A}{\rightleftarrows}} \mathcal{H}_{\widehat{A}} = \mathbb{C}\widehat{A}$$

with

$$\mathbb{C}A = \mathcal{H}_A \underset{\mathcal{F}_A^{-1}}{\overset{\mathcal{F}_A}{\rightleftarrows}} \mathcal{H}_A = \mathbb{C}A$$

Quantum Subroutine $\mathrm{QRAND}_\varphi()$

Step 0. Initialization

$$|\psi_0\rangle = |0\rangle |s_0\rangle$$

Step 1. Application of the inverse Fourier transform $\mathcal{F}_A^{-1}$ of A

$$|\psi_1\rangle = \left(\mathcal{F}_A^{-1} \otimes 1_S\right) |\psi_0\rangle = \frac{1}{\sqrt{|A|}} \sum_{a \in A} |a\rangle |s_0\rangle$$

where $|A|$ denotes the cardinality of the group A.

Step 2. Application of the unitary transformation U_φ

$$|\psi_2\rangle = U_\varphi |\psi_1\rangle = \frac{1}{\sqrt{|A|}} \sum_{a \in A} |a\rangle |\varphi(a)\rangle$$

Step 3. Application of the Fourier transform $\mathcal{F}_A$ of A

$$|\psi_3\rangle = \left(\mathcal{F}_A^{-1} \otimes 1_S\right)|\psi_2\rangle = \frac{1}{|A|}\sum_{a\in A}\sum_{\chi\in \widehat{A}}\chi(a)|\chi\rangle|\varphi(a)\rangle$$

$$= \sum_{\chi\in \widehat{A}}\frac{\||\varphi(\chi^\bullet)\rangle\|}{|A|}|\chi\rangle\frac{|\varphi(\chi^\bullet)\rangle}{\||\varphi(\chi^\bullet)\rangle\|}$$

where

$$|\varphi(\chi^\bullet)\rangle = \sum_{a\in A}\chi(a)|\varphi(a)\rangle$$

REMARK 5. *This notation is meant to be suggestive, since under the identification $\mathcal{H}_A = \mathbb{C}A$ we have*

$$|\varphi(\chi^\bullet)\rangle = \sum_{a\in A}\chi(a)\,\varphi(a) = \varphi\left(\sum_{a\in A}\chi(a)\,a\right) = \varphi(\chi^\bullet)$$

Step 4. Measurement of the left quantum register. Thus, with probability

$$Prob_\varphi(\chi) = \frac{\||\varphi(\chi^\bullet)\rangle\|^2}{|A|^2}$$

the character χ is the resulting measured value, and the quantum system "collapses" to the state

$$|\psi_4\rangle = |\chi\rangle\frac{|\varphi(\chi^\bullet)\rangle}{\||\varphi(\chi^\bullet)\rangle\|}$$

Step 5. OUTPUT the character χ, and STOP.

REMARK 6. *The quantum subroutine* $\mathrm{QRAND}_\varphi()$ *can also be viewed as a subroutine with the state $|\chi\rangle|\varphi(\chi^\bullet)\rangle$ as a side effect.*

As a result of the above description of $\mathrm{QRAND}_\varphi()$, we have the following theorem:

THEOREM 5. *Let*

$$\varphi : A \longrightarrow S$$

be a map from a finite abelian group A into a finite set S. Then the quantum subroutine $\text{QRAND}_\varphi()$ is an implementation of the probability distribution $Prob_\varphi(\chi)$ on the group $\widehat{A}$ of characters of A given by

$$Prob_\varphi(\chi) = \frac{\|\varphi(\chi^\bullet)\|^2}{|A|^2},$$

for all $\chi \in \widehat{A}$, where $\chi^\bullet$ denotes

$$\chi^\bullet = \sum_{a \in A} \chi(a) a \in \mathbb{C}A$$

REMARK 7. *Please note that the above theorem is true whether or not the map $\varphi : A \longrightarrow S$ has a hidden subgroup.*

We will, on occasion, refer to the probability distribution

$$Prob_\varphi : \widehat{A} \longrightarrow [0, 1]$$

on the character group $\widehat{A}$ as the **stochastic source** $\mathcal{S}_\varphi(\chi)$ which produces a symbol $\chi \in \widehat{A}$ with probability $Prob_\varphi(\chi)$. (See [32].) Thus, $\text{QRAND}_\varphi(\chi)$ is an algorithmic implementation of the stochastic source $\mathcal{S}_\varphi(\chi)$.

Part 4. Vintage Simon Algorithms

We now begin the development of the class of vintage Simon QHSAs. These are QHSAs for which the ambient group A is finite abelian.

8. Properties of the probability distribution $Prob_\varphi(\chi)$ when φ has a hidden subgroup

Let

$$\varphi : A \longrightarrow S$$

be a map from a finite abelian group A to a set S. We now assume that φ has a hidden subgroup K_φ, and hence, a hidden quotient group $H_\varphi = A/K_\varphi$.

Let

$$\nu : A \longrightarrow H_\varphi = A/K_\varphi$$

denote the corresponding natural epimorphism respectively. Then since $Hom_\mathbb{Z}(-, 2\pi\mathbb{R}/2\pi\mathbb{Z})$ is a left exact contravariant functor, the map

$$\begin{array}{rcl} \widehat{\nu} : \widehat{H_\varphi} & \longrightarrow & \widehat{A} \\ \eta & \longmapsto & \eta \circ \nu \end{array}$$

is a monomorphism[10].

Since $\widehat{\nu}$ is a monomorphism, each character η of the hidden quotient group H_φ can be identified with a character χ of A for which $\chi(k) = 1$ for every element of

---

[10]See [7].

K_φ. In other words, $\widehat{H_\varphi}$ can be identified with all characters of A which are trivial on K_φ.

THEOREM 6. *Let*
$$\varphi : A \longrightarrow S$$
be a map from a finite abelian group A into a finite set S. If there exists a hidden subgroup K_φ of φ, and hence a hidden quotient group $H_\varphi = A/K_\varphi$ of φ, then the probability distribution $Prob_\varphi(\chi)$ on $\widehat{A}$ implemented by the quantum subroutine $\mathrm{QRAND}_\varphi()$ is given by
$$Prob_\varphi(\chi) = \begin{cases} \frac{1}{|H_\varphi|} & \text{if } \chi \in \widehat{H_\varphi} \\ 0 & \text{otherwise} \end{cases}$$
In other words, in this particular case, $Prob_\varphi(\chi)$ is nothing more than the uniform probability distribution on the character group $\widehat{H_\varphi}$ of the hidden quotient group H_φ.

PROOF. Since φ has a hidden subgroup K_φ, there exists a hidden injection
$$\iota_\varphi : H_\varphi \longrightarrow S$$
from the hidden quotient group $H_\varphi = A/K_\varphi$ to the set S such that the diagram

$$\begin{array}{ccc} A & \xrightarrow{\varphi} & S \\ \nu \downarrow & \nearrow \iota_\varphi & \\ H_\varphi & & \end{array}$$

is commutative, where $\nu : A \longrightarrow H_\varphi$ denotes the hidden natural epimorphism of A onto the quotient group $H_\varphi = A/K_\varphi$.

Next let
$$\iota_\nu : H_\varphi \longrightarrow A$$
be a transversal map of the subgroup K_φ in A, i.e., a map such that
$$\nu \circ \iota_\nu = id_{H_\varphi} .$$
In other words, ι_ν sends each element h of H_φ to a unique element of the coset $\varphi^{-1}(h)$.

Recalling that
$$\varphi(\chi^\bullet) = \sum_{a \in A} \chi(a) \varphi(a) ,$$
we have
$$\varphi(\chi^\bullet) = \sum_{a \in A} \chi(a) \iota_\varphi \nu a = \sum_{h \in H_\varphi} \left(\sum_{k \in K_\varphi} \chi(\iota_\nu h + k) \right) \iota_\varphi h$$

$$= \sum_{h \in H_\varphi} \chi(\iota_\nu h) \left(\sum_{k \in K_\varphi} \chi(k) \right) \iota_\varphi h = \left(\sum_{k \in K_\varphi} \chi(k) \right) \left(\sum_{h \in H_\varphi} \chi(\iota_\nu h) \iota_\varphi h \right)$$

Thus,

$$\|\varphi(\chi^\bullet)\|^2 = \left|\sum_{k \in K_\varphi} \chi(k)\right|^2 \left\|\sum_{h \in H_\varphi} \chi(\iota_v h) \iota_\varphi h\right\|^2$$

$$= \left|\sum_{k \in K_\varphi} \chi(k)\right|^2 \sum_{h \in H_\varphi} |\chi(\iota_v h)|^2 = \left|\sum_{k \in K_\varphi} \chi(k)\right|^2 |H_\varphi|$$

But by a standard character identity[11], we have

$$\sum_{k \in K_\varphi} \chi(k) = \begin{cases} |K_\varphi| = |A|/|H_\varphi| & \text{if } \chi \in \widehat{H_\varphi} \\ 0 & \text{otherwise} \end{cases}$$

Hence, it follows that

$$Prob_\varphi(\chi) = \frac{\|\varphi(\chi^\bullet)\|^2}{|A|^2} = \begin{cases} \frac{1}{|H_\varphi|} & \text{if } \chi \in \widehat{H_\varphi} \\ 0 & \text{otherwise} \end{cases}$$

□

9. A Markov process $\mathcal{M}_\varphi$ induced by $Prob_\varphi$

Before we can discuss the class of vintage Simon quantum hidden subgroup algorithms, we need to develop the mathematical machinery to deal with the following question:

Question. Let $\varphi : A \longrightarrow S$ be a map from a finite abelian group A to a finite set S. Assume that the map φ has a hidden group K_φ, and hence a hidden quotient group H_φ. From theorem 6 of the previous section, we know that the probability distribution

$$Prob_\varphi : \widehat{A} \longrightarrow [0, 1]$$

is effectively the uniform probability distribution on the character group $\widehat{H_\varphi}$ of the hidden quotient group H_φ. How many times do we need to query the probability distribution $Prob_\varphi$ to obtain enough characters of H_φ to generate the entire character group $\widehat{H_\varphi}$?

We begin with a definition:

DEFINITION 4. *Let*

$$Prob_G : G \longrightarrow [0, 1]$$

*be a probability distribution on a finite abelian group G, and let G_+ denote the subgroup of G generated by all elements g of G such that $Prob_G(g) > 0$. The **Markov***

---

[11] See [**17**].

process $\mathcal{M}_G$ associated with a probability distribution $Prob_G$ is the Markov process with the subgroups G_α of G_+ as states, and with transition probabilities given by

$$Prob\left(G_a \rightsquigarrow G_\beta\right) = Prob_G\left\{g \in G_+ \mid G_\beta \text{ is generated by } g \text{ and the elements of } G_\alpha\right\},$$

where $G_a \rightsquigarrow G_\beta$ denotes the transition from state G_α to state G_β. The **initial state** of the Markov process $\mathcal{M}_G$ is the trivial subgroup G_0. The subgroup G_+ is called the **absorbing subgroup** of G. The **transition matrix** T of the Markov process is the matrix indexed on the states according to some chosen fixed linear ordering with (G_α, G_β)-th entry $T_{\alpha\beta}$ given by $Prob\left(G_a \rightsquigarrow G_\beta\right)$.

The following two propositions are immediate consequences of the above definition:

PROPOSITION 2. *Let*

$$Prob_G : G \longrightarrow [0,1]$$

be a probability distribution on a finite abelian group G. Then the Markov process $\mathcal{M}_G$ is an absorbing Markov process with sole absorbing state G_+, a state which once entered can never be left. The remaining states are transient states, i.e., states once left can never again be entered. Hence,

$$\lim_{n\to\infty} Prob_G\left(G_0 \underset{n}{\rightsquigarrow} G_\alpha\right) = \begin{cases} 1 & \text{if } G_\alpha = G_+ \\ 0 & \text{if } G_\alpha \neq G_+ \end{cases}$$

In other words, if the Markov process $\mathcal{M}_G$ starts in state G_0, it will eventually end up permanently in the absorbing state G_+.

PROPOSITION 3. *Let T be the transition matrix of the Markov process associated with the probability distribution*

$$Prob_G : G \longrightarrow [0,1]$$

Then the probability $Prob\left(G_\alpha \underset{n}{\rightsquigarrow} G_\beta\right)$ that the Markov process $\mathcal{M}_G$ starting in state G_α is in state G_β after n transitions is equal to the (G_α, G_β)-th entry of the matrix T^n, i.e.,

$$Prob\left(G_\alpha \underset{n}{\rightsquigarrow} G_\beta\right) = (T^n)_{\alpha\beta}$$

Under certain circumstances, we can work with a much simpler Markov process.

PROPOSITION 4. *Let G be a finite abelian group with probability distribution*

$$Prob_G : G \longrightarrow [0,1]$$

such that $Prob_G$ is the uniform probability distribution on the absorbing group G_+. Partition the states of the associated Markov process $\mathcal{M}_G$ into the collection of sets

$$\{\mathcal{G}_j \mid j \text{ divides } |G_+|\},$$

where $\mathcal{G}_j$ is the set of all states G_α of $\mathcal{M}_G$ of group order j.

If
$$Prob(G_i \rightsquigarrow \mathcal{G}_j) = \sum_{G_j \in \mathcal{G}_j} Prob(G_i \rightsquigarrow G_j)$$

has the same value for all $G_i \in \mathcal{G}_i$, then the states of $\mathcal{M}_G$ can be **combined (lumped)** to form a Markov process $\mathcal{M}_G^{Lumped}$ with states $\{\mathcal{G}_j \mid j \text{ divides } |G_+|\}$, and with transition probabilities given by

$$Prob^{Lumped}(\mathcal{G}_i \rightsquigarrow \mathcal{G}_j) = Prob(G_i \rightsquigarrow \mathcal{G}_j) \ ,$$

where G_i is an arbitrarily chosen element of $\mathcal{G}_i$, and with initial state $\mathcal{G}_1 = \{G_0\}$.

Moreover, the resulting $\mathcal{M}_G^{Lumped}$ is also an absorbing Markov process with sole absorbing state $\mathcal{G}_{|G_+|} = \{G_+\}$, with all other states transient, and such that

$$Prob\left(G_0 \underset{k}{\rightsquigarrow} G_+\right) = Prob\left(\mathcal{G}_1 \underset{k}{\rightsquigarrow} \mathcal{G}_{|G_+|}\right)$$

As a consequence of the above proposition and theorem 6, we have:

COROLLARY 2. *Let $\varphi : A \longrightarrow S$ be a map from a finite abelian group A to a finite set S, which has a hidden subgroup K_φ, and hence a hidden quotient group H_φ. Moreover, let the ambient group A be the direct sum of cyclic groups of the same prime order p, i.e., let*

$$A = \overset{n}{\underset{1}{\oplus}} \mathbb{Z}_p \ .$$

Then the combined (lumped) process $\mathcal{M}_{\widehat{A}}^{Lumped}$ is a Markov process such that

$$Prob\left(\widehat{A}_0 \underset{k}{\rightsquigarrow} \widehat{H_\varphi}\right) = Prob\left(\mathcal{G}_1 \underset{k}{\rightsquigarrow} \mathcal{G}_{|\widehat{H_\varphi}|}\right)$$

Moreover, if the states of $\mathcal{M}_{\widehat{A}}^{Lumped}$ are linearly ordered as

$$\mathcal{G}_i < \mathcal{G}_j \text{ if and only if } i \text{ divides } j \ ,$$

then the transition matrix T of $\mathcal{M}_{\widehat{A}}^{Lumped}$ is given by

$$T = \begin{pmatrix} 1 & 0 & 0 & 0 & \cdots & 0 & 0 \\ 1-\frac{1}{p} & \frac{1}{p} & 0 & 0 & \cdots & 0 & 0 \\ 0 & 1-\frac{1}{p^2} & \frac{1}{p^2} & 0 & \cdots & 0 & 0 \\ 0 & 0 & 1-\frac{1}{p^3} & \frac{1}{p^3} & \cdots & 0 & 0 \\ \vdots & \vdots & \vdots & \vdots & \ddots & \vdots & \vdots \\ 0 & 0 & 0 & 0 & \cdots & \frac{1}{p^{n-1}} & 0 \\ 0 & 0 & 0 & 0 & \cdots & 1-\frac{1}{p^n} & \frac{1}{p^n} \end{pmatrix}$$

Hence,

$$Prob\left(\widehat{A}_0 \underset{k}{\rightsquigarrow} \widehat{H_\varphi}\right) = \left(T^k\right)_{n1} \ ,$$

from which it easily follows that

$$Prob\left(\widehat{A}_0 \underset{k}{\rightsquigarrow} \widehat{H_\varphi}\right) > 1 - \frac{1}{p-1}\left(\frac{1}{p}\right)^{k-n} S \geq 1 - \frac{1}{(p-1)p^2}$$

for $k \geq n+2$.

10. Vintage Simon quantum hidden subgroup algorithms (QHSAs)

We are now prepared to extend Simon's quantum algorithm to an entire class of QHSAs on finite abelian groups.

Let
$$\varphi : A \longrightarrow S$$
be a map from a finite abelian group A to a finite set S for which there exists a hidden subgroup K_φ, and hence, a hidden quotient group $H_\varphi = A/K_\varphi$.

Following our usual convention, we use additive notation for the ambient group A and multiplicative notation for the hidden quotient group H_φ.

As mentioned in section 2 of this paper, it follows from the standard theory of abelian groups (i.e., Theorem 1) that the ambient group A can be decomposed into the finite direct sum of cyclic groups $\mathbb{Z}_{m_0}, \mathbb{Z}_{m_1}, \ldots, \mathbb{Z}_{m_{\ell-1}}$, i.e.,
$$A = \mathbb{Z}_{m_0} \oplus \mathbb{Z}_{m_1} \oplus \ldots \oplus \mathbb{Z}_{m_{\ell-1}} ,$$
We denote respective generators of the above cyclic groups by
$$a_0, a_1, \ldots, a_{\ell-1} .$$

Consequently, each character χ of the ambient group A can be uniquely expressed as
$$\chi : \sum_{j=0}^{\ell-1} \alpha_j a_j \longmapsto \exp\left(2\pi i \sum_{j=0}^{\ell-1} \alpha_j \frac{y_j}{m_j}\right) ,$$
where $0 \leq y_j < m_j$ for $j = 0, 1, \ldots, \ell - 1$. Thus, we have a one-to-one correspondence between the characters χ of A and ℓ-tuples of rationals (modulo 1) of the form
$$\left(\frac{y_0}{m_0}, \frac{y_1}{m_1}, \ldots, \frac{y_{\ell-1}}{m_{\ell-1}}\right) ,$$
where
$$0 \leq y_j < m_j, \quad j = 0, 1, \ldots, \ell - 1 .$$
As a result, we can and do use the following notation to refer uniquely to each and every character χ of A
$$\chi = \chi_{\left(\frac{y_0}{m_0}, \frac{y_1}{m_1}, \ldots, \frac{y_{\ell-1}}{m_{\ell-1}}\right)} .$$

DEFINITION 5. *Let*
$$A = \mathbb{Z}_{m_0} \oplus \mathbb{Z}_{m_1} \oplus \cdots \oplus \mathbb{Z}_{m_{\ell-1}}$$
be a direct sum decomposition of a finite abelian group A into finite cyclic groups. Let
$$a_0, a_1, \ldots, a_{n-1}$$
denote respective generators of the cyclic groups in this direct sum decomposition.

Then an integer matrix

$$\mathfrak{G} = [\alpha_{ij}]_{k \times n} \qquad \mod(m_0, m_1, \ldots, m_{n-1})$$

*is said to be a **generator matrix** of a subgroup K of A provided*

$$\left\{ \sum_{j=0}^{n-1} \alpha_{ij} a_j \mid 0 \leq i < k \right\}$$

is a complete set of generators of the subgroup K.

A matrix of rationals $\mod 1$

$$\mathfrak{H} = \left[\frac{y_{ij}}{m_j} \right]_{\ell \times n} \qquad \mod 1$$

*is said to be a **dual generator matrix** of a subgroup K of A provided*

$$\left\{ \chi_{(\frac{y_{i0}}{m_0}, \frac{y_{i1}}{m_1}, \ldots, \frac{y_{i(n-1)}}{m_{n-1}})} \mid 0 \leq i < \ell \right\}$$

is a complete set of generators of the character group $\widehat{H}$ of the quotient group $H = A/K$.

Let $\mathcal{M}_\varphi$ be the Markov process associated with the probability distribution

$$Prob_\varphi : \widehat{A} \longrightarrow [0, 1]$$

on the character group $\widehat{A}$ of the ambient group A.

Let $0 \leq \epsilon \ll 1$ be a chosen threshold.

Then a vintage Simon algorithm is given below:

VINTAGE SIMON(φ, ϵ)

Step 1. Select a positive integer ℓ such that

$$Prob_{\widehat{A}} \left(\dot{A}_0 \underset{\ell}{\leadsto} \dot{H}_\varphi \right) < 1 - \epsilon$$

Step 2. Initialize running dual generator matrix

$$\mathfrak{H} = [\ ]$$

Step 3. Query the probability distribution $Prob_\varphi$ ℓ times to obtain ℓ characters (not necessarily distinct) of the hidden quotient group H_φ, while incrementing the running dual generator matrix $\mathfrak{H}$.

LOOP i FROM 0 TO $\ell - 1$ DO

$$\chi_{(\frac{y_{i0}}{m_0}, \frac{y_{i1}}{m_1}, \ldots \frac{y_{i(n-1)}}{m_{n-1}})} = \text{QRAND}_\varphi()$$

$$\mathfrak{H} = \begin{bmatrix} \frac{y_{i0}}{m_0} & \frac{y_{i1}}{m_1} & \cdots & \frac{y_{i(n-1)}}{m_{n-1}} \\ \hline & & \mathfrak{H} & \end{bmatrix}$$

LOOP LOWER BOUNDARY;

Step 4. Compute the generator matrix $\mathfrak{G}$ from the dual generator matrix $\mathfrak{H}$ by using Gaussian elimination to solve the system of equations

$$\sum_{j=0}^{n-1} \frac{y_{ij}}{m_j} x_j = 0 \bmod 1 \qquad 0 \leq i < N_0$$

for unknown $x_j \bmod m_j$.

Step 5. OUTPUT $\mathfrak{G}$ and STOP.

Part 5. Vintage Shor Algorithms

11. Vintage Shor quantum hidden subgroup algorithms(QHSAs)

Let $\varphi : A \longrightarrow S$ be a map with hidden subgroup structure. We now consider QHSPs for which the ambient group A is **free abelian of finite rank** n.

Since the ambient group A is infinite, at least two difficulties naturally arise. One is that the associated complex vector space $\mathcal{H}_A$ is now infinite dimensional, thereby causing some implementation problems. The other is that the Fourier transform of a periodic function on A does not exist as a function[12], but as a generalized function!

Following Shor's lead, we side-step these annoying obstacles by choosing not to work with the ambient group A and the map φ at all. Instead, we work with a group $\widetilde{A}$ and a map $\widetilde{\varphi} : \widetilde{A} \longrightarrow S$ which are "approximations" of A and $\varphi : A \longrightarrow S$, respectively.

The group $\widetilde{A}$ and the **approximating map** $\widetilde{\varphi}$ are constructed as follows:

---

[12] As a clarifying note, let $f : \mathbb{Z} \longrightarrow \mathbb{C}$ be a period P function on $\mathbb{Z}$. Then f on $\mathbb{Z}$ is neither of compact support, nor of bounded L^2 or L^1 norm. So the Fourier transform of f does not exist as a function, but as a generalized function, i.e., as a distribution. However, the function f does induce a function $\widetilde{f} : \mathbb{Z}_P \longrightarrow \mathbb{C}$ which does have a Fourier transform on $\mathbb{Z}_P$ which exists as a function. The problem is that we do not know the period of φ, and as a consequence, cannot do Fourier analysis on the corresponding unknown finite cyclic group.

Choose an epimorphism
$$\mu : A \longrightarrow \widetilde{A}$$
of the ambient group A onto a chosen finite group $\widetilde{A}$, called a **group probe**. Next, select a **transversal**
$$\iota_\mu : \widetilde{A} \longrightarrow A$$
of μ, i.e., a map such that
$$\mu \circ \iota_\mu = id_{\widetilde{A}} \ ,$$
where $id_{\widetilde{A}}$, denotes the identity map on the group probe $\widetilde{A}$. [Consequently, ι_μ is an injection, and in most cases not a morphism at all.]

Having chosen μ and ι_μ, the **approximating map** $\widetilde{\varphi}$ is defined as
$$\widetilde{\varphi} = \varphi \circ \iota_\mu : \widetilde{A} \longrightarrow S$$

Although the map $\widetilde{\varphi}$ is not usually a morphism, the quantum subroutine $\text{QRAND}_{\widetilde{\varphi}}()$ is still a well defined quantum procedure which produces a well defined probability distribution $Prob_{\widetilde{\varphi}}(\chi)$ on the character group $\widehat{\widetilde{A}}$ of the group probe $\widetilde{A}$. As we shall see, if the the map $\widetilde{\varphi}$ is a "reasonably good approximation" to the original map φ, then $\text{QRAND}_{\widetilde{\varphi}}()$ will with high probability produce characters χ of the probe group $\widetilde{A}$ which are "sufficiently close" to corresponding characters η of the hidden quotient group H_φ.

Following this basic strategy, we will now use the quantum subroutine $\text{QRAND}_{\widetilde{\varphi}}()$ to build three classes of vintage Shor QHSAs, where the probe group $\widetilde{A}$ is a finite cyclic group $\mathbb{Z}_Q$ of order Q. In this way, we will create three classes of quantum algorithms which form natural extensions of Shor's original quantum factoring algorithm.

12. Direct summand structure

We digress momentarily to discuss the direct sum structure of the ambient group A when it is free abelian of finite rank n.

Since the ambient group A is free abelian of finite rank n, the hidden subgroup K_φ is also free abelian of finite rank. Moreover, there exist compatible direct sum decompositions of A and K_φ into free cyclic groups

$$\begin{cases} K_\varphi & = & P_1 \mathbb{Z} \oplus \cdots \oplus P_n \mathbb{Z} \\ \\ A & = & \underbrace{\mathbb{Z} \oplus \cdots \oplus \mathbb{Z}}_{n \text{ direct summands}} \end{cases},$$

where $P_1, ..., P_n$ are non-negative integers, and where the inclusion morphism

$$K_\varphi = P_1 \mathbb{Z} \oplus \cdots \oplus P_n \mathbb{Z} \hookrightarrow \underbrace{\mathbb{Z} \oplus \cdots \oplus \mathbb{Z}}_{n} = A$$

is the direct sum of the inclusion morphisms

$$P_j \mathbb{Z} \hookrightarrow \mathbb{Z}$$

It should be mentioned that, since the group K_φ is hidden, the above direct sum decompositions are also hidden. Moreover, the selection of a direct sum decomposition of the ambient group A is operationally equivalent to a selection a basis of A. This leads to the following definition:

DEFINITION 6. *A basis*

$$\{a_1, a_2, \ldots, a_n\}$$

*of the ambient group A corresponding to the above hidden direct sum decomposition of A is called a **hidden basis** of A.*

Question. *How is the hidden basis $\{a_1, a_2, \ldots, a_n\}$ of A related to any "visible" basis $\{a'_1, a'_2, \ldots, a'_n\}$ of A that we might choose to work with?*

The group of automorphisms of the free abelian group A of rank n is isomorphic to the group

$$SL_\pm(n, \mathbb{Z})$$

of $n \times n$ invertible integer matrices. This is the same as the group of $n \times n$ integer matrices of determinant ± 1.

PROPOSITION 5. *Let $\{a_1, a_2, \ldots, a_n\}$ be a hidden basis of A, and let $\{a'_1, a'_2, \ldots, a'_n\}$ be any other basis of A. Then there exists a unique element $M \in SL_\pm(n, \mathbb{Z})$ which carries the basis $\{a'_1, a'_2, \ldots, a'_n\}$ into the hidden basis $\{a_1, a_2, \ldots, a_n\}$.*

Since the image of φ is finite, we know that $P_j > 0$, for all j. Thus, the direct sum decomposition of the inclusion morphism becomes

$$\overbrace{(P_1\mathbb{Z} \oplus \cdots \oplus P_{\overline{n}}\mathbb{Z}) \oplus \underbrace{(\mathbb{Z} \oplus \cdots \oplus \mathbb{Z})}_{n-\overline{n}}}^{K_\varphi} \hookrightarrow \overbrace{\underbrace{(\mathbb{Z} \oplus \cdots \oplus \mathbb{Z})}_{\overline{n}} \oplus \underbrace{(\mathbb{Z} \oplus \cdots \oplus \mathbb{Z})}_{n-\overline{n}}}^{A}$$

As a consequence, the hidden quotient group H_φ is the corresponding direct sum of finite cyclic groups

$$H_\varphi = (\mathbb{Z}_{P_1} \oplus \cdots \oplus \mathbb{Z}_{P_{\overline{n}}}) \oplus \underbrace{(0 \oplus \cdots \oplus 0)}_{n-\overline{n}},$$

and the hidden epimorphism

$$\nu : (P_1\mathbb{Z} \oplus \cdots \oplus P_{\overline{n}}\mathbb{Z}) \oplus \underbrace{(\mathbb{Z} \oplus \cdots \oplus \mathbb{Z})}_{n-\overline{n}} \longrightarrow \mathbb{Z}_{P_1} \oplus \cdots \oplus \mathbb{Z}_{P_{\overline{n}}}$$

is the direct sum of of the epimorphisms
$$\begin{cases} \mathbb{Z} \longrightarrow \mathbb{Z}_{P_j} \\ \mathbb{Z} \longrightarrow 0 \end{cases}$$

As a consequence of the above, we have:

DEFINITION 7. *Let*
$$\{a_1, a_2, \ldots, a_n\}$$
be a hidden basis of A. *Then a corresponding* **induced hidden basis** *of the hidden quotient group* H_φ *is defined as*
$$\{b_1 = \nu(a_1), b_2 = \nu(a_2), \ldots, b_{\overline{n}} = \nu(a_{\overline{n}})\} ,$$
where $\nu : A \longrightarrow H_\varphi$ *denotes the hidden epimorphism.*[13]

The above direct sum decompositions are summarized in the following diagram:

$$\overbrace{\left(\bigoplus_{j=1}^{\overline{n}} P_j \mathbb{Z}\right) \oplus \left(\bigoplus_{j=\overline{n}+1}^{n} \mathbb{Z}\right)}^{K_\varphi} \hookrightarrow \overbrace{\left(\bigoplus_{j=1}^{\overline{n}} \mathbb{Z}\right) \oplus \left(\bigoplus_{j=\overline{n}+1}^{n} \mathbb{Z}\right)}^{A} \xrightarrow{\varphi} S$$

with ν going down to $\underbrace{\bigoplus_{j=1}^{\overline{n}} \mathbb{Z}_{P_j}}_{H_\varphi}$ and ι_φ going up to S.

DEFINITION 8. *Let* H *be a finite abelian group. Then a* **maximal cyclic subgroup** *of* H *is a cyclic subgroup of* H *of highest possible order.*

PROPOSITION 6. *Let* $b_1, b_2, \ldots, b_{\overline{n}}$ *be the above defined induced hidden basis of the hidden quotient group* $H_{\overline{n}} = \mathbb{Z}_{P_1} \oplus \mathbb{Z}_{P_2} \oplus \cdots \oplus \mathbb{Z}_{P_{\overline{n}}}$. *Then a maximal cyclic subgroup of* H_φ *is generated by*
$$b_1 \oplus b_2 \oplus \cdots \oplus b_{\overline{n}} ,$$
and is isomorphic to the finite cyclic group $\mathbb{Z}_P$ *of order*
$$P = \mathrm{lcm}(P_1, P_2, \ldots, P_{\overline{n}}) .$$

---

[13] Please note that the hidden basis $\{a_1, a_2, \ldots, a_n\}$ of A is free in the abelian category. However, the induced basis $\{b_1, b_2, \ldots, b_{\overline{n}}\}$ of H_φ is not because H_φ is a torsion group. $\{b_1, b_2, \ldots, b_{\overline{n}}\}$ is a basis in the sense that it is a set of generators of H_φ such that
$$b_1^{k_1} b_2^{k_2} \cdots b_{\overline{n}}^{k_{\overline{n}}} = 1$$
implies that
$$b_j^{k_j} = 1$$
for every j. (For more information, please refer to [20].)

13. Vintage Shor QHSAs with group probe $\widetilde{A} = \mathbb{Z}_Q$.

Choose a positive integer Q and an epimorphism

$$\mu : A \longrightarrow \mathbb{Z}_Q$$

of the free abelian group A onto the finite cyclic group $\widetilde{A} = \mathbb{Z}_Q$ of order Q.

Next we wish to select a transversal ι_μ of the epimorphism μ.

However, at this juncture we must take care. For, not every choice of the transversal ι_μ will produce an efficient vintage Shor algorithm. In fact, most choices probably will produce highly inefficient algorithms[14]. We emphasize that the efficiency of the class of algorithms we are about to define depends heavily on the choice of the transversal ι_μ.

Following Shor's lead once again, we select a very special transversal ι_μ.

DEFINITION 9. *Let $\mu : A \longrightarrow \mathbb{Z}_Q$ be an epimorphism from a free abelian group A of finite rank n onto a finite cyclic group $\mathbb{Z}_Q$ of order Q, and let $\widetilde{a}$ be a chosen generator of the cyclic group $\mathbb{Z}_Q$.*
A transversal

$$\iota_\mu : \mathbb{Z}_Q \longrightarrow A$$

*is said to be a **Shor transversal** provided*

1) $\iota_\mu(k\widetilde{a}) \longmapsto k\iota_\mu(\widetilde{a})$, *for all $0 \leq k < Q$, and*

2) *There exists a basis $\{a'_1, a'_2, \ldots, a'_n\}$ of A such that, when $\iota_\mu(\widetilde{a})$ is expressed in this basis, i.e., when*

$$\iota_\mu(\widetilde{a}) = \sum_{j=1}^{n} \lambda'_j a'_j ,$$

it follows that

$$\gcd(\lambda'_1, \lambda'_2, \ldots, \lambda'_n) = 1$$

PROPOSITION 7. *Let $\lambda'_1, \lambda'_2, \ldots, \lambda'_n$ be n integers, and let M be a non-singular $n \times n$ integral matrix, i.e., an element of $SL_\pm(n, \mathbb{Z})$. If $\lambda_1, \lambda_2, \ldots, \lambda_n$ are n integers defined by*

$$(\lambda_1, \lambda_2, \ldots, \lambda_n) = (\lambda'_1, \lambda'_2, \ldots, \lambda'_n) M ,$$

then

$$\gcd(\lambda_1, \lambda_2, \ldots, \lambda_n) = \gcd(\lambda'_1, \lambda'_2, \ldots, \lambda'_n)$$

As a corollary, we have

---

[14]For example, consider $A = \mathbb{Z}$, $P = 6$, $Q = 64$, and the transversal defined by $\iota_\mu : 6n+k \longmapsto 6n + k + 64 \lfloor k/2 \rfloor$ for $0 \leq n \leq 10$, where $\begin{cases} 0 \leq k < 6 & \text{if } 0 \leq n < 10 \\ 0 \leq k < 4 & \text{if } n = 10 \end{cases}$. One reason this is a poor choice of transversal is that the image of ι_μ does not contain a representative of every coset of the hidden subgroup $\mathbb{Z}_P$ of the ambient group A.

PROPOSITION 8. *If condition 2) is true with respect to one basis, then it is true with respect to every basis.*

An another immediate consequence of the definition of a Shor traversal, we have the following lemma:

LEMMA 1. *If a Shor transversal*
$$\iota_\mu : \mathbb{Z}_Q \longrightarrow A ,$$
is used to construct the the approximating map
$$\widetilde{\varphi} = \varphi \circ \iota_\mu : \mathbb{Z}_Q \longrightarrow S ,$$
then the approximating map $\widetilde{\varphi}$ has the following property
$$\widetilde{\varphi}(k\widetilde{a}) = [\widetilde{\varphi}(\widetilde{a})]^k ,$$
for all $0 \leq k < Q$, where we have used the hidden injection $\iota_\varphi : H_\varphi \longrightarrow S$ to identify the elements $\widetilde{\varphi}(k\widetilde{a})$ of the set S with corresponding elements of the hidden quotient group H_φ.

14. Finding Shor transversals for vintage $\mathbb{Z}_Q$ Shor algorithms

Surprisingly enough, it is algorithmically simpler to find a Shor transversal $\iota_\mu : \mathbb{Z}_Q \longrightarrow A$ first, and then, as an after thought, to construct a corresponding epimorphism $\mu : A \longrightarrow \mathbb{Z}_Q$.

DEFINITION 10. *Let A be an ambient group, and let $\mathbb{Z}_Q$ be a finite cyclic group of order Q with a selected generator $\widetilde{a}$. Then an injection*
$$\iota : \mathbb{Z}_Q \longrightarrow A$$
*is called a **Shor injection** provided*
1) $\iota(k\widetilde{a}) = k\iota(\widetilde{a})$, *for all $0 \leq k < Q$, and*
2) *There exists a basis $\{a'_1, a'_2, \ldots, a'_n\}$ of the ambient group A such that*
$$\gcd(\lambda'_1, \lambda'_2, \ldots, \lambda'_n) = 1 ,$$
where
$$\iota(\widetilde{a}) = \sum_{j=1}^n \lambda'_j a'_j .$$

PROPOSITION 9. *If condition 2) is true with respect to one basis, it is true with respect to all.*

Next, we need to construct an epimorphism $\mu : A \longrightarrow \mathbb{Z}_Q$ for which $\iota : \mathbb{Z}_Q \longrightarrow A$ is a Shor transversal.

PROPOSITION 10. *Let A be an ambient group, and let $\mathbb{Z}_Q$ be a finite cyclic group of order Q with a selected generator $\tilde{a}$. Given a Shor injection*

$$\iota : \mathbb{Z}_Q \longrightarrow A ,$$

there exists an epimorphism

$$\mu_\iota : A \longrightarrow \mathbb{Z}_Q$$

such that ι is a Shor transversal for μ_ι, i.e., such that

$$\mu_\iota \circ \iota = id_{\mathbb{Z}_Q} ,$$

where $id_{\mathbb{Z}_Q}$ denotes the identity morphism on $\mathbb{Z}_Q$.

PROOF. Select an arbitrary basis $\{a'_1, a'_2, \ldots, a'_n\}$ of A. Then

$$\iota(\tilde{a}) = \sum_{j=1}^{n} \lambda'_j a'_j ,$$

where

$$\gcd(\lambda'_1, \lambda'_2, \ldots, \lambda'_n) = 1 .$$

Hence, from the extended Euclidean algorithm, we can find integers

$$\alpha_1, \alpha_2, \ldots, \alpha_n$$

for which

$$\sum_{j=1}^{n} \alpha_j \lambda'_j = 1 .$$

Define

$$\mu : \{a'_1, a'_2, \ldots, a'_n\} \longrightarrow \mathbb{Z}_Q$$

by

$$\mu(a'_j) = \alpha_j \tilde{a} , \quad j = 1, 2, \ldots, n.$$

Since $a'_1, a'_2, \ldots, a'_n$ is a free abelian basis of the ambient group A, it uniquely extends to a morphism

$$\mu : A \longrightarrow \mathbb{Z}_Q .$$

It immediately follows that μ is an epimorphism because

$$\mu\left(\sum_{j=1}^{n} \lambda'_j a'_j\right) = \sum_{j=1}^{n} \alpha_j \lambda'_j \tilde{a} = \tilde{a} .$$

□

Thus the task of finding an epimorphism $\mu : A \longrightarrow \mathbb{Z}_Q$ and a corresponding Shor transversal reduces to the task of finding n integers $\lambda'_1, \lambda'_2, \ldots, \lambda'_n$ such that

$$\gcd(\lambda'_1, \lambda'_2, \ldots, \lambda'_n) = 1 .$$

This leads to the following probabilistic subroutine which finds a random Shor traversal:

RANDOM_SHOR_TRANSVERSAL($\{a'_1, a'_2, \ldots, a'_n\}$, Q, $\tilde{a}$, n)

```
# INPUT:     A basis {a'_1, a'_2, ..., a'_n} of A, a positive integer Q,
#            a selected generator ã of A, and the rank n of A
# OUTPUT:    Shor transversal ι_μ : Z_Q ⟶ A
# SIDE EFFECT: Epimorphism μ : A ⟶ Z_Q
# SIDE EFFECT: Random integers λ'_1, λ'_2, ..., λ'_n
```

GLOBAL: $\mu : A \longrightarrow \mathbb{Z}_Q$
GLOBAL: $\lambda'_1, \lambda'_2, \ldots, \lambda'_n$

Step 0 IF $n = 1$ THEN (SET $\lambda'_1 = 1$ AND GOTO **Step 4**)

Step 1 Select with replacement n random $\lambda'_1, \lambda'_2, \ldots, \lambda'_n$ from $\{1, 2, \ldots, Q\}$.

Step 2 Use the extended Euclidean algorithm to determine
$$d = \gcd(\lambda'_1, \lambda'_2, \ldots, \lambda'_n)$$
and integers $\alpha_1, \alpha_2, \ldots, \alpha_n$ such that $\sum_{j=1}^n \alpha_j \lambda'_j = d$

Step 3 IF $d \neq 1$ THEN GOTO **Step 1** ELSE GOTO **Step 4**

Step 4 Construct Shor transversal $\iota_\mu : \mathbb{Z}_Q \longrightarrow A$ as $\iota_\mu(k\widetilde{a}) = k \sum_{j=1}^n \lambda'_j a'_j$, for $0 \leq k < Q$

Step 5 Construct epimorphism $\mu : A \longrightarrow \mathbb{Z}_Q$ as
$$\mu(a'_j) = \alpha_j \widetilde{a} \text{ for all } j = 1, 2, \ldots, n$$

Step 6 OUTPUT transversal $\iota_\mu : \mathbb{Z}_Q \longrightarrow A$ and STOP

THEOREM 7. *For $n > 1$, the average case complexity of the* RANDOM_SHOR_TRANSVERSAL *subroutine is*
$$O\left(n (\lg Q)^3\right) .$$

PROOF. The computationally dominant part of this subroutine is the main loop Steps 1 through 3.

Each iteration of the main loop executes the extended Euclidean algorithm n times to find the gcd d. Since the computational complexity of the extended Euclidean algorithm[15] is $O\left((\lg Q)^3\right)$, it follows that the computational cost of one iteration of steps 1 through 3 is
$$O\left(n (\lg Q)^3\right) .$$

But by Corollary 7 of Appendix B,
$$Prob_Q(\gcd(\lambda'_1, \lambda'_2, \ldots, \lambda'_n) = 1) = \Omega(1) .$$

---

[15] See [**11**, Chap. 31].

Thus, the average number of iterations before a successful exit to Step 4 is $O(1)$.
Hence, the average case complexity of steps 1 through 4 is

$$O\left(n(\lg Q)^3\right).$$

□

REMARK 8. *Our objective in this paper is to find reasonable asymptotic bounds, not the tightest possible bounds. For example, the above bound is by no means the tightest possible. For a tighter bound for the Euclidean algorithm is $O\left((\lg Q)^2\right)$ which can be found in [11]. Thus, the bound found in the above theorem can be tightened to at least $O\left(n(\lg Q)^2\right)$.*

15. Maximal Shor transversals

Unfortunately, the definition of a Shor transversal is in some instances not strong enough to extend Shor's quantum factoring algorithm to ambient groups which are free abelian groups of finite rank. From necessity, we are forced to make the following definition.

DEFINITION 11. *Let $a_1, a_2, \ldots, a_n$ be a hidden basis of the ambient group A, let $\widetilde{a}$ be a chosen generator the cyclic group probe $\mathbb{Z}_Q$, and let $H_\varphi = \mathbb{Z}_{P_1} \oplus \mathbb{Z}_{P_2} \oplus \cdots \oplus \mathbb{Z}_{P_n}$ be the corresponding hidden direct sum decomposition. A **maximal Shor transversal** is a Shor transversal $\iota_\mu : \mathbb{Z}_Q \longrightarrow A$ such that*

$$\gcd(\lambda_j, P_j) = 1, \quad \text{for } 0 \leq j < n,$$

where the integers $\lambda_1, \lambda_2, \ldots, \lambda_n$ are defined by

$$\iota_\mu(\widetilde{a}) = \lambda_1 a_1 + \lambda_2 a_2 + \ldots, +\lambda_n a_n$$

REMARK 9. *Thus, for maximal Shor traversals, $\iota_\mu(\widetilde{a})$ maps via the hidden epimorphism $\nu : A \longrightarrow A/K_\varphi$ to a maximum order element of the hidden quotient group H_φ.*

One of the difficulties of the above definition is that it does not appear to be possible to determine whether or not a Shor transversal is maximal without first knowing the hidden direct sum decomposition of the hidden quotient group H_φ. We address this important issue in the following corollary, which is an immediate consequence of corollary 8 (found in Appendix B):

COROLLARY 3. *Let*
$$P_1, P_2, \ldots, P_n$$
be n fixed positive integers, and let Q be an integer such that
$$Q \geq \operatorname{lcm}(P_1, P_2, \ldots, P_n) \ ,$$
where $n > 1$.

If Conjecture 1 (found in Appendix B) is true, then the probability that the subroutine RANDOM_SHOR_TRANSVERSAL *produces a maximal Shor transversal is*
$$\Omega \left(\frac{1}{\prod_{j=1}^{n} \lg \lg P_j} \right) = \Omega \left(\frac{1}{(\lg \lg Q)^n} \right) \ .$$

16. Identifying characters of cyclic groups with points on the unit circle $\mathbb{S}^1$ in the complex plane $\mathbb{C}$.

We will now begin to develop an answer to the following question:

Question. 1. *Are the characters of the group probe $\mathbb{Z}_Q$ produced by the quantum subroutine* QRAND$_{\widetilde{\varphi}}()$ *"close enough" to the characters of a maximal cyclic subgroup of the hidden quotient group H_φ?*

If QRAND$_{\widetilde{\varphi}}()$ produces a character χ of $\mathbb{Z}_Q$ which is "close enough" to some character η of a maximal cyclic subgroup $\mathbb{Z}_P$ of the hidden quotient group H_φ, then the character χ can be used to find the corresponding closest character η of $\mathbb{Z}_P$. Each time such a character η is found, something more is known about the hidden quotient group H_φ and the hidden subgroup K_φ. In this way, we have the conceptual genesis of a class of vintage $\mathbb{Z}_Q$ Shor algorithms.

But before we can answer the above question, we need to answer a more fundamental question, namely:

Question. 2. *What do we mean by "close enough"? I.e, what do we mean by saying that a character χ of $\widetilde{A} = \mathbb{Z}_Q$ is "close enough" to some character η of $\mathbb{Z}_P$?*

To answer this last question, we need to introduce two additional concepts:

1) The concept of a common domain for the characters χ of $\mathbb{Z}_Q$ and the characters η of $\mathbb{Z}_P$.
2) The concept of a group norm which is to be used to define when two characters are "close."

In this section, we address item **1)**. In the next, item **2)**.

We begin by noting that the character group $\widehat{\mathbb{Z}}$ of the infinite cyclic group $\mathbb{Z}$ is simply the group $\mathbb{S}^1$, i.e.,
$$\widehat{\mathbb{Z}} = \mathbb{S}^1 = \left\{ \chi_\theta : n \longmapsto e^{2\pi i \theta n} \mid 0 \leq \theta < 1 \right\}$$
In other words, the characters of $\mathbb{Z}$ can be identified with the points on the unit radius circle in the complex plane $\mathbb{C}$.

Moreover, given an arbitrary epimorphism
$$\tau : \mathbb{Z} \longrightarrow \mathbb{Z}_m$$
of the infinite cyclic group $\mathbb{Z}$ onto a finite cyclic group $\mathbb{Z}_m$, the left exact contravariant functor[16]
$$Hom_{\mathbb{Z}}(-, 2\pi\mathbb{R}/2\pi\mathbb{Z})$$
transforms τ into the monomorphism
$$\begin{array}{rcl} \widehat{\tau} : \widehat{\mathbb{Z}_m} & \longrightarrow & \widehat{\mathbb{Z}} \\ \eta & \longmapsto & \eta \circ \tau \end{array}$$
In this way the characters of $\mathbb{Z}_m$ can be identified with the points of $\widehat{\mathbb{Z}} = \mathbb{S}^1$.

Thus, to find a common domain $\mathbb{S}^1$ for the characters of the group probe $\mathbb{Z}_Q$ and the maximal cyclic group $\mathbb{Z}_P$, all that need be done is to find epimorphisms $\widetilde{\mu} : \mathbb{Z} \longrightarrow \mathbb{Z}_Q$ and $\widetilde{\tau} : \mathbb{Z} \longrightarrow \mathbb{Z}_P$. This is accomplished as follows:

Let a be a generator of the infinite cyclic group $\mathbb{Z}$, and let $a_1, a_2, \ldots, a_n$ be a hidden basis of the ambient group A. Then the epimorphisms $\widetilde{\mu}$ and $\widetilde{\tau}$ are defined as
$$\begin{array}{rcl} \widetilde{\mu} : \mathbb{Z} & \longrightarrow & \mathbb{Z}_Q \\ ka & \longmapsto & k\widetilde{a} \end{array} \quad \text{and} \quad \begin{array}{rcl} \widetilde{\tau} : \mathbb{Z} & \longrightarrow & \mathbb{Z}_P \\ ka & \longmapsto & \nu\left[k\left(a_1 + a_2 + \ldots + a_n\right)\right] \end{array},$$
where $\widetilde{a}$ is the selected generator of the group probe $\mathbb{Z}_Q$, and where $\nu : A \longrightarrow H_\varphi$ is the hidden epimorphism.

Thus, as a partial answer to **Question 2** of Section 16, a character χ of the probe group $\widetilde{A} = \mathbb{Z}_Q$ is "close" to some character η of the maximal cyclic subgroup $\mathbb{Z}_P$ $H_\varphi = \bigoplus_{j=1}^{\overline{n}} \mathbb{Z}_{P_j}$, if the corresponding points $\widehat{\widetilde{\mu}}(\chi)$ and $\widehat{\widetilde{\tau}}(\eta)$ on the circle $\mathbb{S}^1$ are "close."

But precisely what do we mean by two points of $\mathbb{S}^1$ being "close" to one another? To answer this question, we need to observe that Shor's algorithm uses, in addition to the group structure of $\mathbb{S}^1$, also the metric structure of $\mathbb{S}^1$.

17. Group norms

We proceed to define a metric structure on the circle group $\mathbb{S}^1$. To do so, we need to define what is meant by a group norm.

DEFINITION 12. *A **(group theoretic) norm** on a group G is a map*
$$|||-||| : G \longrightarrow \mathbb{R}$$
such that

1) $|||x||| \geq 0$, *for all x, and $|||x||| = 0$ if and only if x is the group identity (which is 1 if we think of G as a multiplicative group, or 0 if we think of G as an additive group).*

---

[16] For the definition of a left exact contravariant functor, please refer to, for example, [7].

2) $|||x \cdot y||| \leq |||x||| + |||y|||$ *or* $|||x+y||| \leq |||x||| + |||y|||$, *depending respectively on whether we think of G as a multiplicative or as an additive group.*

Caveat. *The group norms defined in this section are different from the group algebra norms defined in Section 5.*

REMARK 10. *Such a norm induces a metric*
$$G \times G \longrightarrow \mathbb{R}$$
$$(x,y) \longmapsto |||x \cdot y^{-1}||| \ \text{or} \ |||x - y|||$$
depending on whether multiplicative or additive notation is used.

As mentioned in Section 4, we think of the 1-sphere $\mathbb{S}^1$ interchangeably as the multiplicative group
$$\mathbb{S}^1 = \left\{ e^{2\pi i \alpha} \mid 0 \leq \alpha < 1 \right\} \subset \mathbb{C}$$
with multiplication defined as
$$e^{2\pi i \alpha} \cdot e^{2\pi i \beta} = e^{2\pi i (\alpha + \beta)}$$
or as the additive group of reals $\mathbb{R}$ modulo 2π, i.e., as
$$\mathbb{S}^1 = 2\pi \mathbb{R}/2\pi \mathbb{Z} = \{ 2\pi \alpha \mid 0 \leq \alpha < 1 \}$$
with addition defined as
$$2\pi \alpha + 2\pi \beta = (2\pi \alpha + 2\pi \beta) \bmod 2\pi = 2\pi \left(\alpha + \beta \bmod 1 \right)$$
It should be clear from context which of the two representation of the group $\mathbb{S}^1$ is being used.

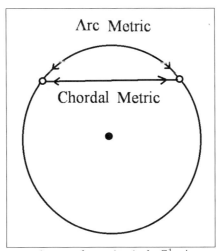

Figure 2. Two metrics on the unit circle $\mathbb{S}^1$, $\text{ARC}_{2\pi}$ and $\text{CHORD}_{2\pi}$.

There are two different norms on the 1-sphere $\mathbb{S}^1$ that we will be of use to us. The first is the **arclength norm**, written $\text{ARC}_{2\pi}$, defined by

$$\text{ARC}_{2\pi}(\alpha) = 2\pi \min\{|\alpha| - \lfloor|\alpha|\rfloor,\ \lceil|\alpha|\rceil - |\alpha|\},$$

which is simply the length of the shortest arc in the 1-sphere $\mathbb{S}^1$ connecting the point $e^{2\pi i \alpha}$ to the point 1.

The second norm is the **chordal length norm**, written $\text{CHORD}_{2\pi}$, defined by

$$\text{CHORD}_{2\pi}(\alpha) = 2|\sin(\pi\alpha)|,$$

which is simply the length of the chord in the complex plane connecting the point $e^{2\pi i \alpha}$ to the point 1.

Shor's algorithm depends heavily on the interrelationship of these two norms. We summarize these interrelationships in the following proposition:

PROPOSITION 11. *The the norms $\text{ARC}_{2\pi}$ and $\text{CHORD}_{2\pi}$ satisfy the following conditions:*

1) $\text{CHORD}_{2\pi}(\alpha) = 2\sin\left(\frac{1}{2}\text{ARC}_{2\pi}(\alpha)\right)$

2) $\frac{2}{\pi}\text{ARC}_{2\pi}(\alpha) \leq \text{CHORD}_{2\pi}(\alpha) \leq \text{ARC}_{2\pi}(\alpha)$

We need the following property of the arclength norm $\text{ARC}_{2\pi}$:

PROPOSITION 12. *Let n be a nonzero integer. If $\text{ARC}_{2\pi}(\alpha) \leq \frac{\pi}{|n|}$, then $\text{ARC}_{2\pi}(n\alpha) = |n|\text{ARC}_{2\pi}(\alpha)$*

18. Vintage $\mathbb{Z}_Q$ Shor QHSAs (Cont.)

Our next step is to look more closely at the probability distribution

$$Prob_{\widetilde{\varphi}} : \widehat{\widetilde{A}} \longrightarrow [0,1].$$

We seek first to use this probability distribution to determine the maximal cyclic subgroup $\mathbb{Z}_P$ of the hidden quotient group H_φ. However, as indicated by the following lemma, there are a number of obstacles to finding the subgroup $\mathbb{Z}_P$.

LEMMA 2. *Let $a_1, \ldots, a_n$ denote a hidden basis of the ambient group $A = \bigoplus_{j=1}^{n} \mathbb{Z}$, and let $P_1, P_2, \ldots, P_{\overline{n}}$ denote the respective orders of the corresponding cyclic direct summands of the hidden quotient group $H_\varphi = \bigoplus_{j=1}^{\overline{n}} \mathbb{Z}_{P_j}$.*

Let $\widetilde{a}$ denote a chosen generator of the group probe $\widetilde{A} = \mathbb{Z}_Q$, and let $\lambda_1, \lambda_2, \ldots, \lambda_n$ denote the unknown integers such that

$$\iota_\mu(\widetilde{a}) = \sum_{j=1}^{n} \lambda_j a_j \in A.$$

Finally, use the hidden injection $\iota_\varphi : H_\varphi \longrightarrow S$ to identify the elements of the hidden quotient group H_φ with the corresponding elements of the set S.

If the approximating map $\widetilde{\varphi}$ is constructed from a Shor transversal, then the order of $\widetilde{\varphi}(\widetilde{a}) \in H_\varphi$ is $\overline{P}$, i.e.,

$$\mathrm{order}\,(\widetilde{\varphi}(\widetilde{a})) = \overline{P},$$

where $\overline{P} = \mathrm{lcm}\left(\overline{P}_1, \overline{P}_2, \ldots, \overline{P}_n\right)$, and where $\overline{P}_j = P_j/\gcd(\lambda_j, P_j)$ for $j = 1, 2, \ldots, n$. Hence,

$$\left\{\widetilde{\varphi}(k\widetilde{a}) = \widetilde{\varphi}(\widetilde{a})^k \mid 0 \leq k < \overline{P}\right\}$$

are all distinct elements of S.

Moreover, if the approximating map $\widetilde{\varphi}$ is constructed from a maximal Shor transversal, $\overline{P} = P = \mathrm{lcm}(P_1, P_2, \ldots, P_n)$.

PROOF.

$$\widetilde{\varphi}(\widetilde{a}) = \varphi \circ \iota_\mu(\widetilde{a}) = \varphi\left(\sum_j \lambda_j a_j\right) = \prod_j \varphi(a_j)^{\lambda_j} = \prod_j b_j^{\lambda_j},$$

where we have used the hidden injection $\iota_\varphi : H_\varphi \longrightarrow S$ to identify the hidden basis element b_j of H_φ with the element $\varphi(a_j)$ of the set S.

Since the order of each b_j is P_j, it follows from elementary group theory that the order of $\prod_j b_j^{\lambda_j}$ must be $\overline{P}$. □

LEMMA 3. *Let $a_1, \ldots, a_n$ be a hidden basis of the ambient group $A = \bigoplus_{j=1}^{n} \mathbb{Z}$, and let $P_1, P_2, \ldots, P_n$ denote the respective orders of the corresponding cyclic direct summands of the hidden quotient group $H_\varphi = \bigoplus_{j=1}^{\overline{n}} \mathbb{Z}_{P_j}$.*

Let $\widetilde{a}$ denote a chosen generator of the group probe $\widetilde{A} = \mathbb{Z}_Q$, let $\lambda_1, \lambda_2, \ldots, \lambda_n$ denote the unknown integers such that

$$\iota_\mu(\widetilde{a}) = \sum_{j=1}^{n} \lambda_j a_j \in A,$$

and let $\chi_{\frac{y}{Q}}$ be a character of $\mathbb{Z}_Q$.

Finally, identify the elements of the hidden quotient group H_φ with the corresponding elements of the set S via the hidden injection $\iota_\varphi : H_\varphi \longrightarrow S$.

If the approximating map $\widetilde{\varphi}$ is constructed from a Shor transversal, then

- When $\overline{P}y \neq 0 \bmod Q$, we have

$$\widetilde{\varphi}\left(\chi^{\bullet}_{\frac{y}{Q}}\right) = \pm e^{i\pi \frac{\overline{P}y}{Q} q} \frac{\text{CHORD}_{2\pi}\left(\frac{\overline{P}y}{Q}(q+1)\right)}{\text{CHORD}_{2\pi}\left(\frac{\overline{P}y}{Q}\right)} \sum_{k_0=0}^{r-1} \chi_{\frac{y}{Q}}(k_0 \widetilde{a}) \widetilde{\varphi}(k_0 \widetilde{a})$$

$$\pm e^{i\pi \frac{\overline{P}y}{Q}(q-1)} \frac{\text{CHORD}_{2\pi}\left(\frac{\overline{P}y}{Q} q\right)}{\text{CHORD}_{2\pi}\left(\frac{\overline{P}y}{Q}\right)} \sum_{k_0=r}^{\overline{P}-1} \chi_{\frac{y}{Q}}(k_0 \widetilde{a}) \widetilde{\varphi}(k_0 \widetilde{a})$$

where $\overline{P} = \text{lcm}\left(\overline{P}_1, \overline{P}_2, \ldots, \overline{P}_n\right)$, where $\overline{P}_j = P_j / \gcd(\lambda_j, P_j)$ for $j = 1, 2, \ldots, n$, and where

$$Q = q\overline{P} + r, \text{ with } 0 \leq r < \overline{P}.$$

- And when $\overline{P}y = 0 \bmod Q$, we have

$$\widetilde{\varphi}\left(\chi^{\bullet}_{\frac{y}{Q}}\right) = (q+1) \sum_{k_0=0}^{r-1} \chi_{\frac{y}{Q}}(k_0 \widetilde{a}) \widetilde{\varphi}(k_0 \widetilde{a}) + q \sum_{k_0=r}^{\overline{P}-1} \chi_{\frac{y}{Q}}(k_0 \widetilde{a}) \widetilde{\varphi}(k_0 \widetilde{a})$$

Moreover, if the approximating map $\widetilde{\varphi}$ is constructed from a maximal Shor transversal, then $\overline{P} = P = \text{lcm}(P_1, P_2, \ldots, P_n)$.

PROOF. We begin by identifying the elements of the hidden quotient group H_φ with the corresponding elements of the set S via injection $\iota_\varphi : H_\varphi \longrightarrow S$.

We first consider the case when $\overline{P}y \neq 0 \bmod Q$.

Then

$$\widetilde{\varphi}\left(\chi^{\bullet}_{\frac{y}{Q}}\right) = \widetilde{\varphi}\left(\sum_{k=0}^{Q-1} \chi_{\frac{y}{Q}}(k\widetilde{a}) k\widetilde{a}\right) = \sum_{k=0}^{Q-1} \chi_{\frac{y}{Q}}(k\widetilde{a}) \widetilde{\varphi}(k\widetilde{a})$$

$$= \sum_{k=0}^{q\overline{P}-1} \chi_{\frac{y}{Q}}(k\widetilde{a}) \widetilde{\varphi}(\widetilde{a})^k + \sum_{k=q\overline{P}}^{Q-1} \chi_{\frac{y}{Q}}(k\widetilde{a}) \widetilde{\varphi}(\widetilde{a})^k$$

$$= \sum_{k_1=0}^{q-1} \sum_{k_0=0}^{\overline{P}-1} \chi_{\frac{y}{Q}}\left[(k_1 \overline{P} + k_0) \widetilde{a}\right] \widetilde{\varphi}[\widetilde{a}]^{k_1 \overline{P} + k_0} + \sum_{n_0=0}^{r-1} \chi_{\frac{y}{Q}}\left[(k_1 \overline{P} + k_0) \widetilde{a}\right] \widetilde{\varphi}[\widetilde{a}]^{k_1 \overline{P} + k_0}$$

From Lemma 2 we have

$$\widetilde{\varphi}[\widetilde{a}]^{k_1 \overline{P} + k_0} = \widetilde{\varphi}[\widetilde{a}]^{k_0} \ .$$

So,

$$\widetilde{\varphi}\left(\chi_{\frac{y}{Q}}^{\bullet}\right) = \left(\sum_{k_1=0}^{q-1} \chi_{\frac{y}{Q}}\left[\left(k_1\overline{P}\right)\widetilde{a}\right]\right) \sum_{k_0=0}^{\overline{P}-1} \chi_{\frac{y}{Q}}\left(k_0\widetilde{a}\right) \widetilde{\varphi}\left(\widetilde{a}\right)^{k_0}$$

$$+ \chi_{\frac{y}{Q}}\left(q\overline{P}\widetilde{a}\right) \sum_{k_0=0}^{r-1} \chi_{\frac{y}{Q}}\left(k_0\widetilde{a}\right) \widetilde{\varphi}\left(\widetilde{a}\right)^{k_0}$$

$$= \left(\sum_{k_1=0}^{q} \chi_{\frac{y}{Q}}\left[\left(k_1\overline{P}\right)\widetilde{a}\right]\right) \sum_{k_0=0}^{r-1} \chi_{\frac{y}{Q}}\left(k_0\widetilde{a}\right) \widetilde{\varphi}\left(\widetilde{a}\right)^{k_0}$$

$$+ \left(\sum_{k_1=0}^{q-1} \chi_{\frac{y}{Q}}\left[\left(k_1\overline{P}\right)\widetilde{a}\right]\right) \sum_{k_0=r}^{\overline{P}-1} \chi_{\frac{y}{Q}}\left(k_0\widetilde{a}\right) \widetilde{\varphi}\left(\widetilde{a}\right)^{k_0}$$

$$= \left(\frac{e^{2\pi i \frac{\overline{P}y}{Q}(q+1)} - 1}{e^{2\pi i \frac{\overline{P}y}{Q}} - 1}\right) \sum_{k_0=0}^{r-1} \chi_{\frac{y}{Q}}\left(k_0\widetilde{a}\right) \widetilde{\varphi}\left(k_0\widetilde{a}\right)$$

$$+ \left(\frac{e^{2\pi i \frac{\overline{P}y}{Q}q} - 1}{e^{2\pi i \frac{\overline{P}y}{Q}} - 1}\right) \sum_{k_0=r}^{\overline{P}-1} \chi_{\frac{y}{Q}}\left(k_0\widetilde{a}\right) \widetilde{\varphi}\left(k_0\widetilde{a}\right)$$

$$= e^{i\pi \frac{\overline{P}y}{Q}q} \left(\frac{e^{\pi i \frac{\overline{P}y}{Q}(q+1)} - e^{-\pi i \frac{\overline{P}y}{Q}(q+1)}}{e^{\pi i \frac{\overline{P}y}{Q}} - e^{-\pi i \frac{\overline{P}y}{Q}}}\right) \sum_{k_0=0}^{r-1} \chi_{\frac{y}{Q}}\left(k_0\widetilde{a}\right) \widetilde{\varphi}\left(k_0\widetilde{a}\right)$$

$$+ e^{i\pi \frac{\overline{P}y}{Q}(q-1)} \left(\frac{e^{\pi i \frac{\overline{P}y}{Q}q} - e^{-\pi i \frac{\overline{P}y}{Q}q}}{e^{\pi i \frac{\overline{P}y}{Q}} - e^{-\pi i \frac{\overline{P}y}{Q}}}\right) \sum_{k_0=r}^{\overline{P}-1} \chi_{\frac{y}{Q}}\left(k_0\widetilde{a}\right) \widetilde{\varphi}\left(k_0\widetilde{a}\right)$$

$$= e^{i\pi \frac{\overline{P}y}{Q}q} \left(\frac{\sin\left(\pi \frac{\overline{P}y}{Q}(q+1)\right)}{\sin\left(\pi \frac{\overline{P}y}{Q}\right)}\right) \sum_{k_0=0}^{r-1} \chi_{\frac{y}{Q}}\left(k_0\widetilde{a}\right) \widetilde{\varphi}\left(k_0\widetilde{a}\right)$$

$$+ e^{i\pi \frac{\overline{P}y}{Q}(q-1)} \left(\frac{\sin\left(\pi \frac{\overline{P}y}{Q}q\right)}{\sin\left(\pi \frac{\overline{P}y}{Q}\right)}\right) \sum_{k_0=r}^{\overline{P}-1} \chi_{\frac{y}{Q}}\left(k_0\widetilde{a}\right) \widetilde{\varphi}\left(k_0\widetilde{a}\right)$$

For the exceptional case when $\overline{P}y = 0 \bmod Q$, we need only observe that

$$\sum_{k_1=0}^{q} \chi_{\frac{y}{Q}}\left[\left(k_1\overline{P}\right)\widetilde{a}\right] = q+1 \text{ and } \sum_{k_1=0}^{q-1} \chi_{\frac{y}{Q}}\left(k_1\overline{P}\right)\widetilde{a} = q \ .$$

□

As an immediate consequence of above lemmas 2 and 3, we have:

COROLLARY 4. *If the approximating map $\widetilde{\varphi}$ is constructed from a Shor transversal, then*

$$\left\| \widetilde{\varphi}\left(\chi_{\frac{y}{Q}}^{\bullet}\right) \right\|^2 = \begin{cases} \dfrac{r\mathrm{CHORD}_{2\pi}^2\left[\frac{\overline{P}y}{Q}y(q+1)\right] + (\overline{P}-r)\mathrm{CHORD}_{2\pi}^2\left[\frac{\overline{P}y}{Q}yq\right]}{\mathrm{CHORD}_{2\pi}^2\left(\frac{\overline{P}y}{Q}\right)} & \text{if } \overline{P}y \neq 0 \bmod Q \\ r(q+1)^2 + (\overline{P}-r)q^2 & \text{if } \overline{P}y = 0 \bmod Q \end{cases}$$

Moreover, if the approximating map $\widetilde{\varphi}$ is constructed from a maximal Shor transversal, then $\overline{P} = P = \mathrm{lcm}(P_1, P_2, \ldots, P_n)$.

As a consequence of the inequalities found in **Proposition 11**, we have:

COROLLARY 5. *If the approximating map $\widetilde{\varphi}$ is constructed from a Shor transversal, then when $\overline{P}y \neq 0 \bmod Q$ we have*

$$\left\| \widetilde{\varphi}\left(\chi_{\frac{y}{Q}}^{\bullet}\right) \right\|^2 \geq \frac{4}{\pi^2}\left(\frac{r\mathrm{ARC}_{2\pi}^2\left[\frac{\overline{P}y}{Q}y(q+1)\right] + (\overline{P}-r)\mathrm{ARC}_{2\pi}^2\left[\frac{\overline{P}y}{Q}yq\right]}{\mathrm{ARC}_{2\pi}^2\left(\frac{\overline{P}y}{Q}\right)}\right)$$

Moreover, if the approximating map $\widetilde{\varphi}$ is constructed from a maximal Shor transversal, then $\overline{P} = P = \mathrm{lcm}(P_1, P_2, \ldots, P_n)$.

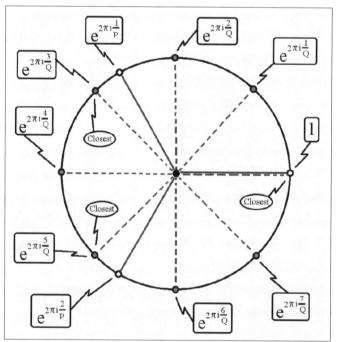

Figure 3. The characters of $\mathbb{Z}_P$ and $\mathbb{Z}_Q$ as points on the circle $\mathbb{S}^1$ of radius 1, with $P = 3$ and $Q = 8$. The characters $\chi_1, \chi_{3/8}, \chi_{5/8}$ of $\mathbb{Z}_Q$ are close respectively to characters $\chi_1, \chi_{\frac{1}{P}}, \chi_{\frac{2}{P}}$ of $\mathbb{Z}_P$. They are the characters of $\mathrm{ARC}_{2\pi}$ distance less than $\frac{\pi}{Q}\left(1 - \frac{P}{Q}\right)$ from some character of $\mathbb{Z}_P$. Also, $\chi_{\frac{1}{P}}$ and $\chi_{\frac{2}{P}}$ are the primitive

characters of $\mathbb{Z}_P$. Unfortunately, since $Q \not\geq P^2$, the characters $\chi_{3/8}$ and $\chi_{5/8}$ of $\mathbb{Z}_Q$ are not sufficiently close respectively to the primitive characters $\chi_{\frac{1}{P}}$ and $\chi_{\frac{2}{P}}$ of $\mathbb{Z}_P$. Hence, the continued fraction algorithm can not be used to find P.

19. When are characters of $\widetilde{A} = \mathbb{Z}_Q$ close to some character of a maximal cyclic subgroup $\mathbb{Z}_P$ of H_φ?

DEFINITION 13. *Let $\mathbb{Z}_{P_1} \oplus \mathbb{Z}_{P_2} \oplus \cdots \oplus \mathbb{Z}_{P_n}$ be the hidden direct sum decomposition of the hidden quotient group H_φ, and let $P = \operatorname{lcm}(P_1, P_2, \ldots, P_n)$. A character $\chi_{\frac{y}{Q}}$ of the group probe $\mathbb{Z}_Q$ is said to be **close** to a character of the maximal cyclic subgroup $\mathbb{Z}_P$ of the hidden quotient group H_φ provided either of the following equivalent conditions are satisfied*

Closeness Condition 1 *There exists an integer d such that*

$$\operatorname{ARC}_{2\pi}\left(\frac{y}{Q} - \frac{d}{P}\right) \leq \frac{\pi}{Q}\left(1 - \frac{P}{Q}\right),$$

or equivalently,

Closeness Condition 1'

$$\operatorname{ARC}_{2\pi}\left(\frac{Py}{Q}\right) \leq \frac{\pi P}{Q}\left(1 - \frac{P}{Q}\right)$$

*If in addition, $Q \geq P^2$, then the the character $\chi_{\frac{y}{Q}}$ of $\mathbb{Z}_Q$ is said to be **sufficiently close** to a character of the maximal cyclic subgroup $\mathbb{Z}_P$.*

It immediately follows from the theory of continued fractions [**21, 33**] that

PROPOSITION 13. *If a character $\chi_{\frac{y}{Q}}$ of $\mathbb{Z}_Q$ is sufficiently close to a character $\chi_{\frac{d}{P}}$ of $\mathbb{Z}_P$, then $\frac{d}{P}$ is a convergent of the continued fraction expansion of $\frac{y}{Q}$.*

However, to determine the sought integer P from the rational $\frac{d}{P}$, the numerator and denominator of $\frac{d}{P}$ must be relatively prime, i.e.,

$$\gcd(d, P) = 1 .$$

This leads to the following definition:

DEFINITION 14. *A character $\chi_{\frac{d}{P}}$ of $\mathbb{Z}_P$ is said to be **primitive** provided that it is a generator of the dual group $\widehat{\mathbb{Z}}_P$.*

PROPOSITION 14. *A character $\chi_{\frac{d}{P}}$ of $\mathbb{Z}_P$ is a primitive character if and only $\gcd(d, P) = 1$. Moreover, the number of primitive characters of $\mathbb{Z}_P$ is $\phi(P)$, where $\phi(P)$ denotes Euler's totient function, i.e., the number of positive integers less than P which are relatively prime to P.*

THEOREM 8. *Assume that $Q \geq P^2$, and that the approximating map $\widetilde{\varphi}$ is constructed from a maximal Shor transversal. Then the probability that $\mathrm{QRAND}_{\widetilde{\varphi}}()$ produces a character of the group probe $\mathbb{Z}_Q$ which is sufficiently close to a primitive character of the maximal cyclic subgroup $\mathbb{Z}_P$ of the hidden quotient group H_φ satisfies the following bound*

$$\mathrm{Prob}_{\widetilde{\varphi}}\left(\begin{array}{c}\chi \text{ sufficiently close to some}\\ \text{primitive character of } \mathbb{Z}_P\end{array}\right) \geq \frac{4}{\pi^2} \frac{\phi(P)}{P}\left(1 - \frac{P}{Q}\right)^2$$

PROOF. Let $\chi_{\frac{y}{Q}}$ be a particular character of the group probe $\mathbb{Z}_Q$ which is sufficiently close to some character of the maximal cyclic subgroup $\mathbb{Z}_P$. We now compute the probability that $\mathrm{QRand}_{\widetilde{\varphi}}()$ will produce this particular character.

First consider the exceptional case when $Py = 0 \bmod Q$. Using the expression for $\left\|\widetilde{\varphi}\left(\chi_{\frac{y}{Q}}^{\bullet}\right)\right\|^2$ given in **Corollary 5**, we have

$$\left\|\widetilde{\varphi}\left(\chi_{\frac{y}{Q}}^{\bullet}\right)\right\|^2 = r(q+1)^2 + (P-r)q^2 \geq Pq^2 = P\left(\frac{Q-r}{P}\right)^2 \geq \frac{1}{P}(Q-P)^2.$$

So

$$\mathrm{Prob}_{\widetilde{\varphi}}\left(\chi_{\frac{y}{Q}}\right) = \frac{\left\|\widetilde{\varphi}\left(\chi_{\frac{y}{Q}}^{\bullet}\right)\right\|^2}{Q^2} \geq \frac{1}{P}\frac{P(Q-P)^2}{Q^2} = \frac{1}{P}\left(1 - \frac{P}{Q}\right)^2 \geq \frac{4}{\pi^2}\frac{1}{P}\left(1 - \frac{P}{Q}\right)^2.$$

Next consider the non-exceptional case when $Py \neq 0 \bmod Q$.

In this case, **Proposition 12** can be applied to both terms in the numerator of the expression given in **Corollary 5**. Hence,

$$\left\|\widetilde{\varphi}\left(\chi_{\frac{y}{Q}}^{\bullet}\right)\right\|^2 \geq \frac{4}{\pi^2}\left(\frac{r\mathrm{ARC}_{2\pi}^2\left[\frac{Py}{Q}y(q+1)\right] + (P-r)\mathrm{ARC}_{2\pi}^2\left[\frac{Py}{Q}yq\right]}{\mathrm{ARC}_{2\pi}^2\left(\frac{Py}{Q}y\right)}\right)$$

$$\geq \frac{4}{\pi^2}\left(\frac{r(q+1)^2\mathrm{ARC}_{2\pi}^2\left[\frac{Py}{Q}y\right] + (P-r)q^2\mathrm{ARC}_{2\pi}^2\left[\frac{Py}{Q}y\right]}{\mathrm{ARC}_{2\pi}^2\left(\frac{Py}{Q}y\right)}\right)$$

$$\geq \frac{4}{\pi^2}r(q+1)^2 + \frac{4}{\pi^2}(P-r)q^2 \geq \frac{4}{\pi^2}rq^2 + \frac{4}{\pi^2}(P-r)q^2$$

$$\geq \frac{4}{\pi^2}Pq^2 = \frac{4}{\pi^2}\frac{1}{P}(Q-r)^2 \geq \frac{4}{\pi^2}\frac{1}{P}(Q-P)^2$$

Thus,

$$\mathrm{Prob}_{\widetilde{\varphi}}\left(\chi_{\frac{y}{Q}}\right) = \frac{\left\|\widetilde{\varphi}\left(\chi_{\frac{y}{Q}}^{\bullet}\right)\right\|^2}{Q^2} \geq \frac{4}{\pi^2}\frac{1}{P}\left(1 - \frac{P}{Q}\right)^2$$

So, in either case we have

$$Prob_{\widetilde{\varphi}}\left(\chi_{\frac{y}{Q}}\right) \geq \frac{4}{\pi^2}\frac{1}{P}\left(1-\frac{P}{Q}\right)^2.$$

We now note that there is one-to-one correspondence between the characters of $\mathbb{Z}_P$ and the sufficiently close characters of $\mathbb{Z}_Q$. Hence, there are exactly $\phi(P)$ characters of the group probe $\mathbb{Z}_Q$ which are sufficiently close some primitive character of the maximal cyclic group $\mathbb{Z}_P$. The theorem follows. $\square$

The following theorem can be found in [**21**, Theorem 328, Section 18.4]:

THEOREM 9.
$$\liminf \frac{\phi(N)}{N/\ln\ln N} = e^{-\gamma},$$
where γ denotes Euler's constant $\gamma = 0.57721566490153286061\ldots$, and where $e^{-\gamma} = 0.5614594836\ldots$.

As a corollary, we have:

COROLLARY 6. $Prob_{\widetilde{\varphi}}(\chi$ sufficiently close to some primitive character of $\mathbb{Z}_P)$ is bounded below by
$$\frac{4}{\pi^2 \ln 2} \cdot \frac{e^{-\gamma} - \epsilon(P)}{\lg\lg Q} \cdot \left(1-\frac{P}{Q}\right)^2,$$
where $\epsilon(P)$ is a monotone decreasing sequence converging to zero. In terms of asymptotic notation,

$$Prob_{\widetilde{\varphi}}(\chi \text{ sufficiently close to some primitive character of } \mathbb{Z}_P) = \Omega\left(\frac{1}{\lg\lg Q}\right).$$

For a proof of the above, please refer to [**33**, **43**].

20. Summary of Vintage $\mathbb{Z}_Q$ Shor QHSAs

Let $\varphi : A \longrightarrow S$ be a map with hidden subgroup structure with ambient group A free abelian of finite rank n, and with image of φ finite. Then as a culmination of the mathematical developments in sections 11 through 19, we have the following **vintage $\mathbb{Z}_Q$ Shor QHSA** for finding the order $P = \text{lcm}(P_1, P_2, \ldots, P_n)$ of the maximum cyclic subgroup Z_P of the hidden quotient group $H_\varphi = \bigoplus_{j=1}^n \mathbb{Z}_{P_j}$. A flowchart of this algorithm is given in Figure 4.

$$\boxed{\text{VINTAGE\_SHOR}(\varphi, Q, n)}$$

\# INPUT: $\varphi : A \longrightarrow S$ and Q and rank n of A

OUTPUT: $P = \text{lcm}(P_1, P_2, \ldots P_n)$ if hidden quotient group
is $H_\varphi = \bigoplus_{j=1}^{n} \mathbb{Z}_{P_j}$

Step 1 Select a basis $a'_1, a'_2, \ldots, a'_n$ of A and a generator $\widetilde{a}$ of $\mathbb{Z}_Q$

Step 2 $(\iota_\mu : \mathbb{Z}_Q \longrightarrow A) = \text{RANDOM\_SHOR\_TRANSVERSL}(\{a'_1, a'_2, \ldots, a'_n\}, Q, \widetilde{a}, n)$

Step 3 Construct $\widetilde{\varphi} = \varphi \circ \iota_\mu : \mathbb{Z}_Q \longrightarrow S$

Step 4 $\chi_{\frac{y}{Q}} = \text{QRAND}_{\widetilde{\varphi}}()$

Step 5 $(d'', P'') = (0, 1)$ # 0-th Cont. Frac. Convergent of $\frac{y}{Q}$
$\qquad (d', P') = \left(1, \left\lfloor \frac{Q}{y} \right\rfloor \right)$ # 1-th Cont. Frac. Convergent of $\frac{y}{Q}$
$\qquad$ INNER LOOP
$\qquad\quad (Save\_d', Save\_P') = (d', P')$
$\qquad\quad (d', P') = \text{NEXT\_CONT\_FRAC\_CONVERGENT}\left(\frac{y}{Q}, (d', P'), (d'', P'')\right)$
$\qquad\quad (d'', P'') = (Save\_d', Save\_P')$
$\qquad\quad$ IF $\varphi(P'a'_j) = \varphi(0)$ for all $j = 1, 2, \ldots, n$ THEN GOTO **Step 6**
$\qquad\quad$ IF $\frac{d'}{P'} = \frac{y}{Q}$ THEN GOTO **Step 2**
$\qquad$ INNER LOOP BOUNDARY

Step 6 OUTPUT P' AND STOP

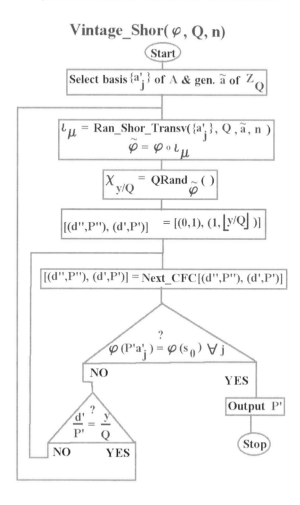

Figure 4. Flowchart for Vintage $\mathbb{Z}_Q$ Shor QHSA. This is a Wandering Shor algorithm.

21. A cursory analysis of complexity

We now make a cursory analysis of the algorithmic complexity of the vintage $\mathbb{Z}_Q$ Shor algorithm. By the word "cursory" we mean that our objective is to find an asymptotic bound which is by no means the tightest possible.

Our analysis is based on the following three assumptions:

- **Assumption 1.** Conjecture 1 (found in Appendix B) is true.

- **Assumption 2.** $U_{\widetilde{\varphi}}$ is of complexity $O\left(n^2 \left(\lg Q\right)^3\right)$.

- **Assumption 3.** The integer Q is chosen so that $Q = 2^L \geq P^2$, where $P = \text{lcm}(P_1, P_2, \ldots, P_n)$.

The following theorem is an immediate consequence of **Assumption 2**.

THEOREM 10. *Let*
$$\widetilde{\varphi} : \mathbb{Z}_Q \longrightarrow S$$
be a map from the cyclic group $\mathbb{Z}_Q$ to a set S, where $Q = 2^L$.
If $U_{\widetilde{\varphi}}$ is of algorithmic complexity
$$O\left(n^2 (\lg Q)^3\right),$$
then the algorithmic complexity of $\text{QRAND}_{\widetilde{\varphi}}()$ is the same, i.e.,
$$O\left(n^2 (\lg Q)^3\right)$$

PROOF. Steps 1 and 3 are each of the same algorithmic complexity as the quantum Fourier transform[17], i.e., of complexity $O\left((\lg Q)^2\right)$. (See [**36**, Chapter 5].). Thus the dominant step in $\text{QRAND}_{\widetilde{\varphi}}()$ is Step 2, which is by assumption of complexity $O\left(n^2 (\lg Q)^3\right)$. □

The complexities of each step of the vintage $\mathbb{Z}_Q$ Shor algorithm are given below. An accompanying abbreviated flow chart of this algorithm is shown in Figure 5.

Step 1 Step 1 is of algorithmic complexity is $O(n)$.

Step 2 By theorem 7 of section 14, Step 2 is of average case complexity $O\left(n^2 (\lg Q)^3\right)$. By corollary 8 of Appendix B, the probability that this step will be successful, i.e., will produce a maximal Shor transversal, is $\Omega\left(\left(\frac{1}{\lg \lg Q}\right)^n\right)$.

Step 3 Step 3 is of algorithmic complexity $O(n)$.

Step 4 By theorem 10 given above, Step 4 is of algorithmic complexity $O\left(n^2 (\lg Q)^3\right)$. By corollary 6 of section 19, the probability (given that Step 2 is successful) that this step will be successful, i.e., will produce a character sufficiently close to a primitive character of the maximal cyclic group $\mathbb{Z}_P$ is $\Omega\left(\frac{1}{\lg \lg Q}\right)$.

Step 5 This step is of algorithmic complexity $O\left(n (\lg Q)^3\right)$. (See, for example, [**30**].)

---

[17]If instead the Hadamard-Walsh transform is used in Step 1, then the complexity of Step 1 is $O(\lg Q)$.

Step 6 For Step 5 to branch to this step, both Steps 2 and 4 must be successful. Thus the probability of branching to step 6 is $Prob_{Success}$ (Step 2) $\cdot$ $Prob_{Success}$ (Step 2) $= \Omega\left(\left(\frac{1}{\lg \lg Q}\right)^{n+1}\right)$

Since the Steps 2 through 5 loop will on average be executed $O\left((\lg \lg Q)^{n+1}\right)$ times, the average algorithmic complexity of the Vintage Z_Q Shor algorithm is $O\left(n^2 (\log Q)^3 (\lg \lg Q)^{n+1}\right)$. (This is, of course, not the tightest possible asymptotic bound.) We formalize this analysis as a theorem:

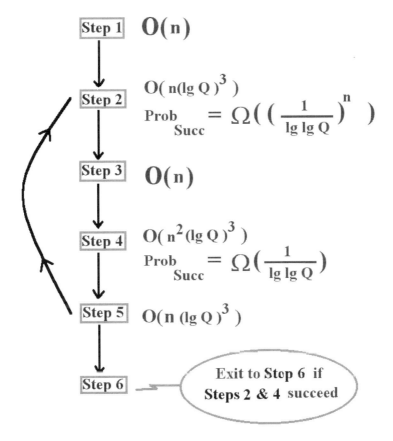

Figure 5. An abbreviated flowchart of the vintage Z_Q Shor Algorithm. The probability of a successful exit to Step 6 is $\Omega\left(\left(\frac{1}{\lg \lg Q}\right)^{n+1}\right)$. Hence, the average number of times Steps 2 through 5 are executed is $O\left((\lg \lg Q)^{n+1}\right)$.

THEOREM 11. *Assuming the three assumptions given in section 21, the average algorithmic complexity of the Vintage $\mathbb{Z}_Q$ Shor algorithm for finding the maximal* cyclic subgroup $\mathbb{Z}_P$ of the hidden quotient group $H_\varphi = \bigoplus_{j=1}^n \mathbb{Z}_{P_j}$ is

$$O\left(n^2 (\lg Q)^3 (\lg \lg Q)^{n+1}\right)$$

22. Two alternative vintage $\mathbb{Z}_Q$ Shor algorithms

As two alternatives to the algorithm described in the last two sections, we give below two other vintage $\mathbb{Z}_Q$ Shor algorithms. Unlike the above described algorithm, these two alternative algorithms do not depend on finding a maximal Shor transversal. The first finds the order of the maximal cyclic subgroup $\mathbb{Z}_P$ of the hidden quotient group H_φ. The second finds the entire hidden subgroup K_φ. Flowcharts for these two quantum algorithms are given in Figures 6 and 7.

An optimal choice for the parameter K of the following algorithm is not known at this time.

| ALTERNATIVE1_VINTAGE_SHOR(φ, Q, n, K) |

\# INPUT: $\varphi : A \longrightarrow S$, Q, rank n of A, and number of
\#   inner loop iterations K
\# OUTPUT: $P = \text{lcm}(P_1, P_2, \ldots P_n)$ if hidden quotient group
\#   is $H_\varphi = \bigoplus_{j=1}^{\overline{n}} \mathbb{Z}_{P_j}$

Step 1   SET $P = 1$

Step 2   Select a basis $a_1', a_2', \ldots, a_n'$ of A and a generator $\widetilde{a}$ of $\mathbb{Z}_Q$

Step 3   OUTER LOOP

    Step 4   INNER LOOP for K iterations

        Step 5 $(\iota_\mu : Z_Q \longrightarrow A) = \text{RAND\_SHOR\_TRANSVR}(\{a_1', a_2', \ldots, a_n'\}, Q, \widetilde{a}, n)$

        Step 6   Construct $\widetilde{\varphi} = \varphi \circ \iota_\mu : \mathbb{Z}_Q \longrightarrow S$

        Step 7 $\chi_{\frac{y}{Q}} = \text{QRAND}_{\widetilde{\varphi}}()$

        Step 8 $(d'', P'') = (0, 1)$ \# 0-th Cont. Frac. Converg. of $\frac{y}{Q}$

                   $(d', P') = \left(1, \left\lfloor \frac{Q}{y} \right\rfloor\right)$ \# 1-th Cont. Frac. Converg. of $\frac{y}{Q}$
                   INNERMOST LOOP
                       $(Save\_d', Save\_P') = (d', P')$
                       $(d', P') = \text{NXT\_CONT\_FRAC\_CONVRG}\left(\frac{y}{Q}, (d', P'), (d'', P'')\right)$
                       $(d'', P'') = (Save\_d', Save\_P')$
                       IF $\varphi(P'\iota_\mu(\widetilde{a})) = \varphi(0)$ THEN GOTO **Step 9**
                       IF $\frac{d'}{P'} = \frac{y}{Q}$ THEN GOTO **Step 4**
                   INNERMOST LOOP BOUNDARY

Step 9	$P = \mathrm{lcm}(P, P')$
Step 10	INNER LOOP LOWER BOUNDARY
Step 11	IF $\varphi(Pa_j) = \varphi(0)$ FOR $j = 1, 2, \ldots, n$ THEN GOTO Step 13
Step 12	OUTER LOOP LOWER BOUNDARY
Step 13	OUTPUT P' AND STOP

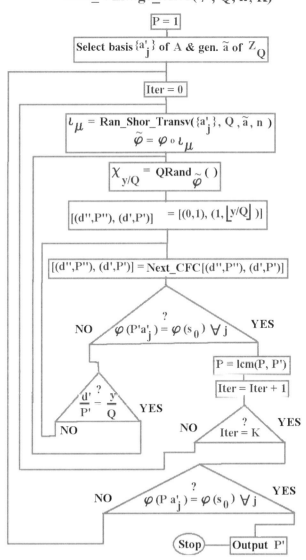

Figure 6. The First Alternate Vintage $\mathbb{Z}_Q$ Shor QHSA. This is a Wandering Shor algorithm.

The following wandering Shor algorithm actually finds the entire hidden quotient group H_φ, and hence the hidden subgroup K_φ:

$$\boxed{\text{Alternative2\_Vintage\_Shor}(\varphi, Q, n)}$$

Input: $\varphi : A \longrightarrow S$, Q, rank n of A
Output: A matrix $\mathfrak{G}$ with row span equal to the
hidden subgroup $K_\varphi = \bigoplus_{j=1}^{n} P_j \mathbb{Z}$

Step 1 Set $\mathfrak{G} = [\ ]$ and NonZeroRows $= 0$

Step 2 Select a basis $a_1', a_2', \ldots, a_n'$ of A and a generator $\widetilde{a}$ of $\mathbb{Z}_Q$

Step 3 Outer Loop Until NonZeroRows $= n$

  Step 4 $(\iota_\mu : \mathbb{Z}_Q \longrightarrow A) = \text{RanShorTransvrsl}(\{a_1', a_2', \ldots, a_n'\}, Q, \widetilde{a}, n)$

  Step 5 Construct $\widetilde{\varphi} = \varphi \circ \iota_\mu : \mathbb{Z}_Q \longrightarrow S$

  Step 6 $\chi_{\frac{y}{Q}} = \text{QRand}_{\widetilde{\varphi}}()$

  Step 7 $(d'', P'') = (0, 1)$ # 0-th Cont. Frac. Converg. of $\frac{y}{Q}$
    $(d', P') = \left(1, \left\lfloor \frac{Q}{y} \right\rfloor\right)$ # 1-th Cont. Frac. Converg. of $\frac{y}{Q}$
    Inner Loop
      $(Save\_d', Save\_P') = (d', P')$
      $(d', P') = \text{NextContFracConverg}\left(\frac{y}{Q}, (d', P'), (d'', P'')\right)$
      $(d'', P'') = (Save\_d', Save\_P')$
      If $\varphi(P' \iota_\mu(\widetilde{a})) = \varphi(0)$ Then Goto $\boxed{\text{Step 8}}$
      If $\frac{d'}{P'} = \frac{y}{Q}$ Then Goto $\boxed{\text{Step 11}}$
    Inner Loop Boundary: Continue

  Step 8 $\mathfrak{G} = \begin{bmatrix} \mathfrak{G} \\ \text{------------} \\ P'\lambda_1' \quad P'\lambda_2' \quad \ldots \quad P'\lambda_n' \end{bmatrix}$

  Step 9 $\mathfrak{G} = \text{Put\_In\_Echelon\_Canonical\_Form}(\mathfrak{G})$

  Step 10 NonZeroRows $= \text{Number\_of\_Non\_Zero\_Rows}(\mathfrak{G})$

Step 11 Outer Loop Lower Boundary: Continue

Step 12 Output matrix $\mathfrak{G}$ and Stop

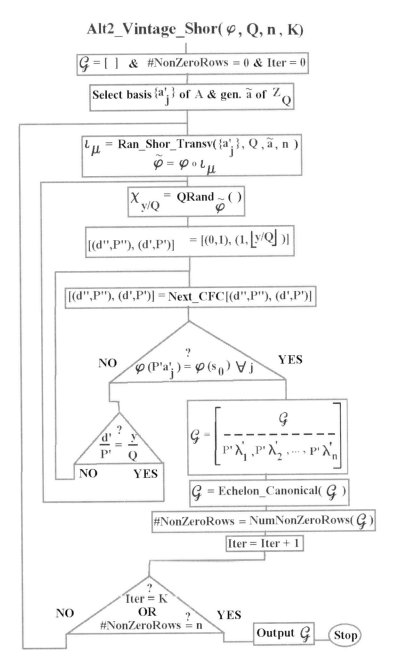

Figure 7. The Second Alternate Vintage $\mathbb{Z}_Q$ Shor QHSA. This is a Wandering Shor algorithm.

Part 6. Epilogue

23. Conclusion

Each of the three vintage $\mathbb{Z}_Q$ Shor QHSAs created in this paper is a natural generalization of Shor's original quantum factoring algorithm to free abelian groups A of finite rank n. The first two of the three find a maximal cyclic subgroup $\mathbb{Z}_P$ of the hidden quotient group H_φ. The last of the three does more. It finds the entire hidden quotient group H_φ.

We also note that these QHSAs can be viewed from yet another perspective as **wandering Shor algorithms** on free abelian groups. By this we mean quantum algorithms which, with each iteration, first select a random cyclic direct summand $\mathbb{Z}$ of the ambient group A and then apply one iteration of the standard Shor algorithm on $\mathbb{Z}$ to produce a random character of the "approximating" group $\widetilde{A} = \mathbb{Z}_Q$.

From this perspective, under the assumptions given in section 21, the algorithmic complexity of the first of these wandering QHSAs is found to be

$$O\left(n^2 (\lg Q)^3 (\lg \lg Q)^{n+1}\right).$$

Obviously, much remains to be accomplished.

It should be possible to extend the vintage $\mathbb{Z}_Q$ Shor algorithms to quantum algorithms with more general group probes of the form

$$\widetilde{A} = \bigoplus_{j=1}^{m} \mathbb{Z}_{Q_j}$$

for $m > 1$. This would be a full generalization of Shor's quantum factoring algorithm to the abelian category.

It is hoped that this paper will provide a useful stepping stone to the construction of QHSAs on non-abelian groups.

24. Acknowledgement

We would like to thank Tom Armstrong, Howard Brandt, Eric Rains, Fernando Souza, Umesh Vazirani, and Yaacov Yesha for some helpful discussions. We would also like to thank the referee for some helpful suggestions.

25. Appendix A. Continued fractions

We give a brief summary of those aspects of the theory of continued fractions that are relevant to this paper. (For a more in-depth explanation of the theory of continued fractions, please refer, for example, to [**21**] and [**31**].)

Every positive rational number ξ can be written as an expression in the form

$$\xi = a_0 + \cfrac{1}{a_1 + \cfrac{1}{a_2 + \cfrac{1}{a_3 + \cfrac{1}{\cdots + \cfrac{1}{a_N}}}}},$$

where a_0 is a non-negative integer, and where $a_1, \ldots, a_N$ are positive integers. Such an expression is called a (finite, simple) **continued fraction**, and is uniquely determined by ξ provided we impose the condition $a_N > 1$. For typographical simplicity, we denote the above continued fraction by

$$[a_0, a_1, \ldots, a_N] \ .$$

The continued fraction expansion of ξ can be computed with the following recurrence relation, which always terminates if ξ is rational:

$$\begin{cases} a_0 = \lfloor \xi \rfloor \\ \xi_0 = \xi - a_0 \end{cases}, \text{ and if } \xi_n \neq 0, \text{ then } \begin{cases} a_{n+1} = \lfloor 1/\xi_n \rfloor \\ \xi_{n+1} = \frac{1}{\xi_n} - a_{n+1} \end{cases}$$

The n-th **convergent** ($0 \leq n \leq N$) of the above continued fraction is defined as the rational number ξ_n given by

$$\xi_n = [a_0, a_1, \ldots, a_n] \ .$$

Each convergent ξ_n can be written in the form, $\xi_n = \frac{p_n}{q_n}$, where p_n and q_n are relatively prime integers ($\gcd(p_n, q_n) = 1$). The integers p_n and q_n are determined by the recurrence relation

$$\begin{aligned} p_0 &= a_0, & p_1 &= a_1 a_0 + 1, & p_n &= a_n p_{n-1} + p_{n-2}, \\ q_0 &= 1, & q_1 &= a_1, & q_n &= a_n q_{n-1} + q_{n-2} \ . \end{aligned}$$

The subroutine

NEXT_CONT_FRAC_CONVERGENT

found in the vintage $\mathbb{Z}_Q$ Shor algorithm given in section 20 is an embodiment of the above recursion.

This recursion is used because of the following theorem which can be found in [**21**, Theorem 184, Secton 10.15]:

THEOREM 12. *Let ξ be a real number, and let d and P be integers with $P > 0$. If*

$$\left|\xi - \frac{d}{P}\right| \leq \frac{1}{2P^2},$$

then the rational number d/P is a convergent of the continued fraction expansion of ξ.

26. Appendix B. Probability Distributions on Integers

Let

$$Prob_Q : \{1, 2, \ldots, Q\} \longrightarrow [0, 1]$$

denote the uniform probability distribution on the finite set of integers $\{1, 2, \ldots, Q\}$. Thus, the probability that a random integer λ from $\{1, 2, \ldots, Q\}$ is divisible by a given prime p is

$$Prob_Q(p|\lambda) = \frac{\lfloor Q/p \rfloor}{Q} \leq \frac{1}{p},$$

where '$\lfloor - \rfloor$' denotes the the floor function.

The limit $Prob_\infty$, should it exist, of the probability distribution $Prob_Q$ as Q approaches infinity, i.e.,

$$Prob_\infty = \lim_{Q \longrightarrow \infty} Prob_\infty,$$

will turn out to be a useful tool. Since $Prob_\infty$ is not a probability distribution, we will call it a **pseudo-probability distribution** on the integers $\mathbb{Z}$. It immediately follows that

$$Prob_\infty(p|\lambda) = \frac{1}{p}.$$

In this sense, we say that the pseudo-probability of a random integer $\lambda \in \mathbb{Z}$ being divisible by a given prime p is $1/p$.

THEOREM 13. *Let n be an integer greater than 1. Let $\lambda'_1, \lambda'_2, \ldots, \lambda'_n$, be n integers selected randomly and independently with replacement from the set $\{1, 2, \ldots, Q\}$ according to the uniform probability distribution. Then the probability that*

$$\gcd(\lambda'_1, \lambda'_2, \ldots, \lambda'_n) = 1$$

is

$$Prob_Q\left(\gcd(\lambda'_1, \lambda'_2, \ldots, \lambda'_n) = 1\right) = \sum_{k=1}^{Q} \mu(k) \left(\frac{\lfloor Q/k \rfloor}{Q}\right)^n,$$

where '$\lfloor - \rfloor$' and '$\mu(-)$' respectively denote the floor and Möbius functions.
Moreover,

$$Prob_\infty\left(\gcd(\lambda'_1, \lambda'_2, \ldots, \lambda'_n) = 1\right) = \zeta(n)^{-1},$$

where $\zeta(n)$ denotes the Riemann zeta function $\zeta(n) = \sum_{k=1}^{\infty} \frac{1}{k^n}$.

PROOF. Let $Primes_Q$ denote the set of primes less than or equal to Q. For each prime p and integer $0 < j \leq n$, let A_{pj} denote the set

$$A_{pj} = \left\{ \vec{\lambda} \in \{1, \ldots, Q\}^n : p \mid \lambda_j \right\}.$$

Since

$$\bigcap_{p \in Primes_Q} \bigcup_{j=1}^{n} \overline{A}_{pj} = \left\{ \vec{\lambda} \in \{1, \ldots, Q\}^n : \forall p \exists j \ p \mid \lambda_j \right\},$$

we have

$$Prob_Q \left(\gcd(\lambda_1, \lambda_2, \ldots, \lambda_n) = 1 \right) = Prob_Q \left(\bigcap_{p \in Primes_Q} \bigcup_{j=1}^{n} \overline{A}_{pj} \right),$$

where $\overline{A}_{pj}$ denotes the complement of A_{pj}.

We proceed to compute $Prob_Q \left(\bigcap_{p \in Primes_Q} \bigcup_{j=1}^{n} \overline{A}_{pj} \right)$ by first noting that:

$$Prob_Q \left(\bigcap_{p \in Primes_Q} \bigcup_{j=1}^{n} \overline{A}_{pj} \right) = 1 - Prob_Q \left(\bigcup_{p \in Primes_Q} \bigcap_{j=1}^{n} A_{pj} \right).$$

So by the inclusion/exclusion principle, we have

$$Prob_Q \left(\bigcup_{p \in Primes_Q} \bigcap_{j=1}^{n} A_{pj} \right) = - \sum_{\substack{S \subseteq Primes_Q \\ S \neq \emptyset}} (-1)^{|S|} Prob_Q \left(\bigcap_{p \in S} \bigcap_{j=1}^{n} A_{pj} \right)$$

$$= - \sum_{\substack{S \subseteq Primes_Q \\ S \neq \emptyset}} (-1)^{|S|} Prob_Q \left(\bigcap_{j=1}^{n} \bigcap_{p \in S} A_{pj} \right)$$

Since the λ_j's are independent random variables, we have

$$Prob_Q \left(\bigcap_{j=1}^{n} \bigcap_{p \in S} A_{pj} \right) = \prod_{j=1}^{n} Prob_Q \left(\bigcap_{p \in S} A_{pj} \right)$$

Moreover, it follows from a straight forward counting argument that

$$Prob_Q \left(\bigcap_{p \in S} A_{pj} \right) = \left\lfloor Q / \prod_{p \in S} p \right\rfloor / Q,$$

from which we obtain

$$Prob_Q \left(\bigcap_{j=1}^{n} \bigcap_{p \in S} A_{pj} \right) = \prod_{j=1}^{n} \left\lfloor Q / \prod_{p \in S} p \right\rfloor / Q = \left(\left\lfloor Q / \prod_{p \in S} p \right\rfloor / Q \right)^n.$$

Thus,

$$Prob_Q \left(\bigcap_{p \in Primes_Q} \bigcup_{j=1}^n \overline{A}_{pj} \right) = \sum_{S \subseteq Primes_Q} (-1)^{|S|} \left(\left\lfloor Q / \prod_{p \in S} p \right\rfloor / Q \right)^n$$

This last expression expands to

$$1 - \sum_{\substack{p \leq Q \\ p \text{ Prime}}} \left(\frac{\lfloor Q/p \rfloor}{Q} \right)^n + \sum_{\substack{p < p' \leq Q \\ p, p' \text{ Prime}}} \left(\frac{\lfloor Q/pq \rfloor}{Q} \right)^n - \sum_{\substack{p < p' < p'' \leq Q \\ p, p', p'' \text{ Prime}}} \left(\frac{\lfloor Q/pp'p'' \rfloor}{Q} \right)^n + \ldots,$$

which can be rewritten as

$$\sum_{k=1}^Q \mu(k) \left(\frac{\lfloor Q/k \rfloor}{Q} \right)^n$$

since $\mu(k) = 0$ for all integers k that are not squarefree.

The last part of this theorem follows immediately from the fact that

$$\lim_{Q \to \infty} \sum_{k=1}^Q \mu(k) \left(\frac{\lfloor Q/k \rfloor}{Q} \right)^n = \sum_{k=1}^\infty \mu(k) \frac{1}{k^n} = \zeta(n)^{-1}.$$

(See [**40**], [**42**], or [**21**].) □

COROLLARY 7. *Let n be an integer greater than 1, and let $\lambda_1', \lambda_2', \ldots, \lambda_n'$ be n integers randomly and independently selected with replacement from the set $\{1, 2, \ldots, Q\}$ according to the uniform probability distribution. Let M be a fixed element of the group $SL_\pm(n, \mathbb{Z})$ of invertible $n \times n$ integer matrices. Finally, let $\lambda_1, \lambda_2, \ldots, \lambda_n$ be n integers given by*

$$(\lambda_1, \lambda_2, \ldots, \lambda_n)^{transpose} = M (\lambda_1', \lambda_2', \ldots, \lambda_n')^{transpose}.$$

Then the probability that

$$\gcd(\lambda_1, \lambda_2, \ldots, \lambda_n) = 1$$

is

$$Prob_Q \left(\gcd(\lambda_1, \lambda_2, \ldots, \lambda_n) = 1 \right) = \sum_{k=1}^Q \mu(k) \left(\frac{\lfloor Q/k \rfloor}{Q} \right)^n,$$

where '$\lfloor - \rfloor$' and '$\mu(-)$' respectively denote the floor and Möbius functions.

Moreover,

$$Prob_\infty \left(\gcd(\lambda_1, \lambda_2, \ldots, \lambda_n) = 1 \right) = \zeta(n)^{-1},$$

where $\zeta(n)$ denotes the Riemann zeta function $\zeta(n) = \sum_{k=1}^\infty \frac{1}{k^n}$. Hence,

$$Prob_Q \left(\gcd(\lambda_1, \lambda_2, \ldots, \lambda_n) = 1 \right) = \Omega \left(\zeta(n)^{-1} \right) = \Omega(1)$$

PROOF. This corollary immediately follows from the fact that the gcd is invariant under the action of $SL_\pm(n, \mathbb{Z})$. □

REMARK 11. *We conjecture that a stronger result holds, namely that the function $\zeta(n)^{-1}$ is actually a lower bound for $Prob_Q\left(\gcd(\lambda_1, \lambda_2, \ldots, \lambda_n) = 1\right)$ for $Q \geq n$.*

We need to make the following conjecture to estimate the algorithmic complexity of Vintage $\mathbb{Z}_Q$ algorithms, also called wandering Shor algorithms.

CONJECTURE 1. *Let n be an integer greater than 1, let $P_1, P_2, \ldots, P_n$ be n fixed positive integers, and let $\lambda_1', \lambda_2', \ldots, \lambda_n'$ be n integers randomly and independently selected with replacement from the set $\{1, 2, \ldots, Q\}$ according to the uniform probability distribution. Let M be a fixed element of the group $SL_\pm(n, \mathbb{Z})$ of invertible $n \times n$ integral matrices, and let*

$$(\lambda_1, \lambda_2, \ldots, \lambda_n) = M(\lambda_1', \lambda_2', \ldots, \lambda_n')$$

Then the conditional pseudo-probability

$$Prob_\infty\left(\gcd(\lambda_j, P_j) = 1 \; \forall j \; \bigg| \; \gcd(\lambda_1, \lambda_2, \ldots, \lambda_n) = 1\right)$$

is given by

$$\frac{\prod_{j=1}^n \frac{\varphi(P_j)}{P_j}}{\prod_{\substack{p \; Prime \\ p | \text{lcm}(P_1, \ldots, P_n)}} (1 - p^{-n})} \geq \prod_{j=1}^n \frac{\varphi(P_j)}{P_j},$$

where $\zeta(-)$ and $\varphi(-)$ denote respectively the Riemann zeta and the Euler totient functions.

Plausibility Argument. (This is not a proof.)

We treat $Prob_\infty$ as if it were a probability distribution on the integers $\mathbb{Z}^n = \{(\lambda_1, \lambda_2, \ldots, \lambda_n)\}$. We assume that M maps this distribution on itself, and that $Prob_\infty(p \mid \lambda_j)$ and $Prob_\infty(q \mid \lambda_j)$ are stochastically independent when p and q are distinct primes.

For fixed j, the probability $Prob_\infty(p \nmid \lambda_j)$ that a given prime divisor p of P_j does not divide λ_j' is

$$1 - \frac{1}{p}.$$

Hence, the probability that P_j and λ_j are relatively prime is

$$Prob_\infty\left(\gcd(P_j, \lambda_j) = 1\right) = \prod_{p | P_j} \left(1 - \frac{1}{p}\right).$$

This can be reexpressed in terms of the Euler totient function as

$$Prob_\infty\left(\gcd(P_j, \lambda_j) = 1\right) = \frac{\varphi(P_j)}{P_j}$$

Since $\lambda_1, \lambda_2, \ldots, \lambda_n$ are independent random variables, we have

$$Prob_\infty \left(\gcd(P_j, \lambda_j) = 1 \forall j \right) = \prod_{j=1}^{n} \frac{\varphi(P_j)}{P_j}.$$

On the other hand, the probability that a given prime p does not divide all the integers $\lambda_1, \lambda_2, \ldots, \lambda_n$ is

$$1 - \frac{1}{p^n}.$$

Thus,

$$Prob_\infty \left(p \nmid \gcd(\lambda_1, \lambda_2, \ldots, \lambda_n) \forall p \text{ s.t. } p \nmid \text{lcm}(P_1, P_2, \ldots, P_n) \right)$$

is given by the expression

$$\prod_{p \nmid \text{lcm}(P_1, P_2, \ldots, P_n)} (1 - p^{-n}) = \frac{\zeta(n)^{-1}}{\prod_{p \mid \text{lcm}(P_1, P_2, \ldots, P_n)} (1 - p^{-n})},$$

where we have used the fact [21] that

$$\zeta(n)^{-1} = \prod_{p \text{ Prime}} \left(1 - \frac{1}{p^n} \right).$$

We next note that the events $\forall j \; \gcd(P_j, \lambda_j) = 1$ and $\forall p \; p \nmid \text{lcm}(P_1, P_2, \ldots, P_n) \longrightarrow p \nmid \gcd(P_1, P_2, \ldots, P_n)$ are stochastically independent since they respectively refer to the disjoint sets of primes $\{p : p \mid \text{lcm}(P_1, P_2, \ldots, P_n)\}$ and $\{p : p \nmid \text{lcm}(P_1, P_2, \ldots, P_n)\}$. Hence, the probability of the joint event

$$Prob_\infty \left(\gcd(P_j, \lambda_j) = 1 \; \forall j \text{ AND } \gcd(\lambda_1, \lambda_2, \ldots, \lambda_n) = 1 \right)$$

is given by the expression

$$\frac{\zeta(n)^{-1} \prod_{j=1}^{n} \frac{\varphi(P_j)}{P_j}}{\prod_{p \mid \text{lcm}(P_1, P_2, \ldots, P_n)} (1 - p^{-n})}.$$

Using exactly the same argument as that used to find an expression for

$$Prob_\infty \left(p \nmid \gcd(\lambda_1, \lambda_2, \ldots, \lambda_n) \forall p \text{ s.t. } p \nmid \text{lcm}(P_1, P_2, \ldots, P_n) \right),$$

we have

$$Prob_\infty \left(\gcd(\lambda_1, \lambda_2, \ldots, \lambda_n) = 1 \right) = \zeta(n)^{-1}.$$

Hence the conditional probability

$$Prob_\infty \left(\gcd(P_j, \lambda_j) = 1 \; \forall j \; \Big| \; \gcd(\lambda_1, \lambda_2, \ldots, \lambda_n) = 1 \right)$$

is given by the expression

$$\frac{\prod_{j=1}^{n} \frac{\varphi(P_j)}{P_j}}{\prod_{p|\text{lcm}(P_1,P_2,\ldots,P_n)} (1-p^{-n})}$$

Finally, since

$$\prod_{p|\text{lcm}(P_1,P_2,\ldots,P_n)} (1-p^{-n}) \leq 1 ,$$

it follows that the conditional probability

$$\text{Prob}_\infty \left(\gcd(P_j, \lambda_j) = 1 \;\forall j \;\Big|\; \gcd(\lambda_1, \lambda_2, \ldots, \lambda_n) = 1 \right)$$

is bounded below by the expression

$$\prod_{j=1}^{n} \frac{\varphi(P_j)}{P_j} .$$

□

The following is an immediate corollary of the above conjecture.

COROLLARY 8. *Let n be an integer greater than 1, and let $P_1, P_2, \ldots, P_n$ be n fixed positive integers. Let $\lambda'_1, \lambda'_2, \ldots, \lambda'_n$ be n integers randomly and independently selected with replacement from the set all integers $\mathbb{Z}$ according to the uniform probability distribution. Let M be a fixed element of the group $SL_\pm(n, \mathbb{Z})$ of invertible $n \times n$ integer matrices. Finally, let $\lambda_1, \lambda_2, \ldots, \lambda_n$ be n integers given by*

$$(\lambda_1, \lambda_2, \ldots, \lambda_n)^{transpose} = M (\lambda'_1, \lambda'_2, \ldots, \lambda'_n)^{transpose} .$$

Then, assuming conjecture 1, we have

$$\text{Prob}_\infty \left(\gcd(\lambda_j, P_j) = 1 \;\forall j \;\Big|\; \gcd(\lambda_1, \lambda_2, \ldots, \lambda_n) = 1 \right) = \Omega \left(\prod_{j=1}^{n} \frac{1}{\lg \lg P_j} \right) ,$$

where $\Omega(\;)$ denotes the asymptotic lower bound 'big-omega.'

Thus, if Q is greater than each P_j, we have

$$\text{Prob}_Q \left(\gcd(\lambda_j, P_j) = 1 \;\forall j \;\Big|\; \gcd(\lambda_1, \lambda_2, \ldots, \lambda_n) = 1 \right) = \Omega \left(\left(\frac{1}{\lg \lg Q}\right)^n \right)$$

PROOF. Since[18]

$$\underline{\lim} \frac{\varphi(n) \ln \ln n}{n} = e^{-\gamma} ,$$

where γ denotes Euler's constant, we have that

$$\frac{\varphi(P_j)}{P_j} = \Omega \left(\frac{1}{\lg \lg P_j} \right)$$

---

[18] See [**21**, Theorem 328, Section 18.4].

Thus, an asymptotic lower bound for the above conditional probability is given by the expression

$$\Omega\left(\prod_{j=1}^{n}\frac{1}{\lg\lg P_j}\right).$$

□

References

[1] Alber, G., T. Beth, M. Horodecki, P. Horodecki, R. Horodecki, M. Rotteler, H. Weinfurther, R. Werner, and A. Zeilinger, **"Quantum Information: An Introduction to Basic Theoretical Concepts and Experiments,"** Springer, (2001).

[2] Bach, Eric, and Jeffrey Shallit, **"Algorithmic Number Theory: Volume I: Efficient Algorithms,"** MIT Press, (1997).

[3] Bernstein, Ethan, and Umesh Vazirani, **Quantum Complexity Theory**, SIAM J. of Computing, Vol. 26, No. 5, (1997), pp 1411-1473.

[4] Boneh, Dan, and Richard J. Lipton, **Quantum cryptanalysis of hidden linear functions**, in **"Lecture Notes in Computer Science – Advances in Cryptology – CRYPTO'95,"** D. Coppersmith (ed.), Springer-Verlag, Berlin, (1995), pp 424-437.

[5] Brassard, Gilles, and Paul Bratley, **"Algorithmics: Theory and Practice,"** Printice-Hall, (1988).

[6] Brassard, Gilles and Peter Hoyer, **An exact quantum polynomial-time algorithm for Simon's problem**, Proceedings of Fifth Israeli Symposium on Theory of Computing and Systems – ISTCS, IEEE Computer Society Press, (1997), pp. 12-23.

[7] Cartan, Henri, and Samuel Eilenberg, **"Homological Algebra,"** Princeton University Press, (1956).

[8] Cheung, Kevin K.H., and Michele Mosca, **Decomposing finite abelian groups**, http://xxx.lanl.gov/abs/cs.DS/0101004.

[9] Cleve, Richard, Artur Ekert, Chiara Macchiavello, and Michele Mosca, **Quantum Algorithms Revisited**, Phil. Trans. Roy. Soc. Lond., A, (1997). http://xxx.lanl.gov/abs/quant-ph/9708016

[10] Coppersmith, D., **An approximate quantum Fourier transform used in quantum factoring,** IBM Research Report RC 19642, (1994).

[11] Cormen, Thomas H., Charles E. Leiserson, and Ronald L. Rivest, **"Introduction to Algorithms,"** McGraw-Hill, (1990).

[12] Cox, David, John Little, and Donal O'Shea, **"Ideals, Varieties, and Algorithms: An Introduction to Computational Algebraic Geometry and Commutative Algebra"** (second edition), Springer-Verlag, (1996).

[13] Cox, David, John Little, and Donal O'Shea, **"Using Algebraic Geometry,"** Springer, (1998).

[14] Ekert, Artur K.and Richard Jozsa, **Quantum computation and Shor's factoring algorithm**, Rev. Mod. Phys., 68,(1996), pp 733-753.

[15] Ettinger, Mark, and Peter Hoyer, **On Quantum Algorithms for Noncommutative Hidden Subgroups**, (1998). http://xxx.lanl.gov/abs/quant-ph/9807029

[16] Ettinger, Mark, Peter Hoyer, Emanuel Knill, **Hidden Subgroup States are Almost Orthogonal**, http://xxx.lanl.gov/abs/quant-ph/9901034.

[17] Fulton, William, and Joe Harris, **"Representation Theory,"** Springer-Verlag, (1991).

[18] Gathen, Joachim von zur, and Jurgen Gerhard, "Modern Computer Algebra," Cambridge University Press, (1999).

[19] Geddes, Keith O., Stephen R. Czapor, and George Labahn, **"Algorithms for Computer Algebra,"** Kluwer Academic Publishers, Boston, (1992).

[20] Hall, Marshall, Jr., **"The Theory of Groups,"** Macmillan, New York, (1959).

[21] Hardy, G.H., and E.M. Wright, **"An Introduction to the Theory of Numbers,"** Oxford Press, (1965).

[22] Hirvensalo, Mika, **"Quantum Computing,"** Springer, (2001).

[23] Hoyer, Peter, **Efficient quantum transforms**, http://xxx.lanl.gov/abs/quant-ph/9702028.
[24] Ivanyos, Gabor, Frederic Magniez, and Miklos Santha, **Efficient quantum algorithms for some instances of the non-Abelian hidden subgroup problem**, (2001). http://xxx.lanl.gov/abs/quant-ph/0102014
[25] Jozsa, Richard, **Quantum algorithms and the Fourier transform**, quant-ph preprint archive 9707033 17 Jul 1997.
[26] Jozsa, Richard, Proc. Roy. Soc. London Soc., Ser. A, 454, (1998), 323 - 337.
[27] Jozsa, Richard, Quantum factoring, discrete logarithms and the hidden subgroup problem, IEEE Computing in Science and Engineering, (to appear). http://xxx.lanl.gov/abs/quant-ph/0012084
[28] Kemeny, John G., and J. Laurie Snell, "**Finite Markov Chains**," Van Nostrand, (1960).
[29] Kitaev, A., **Quantum measurement and the abelian stabiliser problem,** (1995), quant-ph preprint archive 9511026.
[30] Knuth, Donald E., "**The Art of Computer Programming**," (second edition), Addison-Wesley, Reading, Massachusetts, (1981).
[31] LeVeque, , William Judson, "**Topics in Number Theory**," Addison-Wesley, (1956).
[32] Lomonaco, Samuel J., Jr., **A Rosetta Stone for quantum mechanics with an introduction to quantum computation**, in "Quantum Computation: A Grand Mathematical Challenge for the Twenty-First Century and the Millennium" PSAPM/58, American Mathematical Society, Providence, RI, (2002). (http://xxx.lanl.gov/abs/quant-ph/0007045)
[33] Lomonaco, Samuel J., Jr., **Shor's quantum factoring algorithm,** in "Quantum Computation: A Grand Mathematical Challenge for the Twenty-First Century and the Millennium," PSAPM/58, American Mathematical Society, Providence, RI, (2002). (http://xxx.lanl.gov/abs/quant-ph/0010034)
[34] Lomonaco, Samuel J., Jr., **The non-abelian Fourier transform and quantum computation**, MSRI Streaming Video, (2000), http://www.msri.org/publications/ln/msri/2000/qcomputing/lomonaco/1/index.html
[35] Mosca, Michelle, and Artur Ekert, **The Hidden Subgroup Problem and Eigenvalue Estimation on a Quantum Computer**, Proceedings of the 1st NASA International Conference on Quantum Computing and Quantum Communication, Springer-Verlag, (to appear). (http://xxx.lanl.gov/abs/quant-ph/9903071)
[36] Nielsen, Michael A., and Isaac L. Chuang, "**Quantum Computation and Quantum Information**," Cambridge University Press, (2000).
[37] Preskill, John, "**Quantum Computation**," Lecture Notes, http://www.theory.caltech.edu/people/preskill/ph229/#lecture
[38] Pueschel, Markus, Martin Roetteler, and Thomas Beth, **Fast Quantum Fourier Transforms for a Class of Non-abelian Groups**, (1998). http://xxx.lanl.gov/abs/quant-ph/9807064
[39] Roetteler, Martin, and Thomas Beth, **Polynomial-Time Solution to the Hidden Subgroup Problem for a Class of non-abelian Groups**, (1998). http://xxx.lanl.gov/abs/quant-ph/9812070
[40] Rosser, J. Barkley, and Lowell Schoenfeld, **Approximate formulas for some functions of prime numbers**, Illinois J. Math., v.6, (1962), pp. 64-94.
[41] Russell, Alexander, and Ammon Ta-Shma, **Normal Subgroup Reconstruction and Quantum Computation Using Group Representations**, STOC, (2000).
[42] Schoenfeld, Lowell, **Sharper bounds for the Chebyshev functions $\theta(x)$ and $\psi(x)$. II,** Math. Comp., vol. 30, No. 134, (1976), pp. 337-360.
[43] Shor, Peter W., **Polynomial time algorithms for prime factorization and discrete logarithms on a quantum computer**, SIAM J. on Computing, 26(5) (1997), pp 1484 - 1509. (http://xxx.lanl.gov/abs/quant-ph/9508027)
[44] Shor, Peter W., **Introduction to quantum algorithms,** in "Quantum Computation: A Grand Mathematical Challenge for the Twenty-First Century and the Millennium," PSAPM/58, American Mathematical Society, Providence, RI, (2002). (http://xxx.lanl.gov/abs/quant-ph/0005003)
[45] Simon, Daniel R., **On the power of quantum computation**, SIAM J. Comput., Vol. 26, No. 5, (1997), pp 1474-1483.
[46] van Dam, Wim, and Sean Hallgren, **Efficient Quantum Algorithms for Shifted Quadratic Character Problems**, http://xxx.lanl.gov/abs/quant-ph/0011067.

[47] Vazirani, Umesh, **On the power of quantum computation**, Philosophical Tranactions of the Royal Society of London, Series A, 354:1759-1768, August 1998.
[48] van Dam, Wim, and Lawrence Ip, **Quantum Algorithms, for Hidden Coset Problems**, manuscript, http://www.cs.caltech.edu/~hallgren/hcp.pdf

(Samuel J. Lomonaco, Jr.) DEPARTMENT OF COMPUTER SCIENCE AND ELECTRICAL ENGINEERING, UNIVERSITY OF MARYLAND BALTIMORE COUNTY, 1000 HILLTOP CIRCLE, BALTIMORE, MD 21250

E-mail address, Samuel J. Lomonaco, Jr.: `Lomonaco@UMBC.EDU`
URL: WebPage: `http://www.csee.umbc.edu/~lomonaco`

(Louis H. Kauffman) DEPARTMENT OF MATHEMATICS, STATISTICS, AND COMPUTER SCIENCE, UNIVERSITY OF ILLINOIS AT CHICAGO, CHICAGO, IL 60607-7045

E-mail address, Louis H. Kauffman: `kauffman@uic.edu`
URL: WebPage: `http://math.uic.edu/~kauffman`

Improved Two-Party and Multi-Party Purification Protocols

Elitza N. Maneva and John A. Smolin

ABSTRACT. We present an improved protocol for entanglement purification of bipartite mixed states using several states at a time rather than two at a time as in the traditional recurrence method. We also present a generalization of the hashing method to n-partite cat states, which achieves a finite yield of pure cat states for any desired fidelity. Our results are compared to previous protocols.

1. Introduction

Entanglement is a fundamental resource in quantum information. It can be used for secure quantum cryptography [1] and is an essential part of known algorithms for quantum computation [2, 3] (strangely, it is not known that all quantum algorithms which outperform their classical counterparts *require* entanglement. See [4] for a situation in which quantum states display a form of nonlocality, but which involves no entanglement).

Early studies of entanglement purification [5, 6, 7] focused mainly on bipartite entanglement, attempting to distill pure EPR pairs [8] from bipartite mixed states. More recently, Murao, Plenio, Popescu, Vedral and Knight [9] have studied the generalization of such schemes to distilling three-party (GHZ [10]) and multi-party states of the form (sometimes called "cat" states [11])

$$|\Phi^+\rangle = \frac{1}{\sqrt{2}}(|00\ldots 0\rangle + |11\ldots 1\rangle) \tag{1}$$

from three-party and multi-party entangled mixed states. However, they do not study generalizations of the hashing method of [6]. The hashing scheme has the major advantage over the recurrence style scheme of [9] that it achieves a finite yield of pure cat states for any arbitrarily high fidelity, whereas the yield for any recurrence method goes to zero.

In this paper we study an improved purification protocol for two parties, and a generalization of hashing to multiple parties.

2000 *Mathematics Subject Classification.* Primary 81P68.

The authors would like to thank Peter Shor for helpful discussions, and the Summer Undergraduate Research Fellowship program at the California Institute of Technology and IBM for support. J.A. Smolin also thanks the Army Research office for support under contract number DAAG55-98-C-0041.

First, we define some notation: All of our studies will apply to entangled mixed qubit states of N parties (conventionally known as Alice, Bob, etc.), diagonal in the following basis:

$$|\psi_{p,i_1 i_2 \ldots i_{N-1}}\rangle = \frac{|0 i_1 i_2 \ldots i_{N-1}\rangle + (-1)^p |1 \bar{i}_1 \bar{i}_2 \ldots \bar{i}_{N-1}\rangle}{\sqrt{2}} \tag{2}$$

where p and the i's are zero or one, and a bar over a bit value indicates its logical negation. This gives 2^N orthogonal states.

These states correspond to the simultaneous eigenvectors of the following operators (There are N operators in all, one special one of the X form, and $N-1$ involving Z and I):

$$\begin{aligned}
S_0 &= X \otimes X \otimes X \otimes X \ldots X \\
S_1 &= Z \otimes Z \otimes I \otimes I \ldots I \\
S_2 &= Z \otimes I \otimes Z \otimes I \ldots I \\
S_3 &= Z \otimes I \otimes I \otimes Z \ldots I \\
&\quad \vdots \\
S_{N-1} &= Z \otimes I \otimes I \otimes I \ldots Z
\end{aligned} \tag{3}$$

The X, Z, and I operators, along with the Y operator which we don't use here, are members of the Pauli group (for more details, see [12]). The p from Equation (2) corresponds to whether a state is a $+1$ or -1 eigenvector of S_0 ($p=0$ for a $+1$ eigenvector and $p=1$ for a -1 eigenvector). This is called the "phase" bit of the cat state. The i_js correspond to whether a state is a $+1$ or -1 eigenvector of S_j for $j = 1, \ldots, (N-1)$, which we call the amplitude bits. Thus, the following is the set of generators of the stabilizer group of $|\psi_{p,i_1 i_2 \ldots i_{N-1}}\rangle$:

$$\{(-1)^p S_0, (-1)^{i_1} S_1, \ldots, (-1)^{i_{N-1}} S_{N-1}\} \tag{4}$$

It is important to realize at this point that since these operators are all tensor products of operators on the subsystems $1 \ldots N$ each with eigenvalue ± 1, they can be measured using only local quantum operations plus classical communication. If an unknown one of the ψ's is shared among N parties and they wish to determine the eigenvalue corresponding to one of the S_i, each party just measures his or her operator and reports the result to everyone else. The eigenvalue of the whole operator is the product of their individual results. Furthermore, since Z and I commute, it is possible to measure the eigenvalues of *all* the $S_{i>0}$. On the other hand, X and Z do not commute and therefore if S_0 is measured, none of the $S_{i>0}$ can be measured (a random result would occur) and similarly if any of the $S_{i>0}$ are measured the result of an S_0 measurement will be randomized. In other words, the parties can locally measure either all the amplitude bits or the phase bit for an unknown cat state.

The other tool we will need is the multilateral quantum XOR gate, in which each party's bits are XORed together in a quantum-coherent way (see Fig. 1). Following Gottesman [12] we can work out how a tensor product of two cat states behaves under the multilateral XOR. The generators of the stabilizer group behave

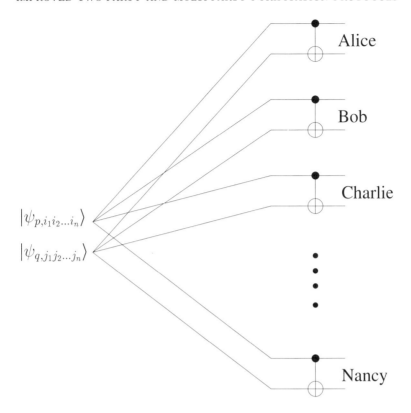

FIGURE 1. The multi-party XOR.

as follows under the quantum XOR operation:

$$\begin{aligned} X \otimes I &\to X \otimes X \\ I \otimes X &\to I \otimes X \\ Z \otimes I &\to Z \otimes I \\ I \otimes Z &\to Z \otimes Z \end{aligned} \tag{5}$$

We work out here the case of three parties, the generalization to n-partite cat states will be apparent. Given $|\psi_{p,i_1i_2}\rangle$ and $|\psi_{q,j_1j_2}\rangle$ with stabilizers as in Eq. (3) the generators of the stabilizers of the tensor product of these are given by:

$$\begin{aligned} &\{(-1)^p XXXIII, \quad (-1)^{i_1} ZZIIII, \quad (-1)^{i_2} ZIZIII, \\ &\phantom{\{}(-1)^q IIIXXX, \quad (-1)^{j_1} IIIZZI, \quad (-1)^{j_2} IIIZIZ\} \end{aligned} \tag{6}$$

(We have omitted the $\otimes$ symbol for brevity.) Now, applying the rule for the XOR operation (5) to corresponding operators (the first and fourth positions correspond to the first party's piece of $|\psi_{p,i_1i_2}\rangle$ and $|\psi_{q,j_1j_2}\rangle$ respectively, etc.) we get:

$$\begin{aligned} &\{(-1)^p XXXXXX, \quad (-1)^{i_1} ZZIIII, \quad (-1)^{i_2} ZIZIII, \\ &\phantom{\{}(-1)^q IIIXXX, \quad (-1)^{j_1} ZZIZZI, \quad (-1)^{j_2} ZIZZIZ\} \end{aligned} \tag{7}$$

We can easily find another set of generators for the same stabilizer group which is again the tensor form of (6):

$$\text{(8)} \quad \begin{aligned} &\{(-1)^{p+q}XXXIII, \quad (-1)^{i_1}ZZIIII, \quad (-1)^{i_2}ZIZIII, \\ &(-1)^{q}IIIXXX, \quad (-1)^{i_1+j_1}IIIZZI, \quad (-1)^{i_2+j_2}IIIZIZ\} \end{aligned}$$

This is simply the set of generators corresponding to $|\psi_{p\oplus q,i_1 i_2}\rangle \otimes |\psi_{q,i_1\oplus j_1\ i_2\oplus j_2}\rangle$. What has happened is that the phase bits have been XORed together with the result put into the phase bit of the source state and the amplitudes are each XORed together and stored in the target state's amplitude bits. This suggests that the action of the multilateral quantum XOR gate (MXOR) can be characterized by its action on the purely classical representation of states as a set of bits, $(p, i_1, i_2, \ldots, i_{N-1})$:

$$\text{(9)} \quad \begin{aligned} &\text{MXOR}[(p, i_1, i_2, \ldots, i_{N-1}), (q, j_1, j_2, \ldots, j_{N-1})] = \\ &(p \oplus q, i_1, i_2, \ldots, i_{N-1}), (q, i_1 \oplus j_1, i_2 \oplus j_2, \ldots, i_{N-1} \oplus j_{N-1}) \end{aligned}$$

Due to its linearity quantum mechanics allows us to think of mixed states as if they are really one of the pure states in the mixture but that we are simply lacking the knowledge of which one (if the states in the mixture come with unequal probability, we are not completely lacking knowledge of which state is in the mixture, but we only know the probabilities, not which state we actually have). Since all the cat states (2) are interconvertible by local operations [**13**] if we had a mixture of cat states and could determine which one we actually had, we would be able to convert it to a $|\Phi^+\rangle$ and would have purified the mixture.

Putting everything we have said up to now together lets us find purification schemes that are essentially classical; only the rules of what we can do are given by quantum mechanics:

- Mixed states diagonal in the cat basis can be thought of as being simply unknown members of the set of cat states.
- The cat states (2) are all interconvertible by local operations, so determining which cat state one has is sufficient to have purified it.
- Either the p or all the i's of an unknown cat state may be measured by local operations plus classical communication.
- The multilateral XOR operation operates classically on the ps and i's of pairs of cat states according to Eq. (10).

Our purification schemes will thus work by treating a set of many mixed states (which are diagonal in the cat basis) as a set of unknown cat states, and attempting to determine the unknown states, discarding them if we cannot.

We are now prepared to analyze the efficiencies of various entanglement purification protocols applied to mixed states. In particular, we concentrate on the generalization of the Werner state [**14**]:

$$\text{(10)} \quad \rho_W = \alpha|\Phi^+\rangle\langle\Phi^+| + \frac{1-\alpha}{2^N}\mathbb{1}, \ 0 \leq \alpha \leq 1$$

The fidelity of ρ_W relative to the desired pure state $|\Phi^+\rangle$ is $F = \langle\Phi^+|\rho_W|\Phi^+\rangle = \alpha + \frac{1-\alpha}{2^N}$. We rewrite ρ_W in the cat basis (2) as

$$\text{(11)} \quad \begin{aligned} \rho_W = &\left(\alpha + \frac{1-\alpha}{2^N}\right)|\psi_{0,00\ldots 0}\rangle\langle\psi_{0,00\ldots 0}| \\ &+ \frac{1-\alpha}{2^N} \sum_{p,i_1 i_2 \ldots i_{N-1} \neq 0,00\ldots 0} |\psi_{p,i_1 i_2 \ldots i_{N-1}}\rangle\langle\psi_{p,i_1 i_2 \ldots i_{N-1}}|. \end{aligned}$$

Thus, the Werner state is diagonal in the cat basis and we can think of it as really being one of the cat states. We write the unknown cat states as N unknown strings of bits: $b_0, b_1, b_2, \ldots, b_{N-1}$, where b_0 is formed by concatenating the (unknown) phase bits of all the cat states, and the b_j for $j > 0$ are formed by concatenating the jth amplitude bits. Together the b_j make up the total bitstring B.

2. Bipartite Protocol

The case of two parties has been studied [5, 6, 7]. The protocols can distill pure entanglement from any Werner state with fidelity $F > 1/2$. The recurrence methods that work on Werner states near $F = 1/2$ involve local quantum operations on two mixed states at a time. For high fidelities, the best known strategy (the *hashing* method) obtains high yields in the limit of arbitrarily large numbers of states. It seemed that operations on an intermediate number of mixed states might give better yield for intermediate fidelities, and this turns out to be the case.

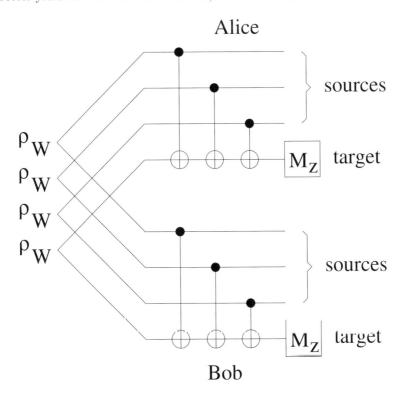

FIGURE 2. The sequence of XOR gates and final measurement in the z basis for our bipartite strategy for $m = 4$.

Our new strategy is to choose a block size m and to take $m - 1$ Werner states and do a bipartite XOR between each one and an mth Werner state, and then to measure the amplitude bits of that target state. The sequence of XORs is illustrated in Fig. 2 for the $m = 4$ case. This is a natural generalization of the recurrence method whose single step is just this method for $m = 2$. If any amplitude bit is nonzero, the two measurements disagree, the source states are discarded. If all amplitude bits are zero the states are said to have "passed" and the hashing method is performed on the $m-1$ source states along with other source states that passed. The advantage over the $m=2$ recurrence is that fewer than half the mixed states are used up inherently just by being measured targets.

In [6] the hashing method was used only on states whose mixture probabilities were *independent*. We note that hashing is a quite general method for extracting entropy from strings of bits, even if there are correlations among the bits. One merely needs to take as many hash bits as there is entropy in the bitstring. Thus, the yield of our method is:

$$(12) \qquad p_{\text{pass}} \frac{m-1}{m} \left(1 - \frac{H(\text{passed source states})}{m-1}\right)$$

The calculation of the entropy of a block of passed states and of p_{pass} is straightforward. One simply keeps track of the probability of each possible string B (of $2mN$ bits corresponding to a block of m Werner states) given the probabilities in the Werner mixture (12), applies the MXOR rule (10) to B to yield a B' and groups the like B's which have passed to yield a final distribution P_i (of the $2(m-1)N$ bits corresponding $m-1$ states left after the MXOR operation). We then have $p_{\text{pass}} = \sum_i P_i$ and the normalized distribution $P'_i = P_i/p_{\text{pass}}$ and the entropy $H(\text{passed source states})$ given by $-\sum_i P'_i \log_2 P'_i$.

Figure 3 compares the yield for our new method for various values of m with the previous recurrence continued by hashing protocol. We have not found a simple way to analyze what happens if our multi-bit step is iterated, rather than passing on immediately to hashing. For the $m=2$ recurrence only one passed source state remains and it is identical in all respects to every other passed source state. For $m>2$ there are multiple correlated passed states and it is not clear just how to treat them. For instance, at $m=3$ there are two passed states from each operation and there is no way to combine the 4 passed states from two operations into another $m=3$ step.

3. Multipartite Hashing

In [9] multi-party recurrence methods are studied, but not multi-party hashing which is needed to achieve finite yields. Here we present a multi-party hashing method.

In the case of two parties it is known [6] how to extract the parity of any random subset of all the bits in B. For more than two parties it is not known now to do this. Instead, we can choose to extract any random subset parity on either the parity bitstring b_0 or on all the amplitude bitstrings b_j, $j>0$ in parallel. This follows immediately from Eq. (10): This is done multilaterally XORing together all the states in the desired subset choosing one of them to be the target. See Fig. 4. Depending on the direction of the XOR gates either the phase bits or all the amplitude bits accumulate in the target state, which can then be measured.

Our multilateral hashing protocol will be to choose a large block size m and then to extract $m\max_{j>0}[\{H(b_j)\}]$ random subsets of each amplitude bitstring in parallel (as shown in Figure 4a), where $H(b_j)$ is the entropy per bit in string b_j. This is sufficient to determine all the bits of all the $b_{j>0}$ as it is just doing the same random hash on each bitstring. Even though the random hashes are all the same, since they are uncorrelated with the bitstrings being determined this many hash bits will be enough to determine all bits of the $b_{j>0}$ [15]. This procedure actually extracts too much information (and thereby uses up too many states as measured

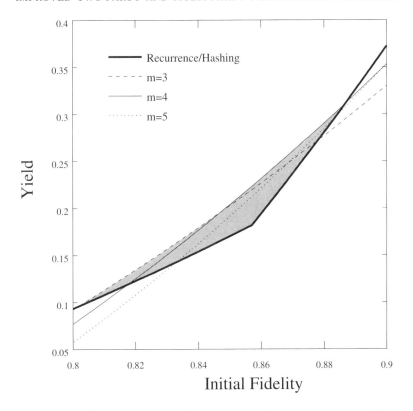

FIGURE 3. Yields of various bipartite purification protocols. The thick line is the recurrence continued by hashing method, the dashed line is our new method for $m = 3$, solid for $m = 4$ and dotted for $m = 5$. The shaded region is the region where our new method improves on the recurrence/hashing method. Note that yield for $m = 5$ is never the best, though for some fidelities it is better than the recurrence/hashing method. The $m = 6$ case (not shown) behaves similarly, while for $m > 6$ the new method is always worse than recurrence/hashing. The sharp "knee" visible in the recurrence/hashing line is the point above which recurrence is never used and hashing is done immediately.

targets), so perhaps a more efficient protocol exists, but this has the virtue of using only the multilateral XOR operation which maps cat states to cat states. After determining the amplitude bits, to find b_0 we use multilateral XORs arranged as in Fig. 4b, and find the hash of the string by measuring another $H(b_0)$ of the states. The yield of this hashing protocol D_h is given by

$$D_h = (1 - \max_{j>0}[\{H(b_j)\}] - H(b_0)) \tag{13}$$

For the case of Werner states all the b's have the same entropy and Eq. (13) reduces to

$$D_W = 1 - 2H_2\left(\frac{(1-f)2^{N-1}}{2^N - 1}\right) \tag{14}$$

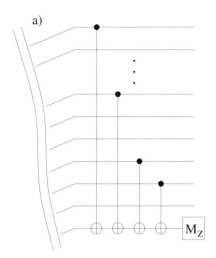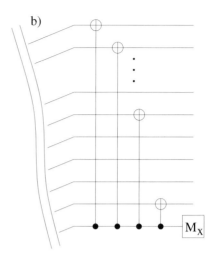

FIGURE 4. Multi-party hashing: These hashes are done on large blocks of bits (indicated by the vertical ellipsis) and are done multilaterally (only one party's operations are shown, the other $N-1$ parties operations are identical).
a) Finding a random subset parity on all the $b_{j>0}$ in parallel. In this case the first, third, sixth and seventh states shown are XORed multilaterally into the last one which is then measured to determine the eigenvalue of the Z operator.
b) Finding a random subset parity on b_0. In this instance the parity of the first, second, fourth and eighth states shown are XORed with the last one, which is then measured in the eigenbasis of the X operator. Note the reversal of the direction of the XOR gates with respect to a).

or in the limit as the number of parties goes to infinity

$$D_W^\infty = 1 - 2H_2\left(\frac{1-f}{2}\right). \tag{15}$$

where $H_2(x) = -x\log_2 x - (1-x)\log_2(1-x)$. Eq. 14 is graphed for several values of N in Figure 5. By using the recurrence method and switching to hashing as soon as it gives better yield, one can obtain positive final yield to arbitrarily high fidelity for any initial fidelity for which the recurrence method of [9] improves the fidelity.

4. Conclusions and Comments

We have found improved bipartite recurrence protocols for the purification of entanglement from mixed quantum states. We have also demonstrated the first finite-yield method for purification of cat states in a multi-party setting. It is worthwhile to note that both of these new procedures were analyzed for mixed state diagonal in the cat basis, but that in fact they will work for *any* mixed state just as well, by considering the state's cat-basis diagonal elements. This is unlikely to be the optimally efficient strategy for non-diagonal states however. In

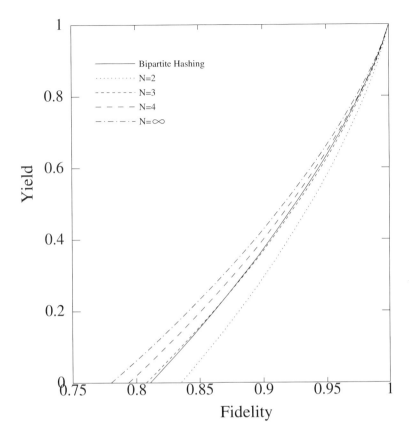

FIGURE 5. Yields for multipartite hashing for various numbers of parties. The solid line is the two-party hashing method of [6]. The dotted line is the corresponding $N = 2$ version of our new hashing method, which has a lower yield since it works on the amplitude and phase bits as separate hash strings even though for the bipartite case it is known how to extract their entropy together, which is more efficient. The lines consisting of dashes, longer dashes and dots with dashes are the $N = 3$, $N = 4$, and $N = \infty$ cases respectively.

[6] there is an example of a state for which the conventional bipartite recurrence and hashing cannot distill any pure entanglement, but which can nevertheless simply be distilled. One expects such examples to exist for our new methods as well (indeed, the example in [6] *is* an example for the bipartite case of new methods which will similarly fail to distill it.

For our bipartite protocol, while clearly not optimal it is not so bad to have passed over to hashing instead of recurring the protocol. Recurrence methods have vanishing yield if one desires arbitrarily high fidelity of the purified states, so hashing needs to be used eventually in any case. Additionally it would likely be best to produce a variable block size protocol that begins as the recurrence method for low fidelity, switches to a larger block size at some higher fidelity and finally is continued by hashing. A calculation of the yield of such a method is cumbersome, and seemingly provides little insight. We hope that our having pointed out that block size $m > 2$ methods can improve over recurrence will stimulate further work in this area to develop a deeper understanding, rather than just a brute-force analysis. Much progress has been made on purification involving only one-way classical communication. Such protocols directly correspond to quantum error-correcting codes (cf. [6]) but recurrence protocols inherently involve two-way classical communication so all parties know which states to discard. So far little of the coding theory has been applied to this case. There does appear to be some relation between these two-way purification protocols and quantum error-detecting codes, and some progress is being made in this area [16].

References

[1] A. K. Ekert, Phys. Rev. Lett. **67**, 661 (1991).
[2] P.W. Shor, "Algorithms for quantum computation: discrete log and factoring," Proceedings of the 35th Annual Symposium on the Foundations of Computer Science (IEEE Computer Society Press, Los Alamitos, CA 1994), p. 124.
[3] L.K. Grover, "A fast quantum mechanical algorithm for database search," Proceedings of the 28th Annual ACM Symposium on Theory of Computing, 1996, pp. 212-219.
[4] C.H. Bennett, D.P. DiVincenzo, C.A. Fuchs, T. Mor, E. Rains, P.W. Shor, J.A. Smolin, and W.K. Wootters, Phys. Rev. A **59**, 1070 (1999).
[5] C.H. Bennett, G. Brassard, S. Popescu, B. Schumacher, J.A. Smolin, and W.K. Wootters, Phys. Rev. Lett. **76**, 722 (1996).
[6] C.H. Bennett, D.P. DiVincenzo, J.A. Smolin, and W.K. Wootters, Phys. Rev. A **54**, 3824 (1996).
[7] D. Deutsch, A. Ekert, R. Jozsa, C. Macchiavello, S. Popescu, A. Sanpera, "Quantum Privacy Amplification and the security of Quantum Cryptography Over Noisy Channels," Phys. Rev. Lett. **77**, 2818 (1996).
[8] A. Einstein, B. Podolsky, N. Rosen, Phys. Rev. **47**, 777 (1935).
[9] M. Murao, M.B. Plenio, S. Popescu, V. Vedral, P.L. Knight "Multi-Particle Entanglement Purification Protocols", Phys. Rev. A **57**, 4075 (1998), also available as LANL preprint quant-ph/9712045.
[10] D.M. Greenberger, M. Horne, A. Zeilinger, Am. J. Phys. **58**, 1131 (1990).
[11] These are known as cat states since they are generalizations of the state of the many particles making up Schröedinger's cat, namely $(|\text{alive alive alive}\dots \text{alive}\rangle + |\text{dead dead dead}\dots \text{dead}\rangle)/\sqrt{2}$.
[12] D. Gottesman, preprint (1998) quant-ph/9807006.
[13] To vary the amplitude bits, the appropriate parties perform X operations, and to change the phase bit any one party performs a Z operation.
[14] R.F. Werner, Phys Rev. A **40**, 4277 (1989).
[15] If the bits of the various $b_{j>0}$ are uncorrelated for different values of j this can be thought of as taking the same random hash on different sets of data. Since the data are independent of the hash this will work just as well as if different hashes were chosen for each j. In the case where the $b_{j>0}$ are correlated the situation can only get better. Consider the case of complete correlation–when all the bits in all the strings are the same the same hashes are clearly sufficient to determine them all.
[16] D. Gottesman, private communication.

(Elitza N. Maneva) CALIFORNIA INSTITUTE OF TECHNOLOGY, PASADENA CA 91126-0671
E-mail address, Elitza N. Maneva: elitza@its.caltech.edu

(John A. Smolin) IBM T.J. WATSON RESEARCH CENTER, YORKTOWN HEIGHTS, NY 10598
E-mail address, John A. Smolin: smolin@watson.ibm.com

Quantum Games and Quantum Algorithms

David A. Meyer

ABSTRACT. A quantum algorithm for an oracle problem can be understood as a quantum strategy for a player in a two-player zero-sum game in which the other player is constrained to play classically. I formalize this correspondence and give examples of games (and hence oracle problems) for which the quantum player can do better than would be possible classically. The most remarkable example is the Bernstein-Vazirani quantum search algorithm which I show creates no entanglement at any timestep.

1. Introduction

Despite the exuberance of people working on quantum computation—and more generally, quantum information theory—we discuss remarkably few quantum algorithms. These include, and are largely limited to, the Deutsch-Jozsa [1], Simon [2], Shor [3] and Grover [4] algorithms. Of course, quantum versions of many other information processing tasks have also been studied: cryptography [5], error correction [6], communication channels [7], distributed computation [8], *etc.* When I was invited to speak about quantum computing at Microsoft Research in January 1998, I decided to try to add game theory to this list.

My motivations were twofold: First, as I explained in that talk, von Neumann was not only the driving force behind the development of modern digital computers [9]—a subject of great interest to Microsoft—but also one of the founders of quantum mechanics [10], and thus someone whose ideas are central to quantum computing. But he had another great interest—shared with Microsoft—economics! Von Neumann essentially invented game theory [11]; his book with Morgenstern, *Theory of Games and Economic Behavior* [12] raises (and in some cases, answers) many of the questions which preoccupy game theorists and economists today. Second, I hoped that something like the argument that identifies which two-person zero-sum games have optimal mixed, rather than pure, classical strategies might provide some insight into which problems are solvable more efficiently by quantum rather than classical algorithms. This hope was probably somewhat naïve, but it brings us to the first question I'll address in this talk: What do quantum games have to do with quantum algorithms?

2000 *Mathematics Subject Classification.* Primary 81P68, 68Q15; Secondary 91A05.
Key words and phrases. quantum strategy, quantum algorithm, query complexity, entanglement.

The quantum game I described originally [13], PQ PENNY FLIP, is perhaps too simple to make the connection with quantum algorithms completely clear. In fact, it is so simple, involving only one qubit, that several people have pointed out that it could be simulated classically [14]. I have argued elsewhere that this misses the point slightly, that the issue is not whether there exists a classical simulation, but how the complexity of that simulation would scale if the size of the game were to increase [15]. To illustrate this point explicitly, in this talk I'll tell a story which involves a game which has instances of arbitrarily large size. My discussion of this game naturally includes a description of Grover's algorithm [4], which not only helps to answer the first question, but raises a second and third.

The second is: Are there sophisticated quantum search algorithms? More explicitly, are there 'databases' which can be 'searched' with better than the square root speedup that Grover's algorithm provides over the best possible classical algorithm? [4] Since Bennett, Bernstein, Brassard and Vazirani [16], Boyer, Brassard, Høyer and Tapp [17], and Zalka [18] have shown that Grover's algorithm is optimal, I will explain the natural changes in the problem which make this question interesting.

Recently Lloyd has argued that Grover's algorithm can be implemented without entanglement [19]. At first glance this may appear surprising: many people have stated that the power of quantum computing derives from entanglement [20,21]. This belief underlies the criticism that NMR experiments do not realize quantum computation because the state of the system at each timestep is separable [22], *i.e.*, a convex combination of product states [23]. Similarly, van Enk observed that the one qubit PQ PENNY FLIP game not only can be simulated classically, but involves no entanglement [14], which would suggest by the same 'reasoning' that such a quantum game would be unrelated to quantum algorithms. In fact, the quantum game in the story to come provides an answer to the third question I'll address: Can quantum-over-classical improvements be achieved without entanglement?

2. PQ games

I'll begin by reviewing briefly the one qubit game PQ PENNY FLIP [13]. In our first episode, the starship Enterprise is facing some imminent catastrophe when the superpowerful being Q appears on the bridge and offers to rescue the ship if Captain Picard[1] can beat him at a simple game: Q produces a penny and asks Picard to place it in a small box, head up. Then Q, followed by Picard, followed by Q, reaches into the box, without looking at the penny, and either flips it over or leaves it as it is. After Q's second turn they open the box and Q wins if the penny is head up. Q wins every time they play, using the following quantum strategy:

$$
\begin{aligned}
|0\rangle &\xmapsto[H]{Q} \frac{1}{\sqrt{2}}(|0\rangle + |1\rangle) \\
&\xmapsto[\sigma_x \text{ or } I_2]{\text{Picard}} \frac{1}{\sqrt{2}}(|0\rangle + |1\rangle) \\
&\xmapsto[H]{Q} |0\rangle
\end{aligned}
$$

---

[1]Captain Picard and Q are characters in the popular American television (and movie) series *Star Trek: The Next Generation* whose initials and abilities are ideal for this illustration. See [24].

Here $|\cdot\rangle$ is Dirac notation [25] for an element of Hilbert space, 0 denotes 'head' and 1 denotes 'tail', $H = \frac{1}{\sqrt{2}}\begin{pmatrix} 1 & 1 \\ 1 & -1 \end{pmatrix}$ is the Hadamard transformation, and $\sigma_x = \begin{pmatrix} 0 & 1 \\ 1 & 0 \end{pmatrix}$ implements Picard's possible action of flipping the penny over. Q's quantum strategy of putting the penny into the equal superposition of 'head' and 'tail' on his first turn means that whether Picard flips the penny over or not, it remains in an equal superposition which Q can rotate back to 'head' by applying H again since $H = H^{-1}$. So when they open the box, Q always wins.

Notice that if Q were restricted to playing classically, *i.e.*, to implementing only σ_x or I_2 on his turns, an optimal strategy for both players would be to flip the penny over or not with equal probability on each turn. In this case Q would win only half the time, so he does substantially better by playing quantum mechanically.

The structure of PQ PENNY FLIP motivates the following definition which formalizes one meaning for quantum game.[2]

Definitions. A *PQ game* consists of
  (i) a Hilbert space $\mathcal{H}$—the possible states of the game—with $N = \dim \mathcal{H}$,
  (ii) an initial state $\psi_0 \in \mathcal{H}$,
  (iii) subsets $Q_i \subset U(N)$, $i \in \{1,\ldots,k+1\}$—the elements of Q_i are the moves Q chooses among on turn i,
  (iv) subsets $P_i \subset S_N$, $i \in \{1,\ldots,k\}$, where S_N is the permutation group on N elements—the elements of P_i are the moves Picard chooses among on turn i, and
  (v) a projection operator Π on $\mathcal{H}$—the subspace W_Q fixed by Π consists of the winning states for Q.

Since only Picard and Q play, these are *two-player* games; they are *zero-sum* since when Q wins, Picard loses, and *vice versa*.

A *pure quantum strategy* for Q is a sequence $(u_i \in Q_i)_{i=1}^{k+1}$. A *pure (classical) strategy* for Picard is a sequence $(s_i \in P_i)_{i=1}^{k}$, while a *mixed (classical) strategy* for Picard is a sequence of probability distributions $(f_i : P_i \to [0,1])_{i=1}^{k}$. If both Q and Picard play pure strategies, the corresponding *evolution* of the PQ game is described by
$$\psi_f = u_{k+1} s_k u_k \ldots u_2 s_1 u_1 \psi_0.$$
After Q's last move the state of the game is measured with Π. According to the rules of quantum mechanics [10], the players observe the eigenvalue 1 with probability $\mathrm{Tr}(\psi^\dagger \Pi \psi)$; this is the probability that the state is projected into W_Q and Q wins. More generally, if Picard plays a mixed strategy, the corresponding *evolution* of the PQ game is described by
$$\rho_f = u_{k+1}\Big(\sum_{s_k \in P_k} f_k(s_k) s_k u_k \ldots u_2 \Big(\sum_{s_1 \in P_1} f_1(s_1) s_1 u_1 \rho_0 u_1^\dagger s_1^\dagger\Big) u_2^\dagger \ldots u_k^\dagger s_k^\dagger\Big) u_{k+1}^\dagger,$$
where $\rho_0 = \psi_0 \otimes \psi_0^\dagger$. Again, after Q's last move ρ_f is measured with Π; the probability that ρ_f is projected into $W_Q \otimes W_Q^\dagger$ and Q wins is $\mathrm{Tr}(\Pi \rho_f)$.

---

[2] Eisert, Wilkens and Lewenstein have proposed a different formalism for quantum games and have applied it to find a (unique) Pareto optimal equilibrium for the Prisoners' Dilemma when the players are allowed to select moves from the same specific subset of unitary transformations [26]. Benjamin and Hayden recently showed that this is not a solution when the players are allowed to make arbitrary unitary moves [27]; in fact, in this case there is no equilibrium, just as in PQ PENNY FLIP [13]. Nevertheless, that formalism still seems likely to be interesting, although more closely related to quantum communication protocols [8,28] than to quantum algorithms.

Finally, an *equilibrium* is a pair of strategies, one for Picard and one for Q, such that neither player can improve his probability of winning by changing his strategy while the other does not.

As I'll show in the next section, the structure of a PQ game specializes to the structure of the known quantum algorithms. In general, unlike the simple case of PQ PENNY FLIP, $W_Q = W_Q((s_i))$ or $W_Q = W_Q((f_i))$, *i.e.*, the conditions for Q's win can depend on Picard's strategy. Each of the three example games I consider here suffices to prove that there are games with mixed/quantum equilibria at which Q does better than he would at any mixed/mixed equilibrium; equivalently, there are some quantum algorithms which outperform classical ones.

3. Guessing a number

In our second episode, Q returns to the Enterprise and challenges Captain Picard again. He boasts that if Picard picks any number between 0 and $N-1$, inclusive, he can guess it. Now, Picard is no slouch; he has been studying up on quantum algorithms since the last episode. In particular, he has studied Grover's algorithm [4] and realizes that for $N = 2^n$, Q can determine the number he picks with high probability by playing the following strategy:

$$|0\ldots 0, 0\rangle \quad \xmapsto[H^{\otimes n} \otimes H\sigma_x]{Q} \quad \frac{1}{\sqrt{N}}\sum_{x=0}^{N-1}|x\rangle \otimes \frac{1}{\sqrt{2}}(|0\rangle - |1\rangle) \quad (u_1)$$

$$\xmapsto[s(f_a)]{\text{Picard}} \quad \frac{1}{\sqrt{N}}\sum_{x=0}^{N-1}(-1)^{\delta_{xa}}|x\rangle \otimes \frac{1}{\sqrt{2}}(|0\rangle - |1\rangle) \quad (s_1)$$

$$\xmapsto[H^{\otimes n}\otimes I_2 \circ s(f_0) \circ H^{\otimes n} \otimes I_2]{Q} \quad \cdots, \quad (u_2)$$

where $a \in [0, N-1]$ is Picard's chosen number, and the moves s_1 and u_2 are repeated a total of $k = \lfloor \frac{\pi}{4}\sqrt{N}\rfloor$ times, *i.e.*, $s_k = \cdots = s_1$ and $u_{k+1} = \cdots = u_2$. For $f : \mathbb{Z}_2^n \to \mathbb{Z}_2$, $s(f)$ is the permutation (and hence unitary transformation) defined by

$$s(f)|x, b\rangle = |x, b \oplus f(x)\rangle,$$

where $\oplus$ denotes addition mod 2. This transformation is often referred to as 'f-controlled-NOT'. Each of Picard's moves s_i can be thought of as the response of an oracle which computes $f_a(x) := \delta_{xa}$ to respond to the quantum query defined by the state after u_i. After $O(\sqrt{N})$ such queries, a measurement by $\Pi = |a\rangle\langle a| \otimes I_2$ returns a win for Q with probability bounded above $\frac{1}{2}$, *i.e.*, Grover's quantum algorithm determines a with high probability. (Here $\langle a| := |a\rangle^\dagger$ and the tensor product is implicit in $|a\rangle\langle a|$ [25].)

Notice that if Q were to play classically, he could query Picard about a specific number at each turn, but on the average it would take $N/2$ turns to guess a. A classical equilibrium is for Picard to choose a uniformly at random, and for Q to choose a permutation of N uniformly at random and guess numbers in the corresponding order. Even when Picard plays such a mixed strategy, Q's quantum strategy is optimal; together they define a mixed/quantum equilibrium.

Knowing all this, Picard responds that he will be happy to play, but that Q should only get 1 guess, not $\lfloor\frac{\pi}{4}\sqrt{N}\rfloor$. Q protests that this is hardly fair, but he will play, as long as Picard tells him how close (mod 2) his guess is to the chosen number. Picard agrees, and they play. Q wins. They play again. Q wins again. Picard doesn't understand what's going on. The problem is that in his studies of quantum algorithms, he overlooked an insufficiently appreciated quantum algorithm, the slightly improved [16,29] Bernstein-Vazirani algorithm [30]: Guess x and answer a are vectors in $\mathbb{Z}_2^n$, so $x \cdot a$ depends on the cosine of the angle between the vectors. Thus it seems reasonable to define "how close (mod 2) a guess is to the answer" to be the oracle response $g_a(x) := x \cdot a \mod 2$. Then Q plays as follows:

$$|0\ldots 0,0\rangle \underset{H^{\otimes n}\otimes H\sigma_x}{\overset{Q}{\longmapsto}} \frac{1}{\sqrt{N}} \sum_{x=0}^{N-1} |x\rangle \otimes \frac{1}{\sqrt{2}}(|0\rangle - |1\rangle) \qquad (u_1)$$

$$\underset{s(g_a)}{\overset{\text{Picard}}{\longmapsto}} \frac{1}{\sqrt{N}} \sum_{x=0}^{N-1} (-1)^{x\cdot a} |x\rangle \otimes \frac{1}{\sqrt{2}}(|0\rangle - |1\rangle) \quad (s_1)$$

$$\underset{H^{\otimes n}\otimes I_2}{\overset{Q}{\longmapsto}} |a\rangle \otimes \frac{1}{\sqrt{2}}(|0\rangle - |1\rangle) \qquad (u_2)$$

For $\Pi = |a\rangle\langle a| \otimes I_2$ again, Q wins with probability 1, having queried Picard only once!

Just as before, Picard makes the game hardest for Q classically if he chooses a uniformly at random. Classically Q requires n queries to determine a with probability 1. The classical to quantum improvement in number of queries is thus n to 1, in some sense greater than the Grover improvement from $O(N)$ to $O(\sqrt{N})$.

4. Entanglement

Most remarkably, Bernstein and Vazirani's 'sophisticated' quantum search algorithm achieves this improvement without creating any entanglement at any timestep! Recall the following definition:

Definition. A *pure state*—a vector in $\mathcal{H}$—is *entangled* if it does not factor relative to a given tensor product decomposition of the Hilbert space [31].

In the two algorithms considered in the previous sections, the Hilbert space decomposes into a tensor product of *qubits* [32], i.e., $\mathbb{C}^2$s. To see that the slightly improved [16,29] Bernstein-Vazirani algorithm [30] creates no entanglement relative to this decomposition, note that ψ_0 has no entanglement, and hence $u_1\psi_0$ has none since u_1 is the tensor product of operations on individual qubits. Also, ψ_f is not entangled since $|a\rangle$ is just the tensor product of qubits in states $|0\rangle$ or $|1\rangle$. But ψ_f is obtained from the intermediate state $s_1 u_1 \psi_0$ by the action of u_2 which, like u_1, is the tensor product of operations on individual qubits. So $s_1 u_1 \psi_0$ also is not entangled.

In contrast, for Grover's algorithm, every state after $u_1\psi_0$ is entangled for $n > 2$. By a natural measure, the entanglement oscillates with period $\frac{\pi}{4}\sqrt{N}$, i.e., after $\lfloor \frac{\pi}{4}\sqrt{N} \rfloor$ queries the state is close to $|a\rangle \otimes \frac{1}{\sqrt{2}}(|0\rangle - |1\rangle)$ and its entanglement is fairly small, while after only $\lfloor \frac{\pi}{8}\sqrt{N} \rfloor$ queries the entanglement is fairly large [33]. So what could be the suggestion of Lloyd, to which I alluded in the Introduction, that Grover's algorithm can be implemented without entanglement [19]? He simply observes that since the first n qubits are never entangled with the last qubit, if they are implemented by a single tensor factor of dimension N, there is no entanglement. This is true, of course, by definition, and has been observed earlier in the general quantum computing setting by Jozsa and Ekert [20]. The cost incurred for realizing such a scheme physically, as Lloyd acknowledges, increases exponentially with n and must be paid with increasing energy, mass, or precision.

5. Conclusion

So what do quantum games have to do with quantum algorithms? The two versions of GUESS A NUMBER illustrate the relation for oracle problems. In these cases, quantum algorithms specialize the PQ game definition to require that $s_1 = \cdots = s_k$, *i.e.*, the oracle always responds the same way once a function has been chosen. Furthermore, the winning states for Q, W_Q, depend on Picard's strategy $(s_i)_{i=1}^k$, *i.e.*, on a. The Deutsch-Jozsa [1], Simon [2] and Shor [3] algorithms can also be described this way.

Are there 'sophisticated' quantum algorithms? Yes. The oracle which responds in the Bernstein-Vazirani scenario with $x \cdot a \bmod 2$ is a 'sophisticated database' by comparison with Grover's 'naïve database' which only responds that a guess is correct or incorrect. The former is closely related to the vector space model for information retrieval in which there is a vector space with basis vectors corresponding to the occurrence of key words: database elements define vectors in this space and are ranked according to their inner product with the vector representing a query [34]. Furthermore, in some sense it improves more over the classical optimum than does Grover's algorithm.

And finally, is entanglement required for quantum-over-classical improvements? No. I've shown that, remarkably, the slightly improved version of the Bernstein-Vazirani algorithm does not create entanglement at any timestep, but still solves this oracle problem with fewer queries than is possible classically. Relative to the oracle, this quantum algorithm has no entanglement, unlike Grover's, which does—at least within the standard model of quantum computing. It illustrates both 'sophisticated quantum search' without entanglement and sophisticated "quantum search without entanglement" [35].

6. Acknowledgements

I thank Mike Freedman, Manny Knill, Raymond Laflamme, John Smolin and Nolan Wallach for useful discussions. This work was partially supported by Microsoft Research and by the National Security Agency (NSA) and Advanced Research and Development Activity (ARDA) under Army Research Office (ARO) contract number DAAG55-98-1-0376.

References

[1] D. Deutsch and R. Jozsa, "Rapid solution of problems by quantum computation", *Proc. Roy. Soc. Lond. A* **439** (1992) 553–558.

[2] D. R. Simon, "On the power of quantum computation", in S. Goldwasser, ed., *Proceedings of the 35th Symposium on Foundations of Computer Science*, Santa Fe, NM, 20–22 November 1994 (Los Alamitos, CA: IEEE Computer Society Press 1994) 116–123; D. R. Simon, "On the power of quantum computation", *SIAM J. Comput.* **26** (1997) 1474–1483.

[3] P. W. Shor, "Algorithms for quantum computation: discrete logarithms and factoring", in S. Goldwasser, ed., *Proceedings of the 35th Symposium on Foundations of Computer Science*, Santa Fe, NM, 20–22 November 1994 (Los Alamitos, CA: IEEE Computer Society Press 1994) 124–134; P. W. Shor, "Polynomial-time algorithms for prime factorization and discrete logarithms on a quantum computer", *SIAM J. Comput.* **26** (1997) 1484–1509.

[4] L. K. Grover, "A fast quantum mechanical algorithm for database search", in *Proceedings of the 28th Annual ACM Symposium on the Theory of Computing*, Philadelphia, PA, 22–24 May 1996 (New York: ACM 1996) 212–219.

[5] S. Wiesner, "Conjugate coding", *SIGACT News* **15** (1983) 78–88; C. H. Bennett and G. Brassard, "Quantum cryptography: Public-key distribution and coin tossing", in *Proceedings of the IEEE International Conference on Computers, Systems and Signal Processing*, Bangalore, India, December 1984 (New York: IEEE 1984) 175–179; A. Ekert, "Quantum cryptography based on Bell's theorem", *Phys. Rev. Lett.* **67** (1991) 661–663.

[6] P. W. Shor, "Scheme for reducing decoherence in quantum computer memory", *Phys. Rev. A* **52** (1995) R2493–R2496; A. M. Steane, "Error correcting codes in quantum theory", *Phys. Rev. Lett.* **77** (1996) 793–797.

[7] A. S. Kholevo, "Bounds for the quantity of information transmitted by a quantum communication channel", *Problemy Peredachi Informatsii* **9** (1973) 3–11; transl. in *Problems Inf. Transmiss.* **9** (1973) 177–183; C. H. Bennett, D. P. DiVincenzo, J. Smolin and W. K. Wootters, "Mixed state entanglement and quantum error correction", *Phys. Rev. A* **54** (1996) 3824–3851; B. Schumacher, M. Westmoreland and W. K. Wootters, "Limitation on the amount of accessible information in a quantum channel", *Phys. Rev. Lett.* **76** (1997) 3452–3455.

[8] R. Cleve and H. Buhrman, "Substituting quantum entanglement for communication", *Phys. Rev. A* **56** (1997) 1201–1204; H. Buhrman, W. van Dam, P. Høyer and A. Tapp, "Multiparty quantum communication complexity", *Phys. Rev. A* **60** (1999) 2737–2741.

[9] J. von Neumann, in A. H. Taub, ed., *Collected Works*, Vol. 5, *Design of Computers, Theory of Automata and Numerical Analysis* (New York: Pergamon Press 1961–1963).

[10] J. von Neumann, *Mathematische Grundlagen der Quantenmechanik* (Berlin: Springer-Verlag 1932); transl. by R. T. Beyer as *Mathematical Foundations of Quantum Mechanics* (Princeton: Princeton University Press 1955).

[11] J. von Neumann, "*Zur theorie der gesellschaftsspiele*", *Math. Ann.* **100** (1928) 295–320.

[12] J. von Neumann and O. Morgenstern, *Theory of Games and Economic Behavior*, third edition (Princeton: Princeton University Press 1953).

[13] D. A. Meyer, "Quantum strategies", *Phys. Rev. Lett.* **82** (1999) 1052–1055.

[14] See, *e.g.*, S. J. van Enk, "Quantum and classical game strategies", *Phys. Rev. Lett.* **84** (2000) 789.

[15] D. A. Meyer, "Why quantum strategies are quantum mechanical", published as "Meyer replies", *Phys. Rev. Lett.* **84** (2000) 790.

[16] C. H. Bennett, E. Bernstein, G. Brassard and U. Vazirani, "Strengths and weaknesses of quantum computing", *SIAM J. Comput.* **26** (1997) 1510–1523.

[17] M. Boyer, G. Brassard, P. Høyer and A. Tapp, "Tight bounds on quantum searching", *Fortsch. Phys.* **46** (1998) 493–506.

[18] C. Zalka, "Grover's quantum searching algorithm is optimal", *Phys. Rev. A* **60** (1999) 2746–2751.

[19] S. Lloyd, "Quantum search without entanglement", *Phys. Rev. A* **61** (1999) 010301.

[20] R. Jozsa, "Entanglement and quantum computation", in S. A. Huggett, L. J. Mason, K. P. Tod, S. T. Tsou and N. M. J. Woodhouse, eds., *The Geometric Universe: Science, Geometry, and the Work of Roger Penrose* (Oxford: Oxford University Press 1998) 369–379; A. Ekert and R. Jozsa, "Quantum algorithms: entanglement-enhanced information processing", *Phil. Trans. Roy. Soc. Lond. A* **356** (1998) 1769–1782.

[21] A. M. Steane, "A quantum computer needs only one universe", `quant-ph/0003084`.

[22] S. L. Braunstein, C. M. Caves, R. Jozsa, N. Linden, S. Popescu and R. Schack, "Separability of very noisy mixed states and implications for NMR quantum computing", *Phys. Rev. Lett.* **83** (1999) 1054–1057; R. Schack and C. M. Caves, "Classical model for bulk-ensemble NMR quantum computation", *Phys. Rev. A* **60** (1999) 4354–4362. For a response to this criticism, see R. Laflamme, "Review of 'Separability of very noisy mixed states and implications for NMR quantum computing'", *Quick Reviews in Quantum Computation and Information*, `http://quickreviews.org/qc/`.

[23] R. F. Werner, "Quantum states with Einstein-Podolsky-Rosen correlations admitting a hidden-variable model", *Phys. Rev. A* **40** (1989) 4277–4281.

[24] L. M. Krauss, *The Physics of Star Trek*, with a foreword by Stephen Hawking (New York: HarperCollins 1995).

[25] P. A. M. Dirac, *The Principles of Quantum Mechanics*, fourth edition (Oxford: Oxford University Press 1958).

[26] J. Eisert, M. Wilkens and M. Lewenstein, "Quantum games and quantum strategies", *Phys. Rev. Lett.* **83** (1999) 3077–3080.

[27] S. C. Benjamin and P. M. Hayden, "Comment on 'Quantum games and quantum strategies'", *Phys. Rev. Lett.* **87** (2001) 069801/1.

[28] A. M. Steane and W. van Dam, "Physicists triumph at guess my number", *Phys. Today* **53**(2) (February 2000) 35–39.

[29] R. Cleve, A. Ekert, C. Macchiavello and M. Mosca, "Quantum algorithms revisited", *Proc. Roy. Soc. Lond. A* **454** (1998) 339–354.

[30] E. Bernstein and U. Vazirani, "Quantum complexity theory", in *Proceedings of the 25th ACM Symposium on Theory of Computing*, San Diego, CA, 16–18 May 1993 (New York: ACM Press 1993) 11–20; E. Bernstein and U. Vazirani, "Quantum complexity theory", *SIAM J. Comput.* **26** (1997) 1411–1473. A similar algorithm was rediscovered independently in B. M. Terhal and J. A. Smolin, "Single quantum querying of a database", *Phys. Rev. A* **58** (1998) 1822–1826.

[31] E. Schrödinger, "*Die gegenwärtige Situation in der Quantenmechanik*", *Naturwissenschaften* **23** (1935) 807–812; 823–828; 844–849.

[32] B. Schumacher, "Quantum coding (information theory)", *Phys. Rev. A* **51** (1995) 2738–2747.

[33] D. A. Meyer and N. R. Wallach, "Global entanglement in multiparticle systems", `quant-ph/0108104`.

[34] G. Salton and M. McGill, *Introduction to Modern Information Retrieval* (New York: McGraw-Hill 1983); M. W. Berry, Z. Drmač and E. R. Jessup, "Matrices, vector spaces, and information retrieval", *SIAM Rev.* **41** (1999) 335–362.

[35] D. A. Meyer, "Sophisticated quantum search without entanglement", *Phys. Rev. Lett.* **85** (2000) 2014–2017.

(David A. Meyer) PROJECT IN GEOMETRY AND PHYSICS, DEPARTMENT OF MATHEMATICS, UNIVERSITY OF CALIFORNIA/SAN DIEGO, LA JOLLA, CA 92093-0112

E-mail address, David A. Meyer: `dmeyer@chonji.ucsd.edu`

A Proof That Measured Data and Equations of Quantum Mechanics Can Be Linked Only by Guesswork

John M. Myers and F. Hadi Madjid

ABSTRACT. The design and operation of a quantum-mechanical device as a laboratory instrument puts models written in equations of quantum mechanics in contact with instruments. In designing a quantum-mechanical device of high precision, such as a quantum computer, a scientist faces choices of models and of instruments, and the scientist must choose which model to link to which arrangement of instruments. This contact is recordable in files of a Classical Digital Process-control Computer (CPC) used both to calculate with the equations and to manage the instruments. By noticing that equations and instruments make contact in a CPC, we rewrite equations of quantum mechanics to explicitly include functions of CPC-commands to the instruments. This sets up a proof that a scientist's choice in linking mathematical models to instruments is unresolvable without guesswork to narrow the set of models from which one is to be chosen.

The proof presents the challenge of pursuing its implications. Scientists in any investigative endeavor inherit choices from the past and frame choices for the future, choices open to guesswork and visible in CPC files. To picture the framing of choices and relations among them, we adapt colored Petri nets. Constraining the events of the nets to produce output colors defined by definite functions of input colors excludes guesswork from the firing of net events, and by contrast highlights guesses entering a net fragment as colored tokens placed by a scientist or by instruments on input conditions. The availability of these net fragments makes choice and guesswork part and parcel of physics.

Net fragments as a means of expressing guess-demanding choices are applied to portray guesswork needed in testing and calibrating a quantum computer. The sample size required to test a quantum gate in a quantum computer is shown to grow as the inverse square of the error allowed in implementing the gate.

1. Introduction

This paper stems from earlier work [1] and a proof presented here showing that inquiry in quantum physics continually presents a scientist with choices of equations and of instruments, unresolvable by calculation and measurement. Something else is demanded of the scientist, which may as well be called a *guess*.[1] Challenged by the proof to look at its implications, we noticed that people in investigative

---

2000 *Mathematics Subject Classification.* Primary 81P68, 68Q05; Secondary 81P15, 68Q85.
[1]Other words are *hypothesis, Ansatz, assumption, axiom, postulate,* and sometimes *principle*.

endeavors inherit and frame choices open to guesswork, some of which show up clearly in the computers used in the endeavors.

Section 2 introduces the Classical Process-control Computer (CPC) with its special capacity to manipulate abstractions expressed as equations without contaminating them with its own physics.[2] A scientist can use a CPC not only to calculate with equations, but also to mediate the command of laboratory instruments via digital/analog (D/A) converters and to record experimental results returned from the instruments via analog/digital (A/D) converters. By noticing that both equations and instruments make contact in a CPC, we rewrite equations of quantum mechanics to explicitly include functions of CPC-commands to the instruments. This sets up the proof that the scientist's choice in linking equations to instruments is unresolvable without guesswork to narrow the set of models. A lattice of sets of models is defined, two widely used guesses that narrow the set of models are noted, and the concept of statistical distance between probability distributions is applied to quantum-mechanical models.

Section 3 provides language for displaying and analyzing guess-demanding choices visible in CPC files. To this end Turing machines are introduced and adapted to formalize the definition of a CPC. This allows fragments of colored Petri-nets, opened to exogenous influences, to portray the programming and running of programs in a network of CPC's operated by collaborating scientists. Many of these programs incorporate guesses. This general picture of process-control computation shows programs and other guesses as colors on tokens that a scientist enters on a Petri net that acts as a game board. Mechanisms for one scientist to judge programs (and hence guesses) made by another are sketched, leading to the first of many needs for concurrently operating CPC's.

Section 4 describes some examples in which guessing, quantum mechanics and CPC structures must interact in the building of a quantum computer as a laboratory instrument specified by equations of quantum mechanics. We show the need for guesses to link equations and instruments brings with it a need to test the quantum computing instruments and to calibrate them, guided by test results and guesses. Quantum mechanics imposes a peculiar structure on this testing, related to the statistical distance between models. For the measure of precision conventionally used in quantum computing, the sample size needed for testing a quantum gate is shown to increase as the inverse square of the tolerated imprecision. While many questions are left open to future work, the example demonstrates a frame for analysis and experiment broader than any quantum model alone, a frame that includes the testing of the mathematical models by results of the use of instruments, and so distinguishes what the model says from what the instruments do, allowing provision for guesswork as an ingredient in advancing both models and instruments.

2. Quantum-mechanical models and their links to instruments

Proving the necessity of guesses demands language to describe the linking of numbers in mathematical models to numbers pertaining to laboratory instruments,

---

[2]This capacity stems from regenerative amplifiers and clock-gated memory registers, two inventions used to make all computer hardware insensitive to manufacturing variations, so that, like the placing of a chess piece not quite in the center of a square, deviations in performance, within limits, do not matter. Its independence from its own physics distinguishes a classical computer from a quantum computer.

starting with mathematical language to describe a scientist's choosing one arrangement of laboratory instruments rather than another. We shall describe a situation in which a scientist chooses instruments by using a CPC keyboard to type strings of characters, much as Gödel, in mathematical logic, described equations as strings of characters. The scientist at the CPC keyboard writes and executes programs to command the operation of laboratory instruments and record their results. These programs make use of quantum-mechanical equations, which the scientist also writes into the CPC.

Quantum mechanics as a mathematical language expresses different measuring instruments by different operators, and thus has built into it a recognition that phenomena to be described cannot be independent of the instruments used to study them [**2**]. Still, this dependence is emphasized more some times than others. Some modeling merely assumes that instruments can be found, without saying how, to implement various combinations of state vectors and operators. Such models appear in theories of quantum computing to relate the multiplication of unitary operators to the solving of problems of interest. To see the need for another kind of model, suppose a scientist has computer-controlled instruments (such as lasers) with the potential to implement a quantum computer, and faces the question of what commands the CPC should transmit to the instruments and when it should transmit them in order implement one or another quantum gate. Determining the commands and their timing to implement a quantum gate expressed as a unitary matrix U_j takes a model that expresses the gates as unitary transformations in terms of commands that a process-control computer can transmit to the instruments. Curiously, models of this kind have not been much stressed in physics, and it is a merit of efforts to build quantum computers to make the importance of such models apparent.

2.1. Models and instruments make contact in a CPC. Part of a scientist's control of instruments can work through the use of a process-control computer that transmits commands to the (computer-controlled) instruments and records results produced by them. We confine our analysis to this part, excluding from consideration here (but by no means denigrating) hand work beyond the reach of a process-control computer. We shall portray cases in which a scientist chooses arrangements of instruments, chooses models, and puts the two in contact, linking models to instruments, during a CPC session starting after the instruments have been set up and put under control of a CPC and ending before the scientist has to tinker with the instruments in ways unreachable by the CPC. Within the CPC, laboratory instruments and mathematical models make contact when:

(1) a model resident in a CPC file is used to derive commands for the CPC to transmit to the instruments;
(2) instrumental results collected by a CPC are used to narrow down a set of models. (We shall later see feedback as an example of this.)

Such contact does not spring from nothing, but is brought about by design and depends on choices made by a scientist, including choices of what set of models to start with, what model to choose for use by a CPC in generating commands, and what experiments to run. To picture the design and operation of contact between models and instruments, imagine eavesdropping on CPC's used in various investigations. Commands sent to the instruments by the CPC and the results

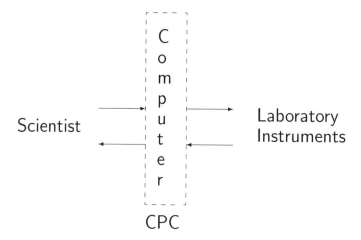

FIGURE 1. Computer mediating contact between scientist and instruments.

received from them, both numerical, are amenable to analysis, as is the scientist's writing of equations, programs, calls for program execution, etc.; we also eavesdrop on displays produced by the CPC for the scientist.

Although the CPC puts instruments in contact with equations involving quantum superposition, the CPC itself is a classical machine, free of quantum superposition, for it needs no quantum behavior within itself, neither to manipulate equations of quantum mechanics nor to manage laboratory instruments. For example, the writing of an expression $|0\rangle + |1\rangle$ for a superposition of quantum states makes use of written characters that themselves exhibit no superposition. And any command to instruments is likewise a character string, including a command to rotate a polarizing filter by 45 degrees to implement the superposed state $|0\rangle + |1\rangle$. Similarly, results of the use of instruments interpreted as demonstrating superposition arrive as bit strings, themselves devoid of superposition.

The CPC is situated between a scientist to its left and laboratory instruments to its right, as shown in figure 1. Working at the CPC, a scientist is limited in action at any moment to the resolution of the choice presented by the CPC at that moment, a choice defined by the files stored in its memory and the state of its processor, and exemplified by a menu displayed by the CPC. Our analysis of the CPC cannot reach beyond its buffers: neither to its left into the scientist, nor to its right where, invisible to eavesdropping, reside digital-to-analog (D/A) and analog-to-digital (A/D) converters and beyond them the laboratory instruments.

2.2. Quantum-mechanical models that recognize commands sent to instruments. For equations of quantum mechanics to model effects of a scientist's choices in arranging instruments, these choices must show up in the equations. To see how this can work, recall that quantum mechanics parses the functioning of instruments into state preparation, transformation, and measurement, three coordinated activities that generate *outcomes*, supposed visible in experimental results by some means unspecified. The three activities are described, respectively, by a state (as a unit vector representing a ray in a Hilbert space), a unitary operator, and a hermitian operator. The only way to make the scientist's choices in arranging instruments show up in quantum-mechanical equations is to make the state vector $|v\rangle$, the transformation operator U, or the measurement operator M, or some combination of them, depend on how these choices are resolved.

A simple and yet, so far as we know, original way to analyze a scientist's choice of arrangements of instruments is to suppose that during a CPC-mediated session the instruments are controlled by CPC-transmitted commands from a set B of possible commands, where $B \subset \mathcal{B}$ and $\mathcal{B}$ is the set of all finite binary strings. We formulate a core set of quantum mechanical models that express the probability of an outcome of instruments in response to a command $b \in B$ sent to the instruments by the CPC, as follows. Let $\mathcal{V}_B$, $\mathcal{U}_B$, and $\mathcal{M}_B$ be the sets of all functions $|v\rangle$, U, and M, respectively, with

$$|v\rangle : B \to \mathcal{H},$$
$$U : B \to \{\text{unitary operators on } \mathcal{H}\}, \text{ and}$$
$$M : B \to \{\text{hermitian operators on } \mathcal{H}\}.$$

The core models exhibit discrete spectra for all $M \in \mathcal{M}$:

PROPERTY 1.

(2.1) $$(\forall b \in B) M(b) = \sum_j m_j(b) M_j(b),$$

where $m_j : B \to \mathcal{R}$ (with $\mathcal{R}$ denoting the real numbers) is the j-th eigenvalue of M, and M_j is the projection onto the j-th eigenspace (so $M_j M_k = \delta_{j,k} M_j$).

Let $\Pr(j|b)$ denote the probability of obtaining the j-th outcome, given transmission by the CPC of a command b. Although not commonly seen in texts, this probability of an outcome given a command is the hinge pin for focusing on quantum mechanical modeling of uses of instruments. Quantum mechanics constrains the models to satisfy:

PROPERTY 2.

(2.2) $$\Pr(j|b) = \langle v(b)|U^\dagger(b) M_j(b) U(b)|v(b)\rangle,$$

where the $\dagger$ denotes the hermitian adjoint.

(Within this modeling scheme, the Schrödinger equation relates a model at a later time to a model at an earlier time by a certain transformation operator U, dependent on the situation.)

Any choice from the sets $\mathcal{B}, \mathcal{V}_B, \mathcal{U}_B$, and $\mathcal{M}_B$ produces some quantum-mechanical model $(|v\rangle, U, M)_B$. Two models $(|v\rangle, U, M)_B$ and $(|v'\rangle, U', M')_B$ generate the same probabilities $\Pr(j|b)$ if they are unitarily equivalent, meaning there exists a $Q : B \to \{\text{unitary operators on } \mathcal{H}\}$ such that $(\forall b \in B)|v'(b)\rangle = Q(b)|v(b)\rangle$, $U'(b) = Q(b)U(b)Q^\dagger(b)$ and $M'(b) = Q(b)M(b)Q^\dagger(b)$. For this reason, any model $(|v\rangle, U, M)_B$ can be reduced to $(|v'\rangle, \mathbf{1}, M)_B$, where $|v'\rangle = U|v\rangle$ and $M' = M$ or, alternatively to $(|v\rangle, \mathbf{1}, M')_B$ where $M' = U^\dagger M U$.

More models are available in more general formulations. When we show that guesswork is necessary even to resolve choices among models of the core set, it follow that guesswork is necessary also to resolve the choices of among a larger set of models involving positive-operator-valued measures, superoperators, etc.

2.3. From results to quantum-mechanical outcomes.
Before stating and proving the proposition that calculations and measurements cannot by themselves link models to outcomes obtained from instruments, we call to the reader's attention that outcomes themselves, in the sense of quantum mechanics, are produced by instruments only with the help of interpretive guesswork.

CLAIM 1. *To speak of actual instruments in the language of quantum mechanics one needs to associate results of the use of the instruments, recorded in a CPC, with* outcomes *in the sense of quantum mechanics or with averages of outcomes.*

Experimental results of the use of instruments become quantum-mechanical outcomes only by a scientist's act of interpreting the results as outcomes. The interpretation involves judgment and guesswork, not only to sidestep imperfections in the instruments, but as a matter of principle, even for the limiting case of instruments supposed free of imperfections. For example, light detectors used in experiments described by models of quantum optics generate experimental results; typically, each of L detectors reports to the CPC at each of a succession of K time intervals a detection result, consisting of 0 (for no detection) or 1 (for detection), so a record contains LK bits. Depending on judgments made about correlations from time interval to time interval and detector to detector, these LK bits may constitute LK quantum outcomes, or one quantum outcome, or some number in between. The number of outcomes in LK bits is determined neither by the experimental results (which in this case are just these bits) nor by general principles of quantum mechanics; yet the parsing of results into outcomes must occur, at least provisionally, before any comparison between equations and measured outcomes can be made. Henceforth, when we speak of outcomes, we presuppose that this piece of guesswork has been accomplished and a decision made to define the parsing of results into outcomes.

2.4. Calculation and measurement by themselves cannot link quantum models to recorded outcomes.
Could it be that the general properties 1 and 2 suffice to determine a model (up to unitary equivalence) if only one collects enough measured results interpreted as outcomes? The answer is: "no; unless some special properties restrict the model more tightly than the form established by properties 1 and 2 alone, one can always find many unitarily inequivalent models $(|v\rangle, U, M)_B$, all of which produce probabilities that match perfectly the relative frequencies of outcomes."

To prove this we define some things to pose the issue more sharply. Let B denote the set of commands used to generate some set of outcomes interpreted from measured results.[3] For any $b \in B$, let $N(b)$ be the number of times that an outcome has been entered in the record for a run of the experiment for command b, and let $J(b)$ be the number of distinct outcomes for command b. For $j = 1, \ldots, J(b)$, let $\lambda_j(b)$ be the j-th distinct outcome obtained for command b, and let $n(j, b)$ be the number of times this j-th distinct outcome λ_j is recorded in response to command b. For all $j > J(b)$ let $\mu_j(b)$ be arbitrary real numbers, and for all $j \geq 1$ let $\phi(j, b)$ be arbitrary real numbers.

PROPOSITION 2.1. *Given any set of recorded outcomes associated with any set B of commands, the set of models satisfying properties 1 and 2 contains many*

---
[3]Practically speaking, B must be a finite set, but the proof holds also for B denumerably infinite.

unitarily inequivalent models $(|v\rangle, U, M)_B$, each of which has a perfect fit with the set of outcomes, in the sense that

(2.3) $$(\forall b)(\forall 1 \leq j \leq J(b))\Pr(j|b) = n(j,b)/N(b).$$

PROOF. It is instructive to start with the special case in which for some $b \in B$, there exist two or more distinct values of j for which $n(j) > 0$. For this case let the set $\{|j\rangle\}$ be an orthonormal basis of a separable Hilbert space. Define a subset S of models satisfying properties 1 and 2 as all models of the form $(|v\rangle, U, M)_B$, where

(2.4) $$|v(b)\rangle \stackrel{\text{def}}{=} \sum_{j=1}^{J(b)} [n(j,b)/N(b)]^{1/2} \exp(i\phi(j,b)|j\rangle),$$

(2.5) $$U(b) \stackrel{\text{def}}{=} \mathbf{1},$$

(2.6) $$M(b) \stackrel{\text{def}}{=} \sum_{j=1}^{J(b)} \lambda_j(b)|j\rangle\langle j| + \sum_{j=J(b)+1}^{\infty} \mu_j(b)|j\rangle\langle j|,$$

with μ_j and ϕ arbitrary real-valued functions. By invoking property 2, one checks that any such model has the claimed perfect fit; yet the set contains many unitarily inequivalent models, which predict conflicting statistics for some possible quantum measurement.[4] This proves the special case.

For the general case, modify the definitions above to

(2.7) $$|v(b)\rangle \stackrel{\text{def}}{=} \sum_{j=1}^{J(b)} [n(j,b)/N(b)]^{1/2} |w_j\rangle,$$

(2.8) $$U(b) \stackrel{\text{def}}{=} \mathbf{1},$$

(2.9) $$M(b) \stackrel{\text{def}}{=} \sum_{j=1}^{J(b)} \lambda_j(b) P_j + \sum_{j=J(b)+1}^{\infty} \mu_j(b) P_j,$$

where $P_j P_k = \delta_{j,k} P_j$, for all j the projection P_j has dimension greater than 1, and $|w_j\rangle$ ranges over all unit vectors of the eigenspace defined by $P_j|w_j\rangle = |w_j\rangle$. In particular, for any j, $\dim(P_j)$ can be as large as one pleases. Then even if there is only one outcome that is ever recorded, there are still many unitarily inequivalent models that perfectly fit the data. □

Proposition 2.1 implies that the density matrix, often supposed to be determined from measured data [3], is undetermined without assuming special properties shortly to be discussed; this follows by expressing the density matrix as $|v\rangle\langle v|$ and noticing that the phases of the off-diagonal elements are undetermined. We leave to the future the demonstration of additional ambiguity in the link between any set of recorded outcomes and models expressed in the mathematical language of quantum mechanics.

---

[4]This happens *e.g.* for primed and unprimed models if for any b, $n(1,b) \neq 0 \neq n(2,b)$ and $\phi(1,b) - \phi(2,b) \neq \phi'(1,b) - \phi'(2,b)$.

2.5. Statistically significant differences between models.

In practice, a scientist has little interest in a model chosen so that its probabilities exactly fit measured relative frequencies. Rather, the scientist wants a simpler model with some appealing structure that comes reasonably close to fitting. Quantum mechanics encourages this predilection, because on account of statistical variation in the sample mean, functions that perfectly fit outcomes on hand at one time are not apt to fit perfectly outcomes acquired subsequently. We show here that accepting statistics no way takes away from the proof that measurements and equations by themselves cannot link models to instruments.

One needs a criterion for the statistical significance of a difference between two quantum-mechanical models (or between a model and measured relative frequencies). Here we limit our attention to models α and β which have a set B of commands in common and for which the spectra of M_α and M_β are the same. For a single command b, the question is whether the difference between the probability distributions $\Pr_\alpha(\cdot|b)$ and $\Pr_\beta(\cdot|b)$ is bigger than typical fluctuations expected in $N(B)$ trials. An answer is that two distributions are indistinguishable statistically in $N(b)$ trials unless

$$(2.10) \qquad N(b)^{1/2} d(\Pr_\alpha(\cdot|b), \Pr_\beta(\cdot|b)) > 1,$$

where d is the statistical distance defined by Wooters in Eq. (10) of [**4**]. Furthermore, Wooters's Eq. (12) shows for two models α and β that differ only in the function $|v\rangle$,

$$(2.11) \qquad d(\Pr_\alpha(\cdot|b), \Pr_\beta(\cdot|b)) \leq \cos^{-1} |\langle v_\alpha(b)|v_\beta(b)\rangle|.$$

To judge the significance of the difference between two models with respect to a set B of commands common to them, a scientist who chooses some weighting of different commands can define a weighted average of $d(\Pr_\alpha(\cdot|b), \Pr_\beta(\cdot|b))$ over all $b \in B$. The same holds if model β is replaced by relative frequencies of outcomes interpreted from measured results.

It is noteworthy that the set of models statistically indistinguishable from a given model can be much larger than would be the case if the "$\leq$" of (2.11) were an equality, as follows.

PROPOSITION 2.2. *For any set of outcomes, two models α and β of the form $(|v\rangle, \mathbf{1}, M)_B$ can perfectly fit the relative frequencies of the outcomes (Proposition 2.1) and yet be mutually orthogonal in the sense that $\langle v_\alpha | v_\beta \rangle = 0$*

PROOF. For any set of measured outcomes, there exists a perfectly fitting model α of the form in the proof of Proposition 2.1 for the general case, for which $(\forall j, b) dim(|w_j(b)\rangle) > 1)$, and a corresponding perfectly fitting model β such that $(\forall j, b)\langle w_{\alpha,j}(b)|w_{\beta,j}(b)\rangle = 0$. For these two models, $\langle v_(b)\alpha(b)|v_\beta(b)\rangle = \sum_j [n(j,b)/N(b)]\langle w_{\alpha,j}(b)|w_{\beta,j}(b)\rangle = 0$. □

Wooters extended the definition of statistical difference to unit vectors. While for any two unit vectors, there exist measurement operators that maximize the statistical distance between them, for any such operator there exist other vectors, mutually orthogonal, that have zero statistical distance relative to this operator. For this reason, among others, statistics still leaves the scientist needing something beyond calculation and measurement to determine a model, for the set of models closer than ϵ in weighted statistical distance to certain measured results certainly includes all the models that exactly fit the data and, without special restrictions dependent on guesses, this set includes models that are mutually orthogonal. Models close to given measured data are not necessarily close to each other in the predictions they make.

2.6. Lattices of models. Properties 1 and 2 set up a big set of models $(|v\rangle, U, M)_B$, $B \subseteq \mathcal{B}$, $|v\rangle \in \mathcal{V}, U \in \mathcal{U}, M \in \mathcal{M}$. Subsets of models of this set are a lattice under set intersection and union. Each command set B establishes a smaller lattice of sets of models, and these lattices will play a part in the testing and calibrating of quantum computers, discussed in section 4, where a scientist encountering problems with a model chooses a set of possible alternatives, and then tries to narrow it. Often this narrowing is seen as choosing values of parameters within a form of model in order to obtain a best fit, say with a criterion of minimizing statistical distance between frequencies of outcomes interpreted from measured results and probabilities calculated from the model. One is free to think of the estimating of parameters in the language of a lattice of models as the using of measured results to select a model from a set of models.

From Proposition 2.1 that showed that the whole set of models defined by properties 1 and 2 is too big to permit measured results to select a model, we have:

PROPOSITION 2.3. *For measured data to uniquely decide to within unitary equivalence which quantum-mechanical model of a set of models best fits experimental results interpreted as outcomes by a criterion of least statistical distance (or any other plausible criterion), the set of models must first be sufficiently narrowed, and this narrowing is underivable from the results and the basic properties 1 and 2 of quantum mechanics.*

Something beyond measured results and calculations from equations is required to narrow a set of models so that measured results can select a model that is "best" by some criterion. Such an act of choosing undefined by calculation and results of observation is what we have called a *guess*.

2.7. Hidden guesswork in conventional quantum mechanical models. The proof casts in a clear light maneuvers conventionally made to narrow down the set of models. Sometimes a community of physicists is in mutual agreement about guesses deemed appropriate, and this agreement obscures from notice the fact that a guess is invoked. As an example of a widely invoked guess, most modeling in quantum physics supposes that the scientist can vary b so as to vary $U(b)$ while holding $v(b)$ and $M(b)$ constant. Indeed, most models used in quantum physics are restricted to the subset of models having the special

PROPERTY 3. *The command b is the concatenation of separate commands for the three types of operations, so that*

$$(2.12) \qquad b = b_v \parallel b_U \parallel b_M,$$

where here the $\parallel$ denotes concatenation of commands.

According to these models, one can vary any one of the three while holding the other two fixed. This specializes (2.1) to the more restrictive form:

$$(2.13) \qquad \Pr(j|b) = \langle v(b_v)|U^\dagger(b_U)M_j(b_M)U(b_U)|v(b_v)\rangle.$$

An additional constraining guess characterizes models widely used in the analysis of quantum computers, a guess prompted by the desire to generate a unitary transformation as a product of other unitary transformations that serve as "elementary quantum gates." For example, one may want to generate the unitary transformation $U(b_{U,1})U(b_{U,2})$. To generate it one causes the CPC to transmit some b_U. For quantum computing to have an advantage over classical computing, the determination of this b_U in terms of $b_{U,1}$ and $b_{U,2}$ must be of polynomial complexity [5]. It is usually assumed that b_U is the simplest possible function of $b_{U,1}$ and $b_{U,2}$, as follows.

Let $B_U \subset B$ be a set of instrument-controlling commands, thought of as strings that can be concatenated. Suppose the function U has the form $U(b_1 \parallel b_2) = U(b_2)U(b_1)$ for all $b_1 \parallel b_2 \in B_U$ (note reversal of order). Then we say the function U respects concatenation.

PROPERTY 4. *Quantum computation employs a subset of models in which U respects concatenation.*

REMARK 2.1. We present properties 1 through 4 not as properties of laboratory instruments, but as properties that a scientist can choose to demand of models. Whether the instruments act that way is another question. There are reasons, relaxation and other forms of decoherence among them, to expect limits to the precision with which instruments can behave in accord with properties 3 and 4. All four properties are used often enough to be conventions, in the sense that a convention is a guess endorsed by a community.

3. Petri nets to show choices open to guesswork

In orchestrating contact between mathematical models and laboratory instruments, scientists set up chains of cause and effect, expressed in computer programs with their "if-then" structure, not as static propositions but as designs for action. Such designs are implemented in experiments; an example is a feedback loop that adjusts the orientation of a filter according to a rule that tells what adjustment to make in immediate response to a result recorded by a light detector. On a more relaxed time scale, physicists make other connections by analyzing outcomes of one generation of experiment, using the equations of a model, to set up design instruments for a next generation. As remarked above, contact between equations and instruments depends on choices made by scientists, including choices of what set of models to start with, what model to choose for use by a CPC in generating commands, and what experiments to run. If these choices could be resolved by some combination of calculation and measurement, one could argue that they are irrelevant to physics. But the propositions of the preceding section show this is not the case, so the design and operation of contact between equations and instruments, with its ineradicable dependence on guesswork, cries out for attention as part and parcel of physics.

Although widespread in practice, the design of contact between equations and instruments is in its infancy as a topic for theoretical attention. A beginning can be seen in Benioff's analysis of sequences of measurements (described quantum mechanically) in which subsequent measurements are functions of outcomes of preceding measurements [6]. Called decision procedures, these involve classical feedback control equations to control instruments described quantum mechanically, in some cases with proved advantages [**7**]. These efforts dealt with measurements occurring at a single location. Designs that put equations and instruments in contact over a network of cooperating investigators are wide open for future attention.

Logic in experiments, in feedback loops at many time scales, is logic in action. This is the logic of models that relate instrument commands to quantum vectors and operators. Here we adapt Petri nets to provide mathematical language by which to express and analyze designs for contact between equations and instruments, designs that include sequencing of effects, decision rules, and interactions among sequences of effects that scientists implement in their instruments. The nets will highlight choices resolvable only by resort to guesswork; they serve as a language with which one can express formally how guessing works in physics, case by case, within CPC-mediated investigations.

3.1. Requirements of CPC's. In order to adapt Petri nets to showing guess-demanding choices visible in CPC's, we start by clarifying how a CPC differs from a Turing machine, on the way to adapting the Turing machine to process control and to use in a network of collaborating scientists. This lays the groundwork for introducing Petri nets.

3.1.1. Timing in the execution of commands. The first thing that makes process-control computing special is timing. In the context of quantum-mechanical models, each unitary transformation maps states possible in one situation to states possible in another situation; for quantum computing this means mapping states possible at an earlier time to states possible at a later time. Thus a unitary transformation is implemented not all at once, but over a time duration. In practice, that duration depends on how the instruments implement the transformation. A written command b_U acts as a musical score. Like sight reading at a piano, executing a program containing the command b_U requires converting the character string b_U—the score—into precisely timed actions—the music. The piano keys, in this analogy, include the output buffers that control the amplitude, phase, frequency, and polarization of lasers of an ion-trap quantum computer or of radio-frequency transmitters for a nuclear-magnetic-resonance (NMR) quantum computer.

For this reason executing a command b_U requires parsing it into pieces (signals) and implementing each signal at a time, the specification of which is contained in the string b_U. Either the CPC that executes a program in which b_U is written parses the command into signals and transmits each signal at its appointed time, or the instruments receiving the command b_U, unparsed, contain programmable counters operating in conjunction with a clock that do this timed parsing. Such programmable counters themselves constitute a special-purpose CPC. So either the scientist's CPC must execute commands by issuing an appropriately timed sequence of signals, or some other CPC attached to the instruments must do this. Either way, the capacity to execute programmed motion in step with a clock is a requirement for a CPC, distinct from and in addition to requirements to act as a Turing machine.

3.1.2. Firewalls in a network of computers. Just as axioms set up branches of mathematics, guesses set up rules for the conduct of experiments and the interpretation of their results, rules often embedded in CPC's. Collaborating scientists accept guesses from each other, at least provisionally, use these in experimenting and modeling; they evaluate some of them, sometimes refining or replacing them. This poses a problem for CPC-mediated inquiry, where guesses engender computer programs, for a scientist's guess can reprogram a CPC, often for better but sometimes, by malice or accident, for worse. Scientists in a collaboration need to test each other's programs and to limit the influence of any program, making the scope of influence of a CPC program a matter for negotiation among the collaborators.

An easy but narrow case is that of a computer running Gödel's test for validity of a claimed derivation [8]. To think about such testing, one models the computer by a Turing machine designed to start from a tape on which the claimed derivation is written and to halt leaving a "yes" or "no" on the tape, according to whether the claim is or is not valid. Such a Turing machine can be emulated by a universal Turing machine executing a testing program to check a passive (non-executed) file containing the claimed derivation.

Not just derivations, but also programs need to be tested with respect to what they do when they are executed. But what is to keep an executing program under test from infecting the program that tests it? Hardware walls of some kind are needed. By limiting our analysis to exclude remote login and insisting on computers that distinguish physically one interface from another, we can see a basic structure for testing programs and for limiting the reach of guesses of any one scientist in a network of CPC's, based on operating two or more CPC's concurrently with controlled interfaces between them, so the testing program and the program under test execute on separate CPC's, with an interface controlled by the testing CPC. By virtue of concurrent operation of CPC's with controlled interfaces, guesses made by collaborators can set up programs that frame choices open to guessing by any one scientist, and that test the performance of the scientist's programs within that frame of choice, allowing freedom to a scientist to program one part of the investigation while insulating other parts. Hardware walls that limit the reach of one person's guesses at any moment are one many motivations for stressing a network of concurrently operation CPC's.

3.2. Turing machines and Petri nets. Here we provide language for displaying and analyzing guess-demanding choices visible in files of CPC's used by collaborating scientists who on occasion reprogram those choices. As a model of a CPC, we assume that each CPC of a network is a Turing Machine adapted for Process-control (TMP), to be defined. Making sense of networks of TMP's handling equations and controlling instruments calls for a descriptive capacity that allows for various viewpoints at various levels of detail. We introduce a specialized use of fragments of colored Petri-nets, opened to exogenous influences, to portray the programming and running of programs in a network of TMP's operated by collaborating scientists.[5]

Different viewpoints and levels of detail are accommodated by morphisms in the category of nets. Isomorphisms between Petri nets trade net detail for color detail [9]. These will be combined with coarsening maps that suppress detail, for example by mapping colored tokens to black tokens. We will show how the programming of a universal TMP (UTMP) portrayable as a single Petri net can produce any number of patterns of use of instruments and equations, portrayable by a host of different Petri nets. This general picture of process-control computation will show programs and other guesses as colors on tokens that a scientist enters on a game board defined by a fragment of a Petri net, and equations of quantum mechanics

---

[5]Our use of Petri nets is impressionistic and a more technical presentation will doubtless be rewarded by exposing issues here overlooked.

written as guesses by a scientist will be seen as colors on tokens that take part in directing and interpreting the use of laboratory instruments.

3.2.1. *Writing* vs. *executing a program*. Computers rest on the writing of motionless characters on a page to describe something moving, a puzzle solved in music by writing notes on staves, to be read in step with a swinging pendulum that chops time into moments, so that written notes that portray a still picture for each moment direct the motion of the playing of a musical instruments [10]. The logical machinery of a computer moves in response to triggering signals, "tick" and "tock", synchronized to distinct phases of the swinging of a pendulum. Computer designers employ truth tables, each of which specifies the response of a clocked circuit at a tock to a stimulus present as an input at a preceding tick. A row of a truth-table can be drawn as a transition in a Finite State Machine (FSM). By coupling an FSM to a memory of unlimited capacity, one arrives at the theoretical concept of a Turing machine, various special cases of which perform various special tasks [11, 12, 13]. And here is the crux of programming: because a state machine is describable by still writing—a table—a Turing machine can be designed to be universal. By coding into its memory the table that describes any given special Turing machine, one causes the universal Turing machine to emulate the given special Turing machine. So, apart from speed and memory requirements, the single universal Turing machine can be put to doing any of the things that any of the special Turing machines can do, making it potentially convenient, once adapted to process control, to designing and implementing contact between equations and instruments. (But demands for quick response require in some cases devices streamlined to a special task better modeled by a special Turing machine than by a universal one.) The next tasks are to adapt the Turing machines, special and universal, to process control, and after that to express them formally by use of colored Petri net fragments.

3.2.2. *Turing machine for process control (TMP)*. To adapt a Turing machine as a model of a process-control computer, we leave the coupling of the FSM to the memory unchanged but add input and output buffers to the FSM. As for the FSM, at whatever level of detail of description one chooses, the control structure of a program (with its "if-then" statements) can be viewed as an FSM consisting of (classical) states drawn as circles, connected by directed arcs, with each arc labeled by an input I that selects it and by an output O [12]; a fragment of such a picture is shown in figure 2(a). An FSM serves as a game board on which a single token can be placed to mark the "current state." Heading toward the hooking together of FSM's to make a Petri net, we suppose that each arc in the FSM is punctuated by a tick event and a tock event, drawn as small boxes, enlarging the FSM into a special case of a condition-event Petri net fragment, as shown in figure 2(b). Once colors are introduced, states shown as dashed circles pointing into an event of the FSM from outside will become the means to express the entrance of guesses. These states are assumed to receive tokens put into them by scientists and instruments undescribed by events of the net. Similarly, dashed states pointed to by arcs from an event are assumed to have tokens taken from them by agents undescribed by events of the net. Figure 2(c) streamlines the picture to the form we shall use, in which more or less vertical arcs are understood to point downward, the dashed states are left undrawn, as are all states with one input and one output event. To emphasize the input and output arcs with their extra tokens, we often call this an FSM fragment to distinguish it from the FSM form of figure 2(a).

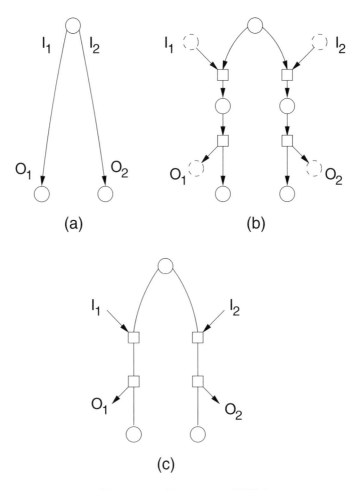

FIGURE 2. Fragment of FSM.

To define a Turing machine for Process-control (TMP), we adapt the FSM of a Turing machine to have for each of its states a cartesian product of states of a set of clocked internal registers and, in addition, input buffers and output buffers, which allow input/output transactions with a scientist, with laboratory instruments, and with other TMP's.

3.2.3. *Colored tokens.* By replacing the black tokens of an FSM fragment by a colored tokens and adjoining to each event a function that defines colors on output tokens in terms of colors on input tokens, any FSM fragment can be mapped one-to-one to the drastic form of figure 3, in which color changes substitute for most of the moves of black tokens on a bigger net. A "fork in the road" for black tokens, turns into a choice between red and green, so to speak, so the descriptive burden is taken up by the functions f_{tick} and f_{tock}; f_{tick} defines the color of a token placed on an internal state depending on a list of colors, one for each input, while f_{tock} defines a list of output colors depending on the color of the token on the internal state. The vertical arc is to be read as directed downward, and the big circles at the top and bottom of a path signify that the path is wrapped around a cylinder,

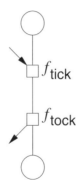

FIGURE 3. FSM with detail pushed to coloring.

so the top is a continuation of the bottom, *i.e.* a loop. An FSM fragment in which the token carries a color will be called a colored FSM.

3.2.4. *Other mappings.* Less drastic mappings are also possible. Any two states of a single FSM can be merged without breaking any arcs by augmenting the color rules in the events that feed them and the events fed by them. If a set of states connected to one another by events is mapped into a single state, the single state then connects to an event that loops back to it; this results in a place-transition Petri net, but not a condition-event net. We restrict the mappings dealt with here to ones that avoid pasting tick and tock events together, thereby avoiding self loops. Two events of an FSM that link the same pair of states can be merged by distinguishing external inputs and outputs by color instead of by place.

The mappings discussed so far are net isomorphisms: they map markings of one net bijectively to markings of the other and preserve the one-step reachability of one marking from another (by the firing of an event). Inverses of these bijections take more richly to less richly colored nets. Going in this direction depends on each state of a colored FSM having a set of possible colors associated with it [9]; then any colored transition corresponds one-to-one to a set of transitions obtained by partitioning sets of colors of input states, as illustrated in figure 4 for a two-in, two-out transition with color sets A, B, C, and D, each partitioned into "+" and "−" subsets. For this to make sense, it must be that an event which has tokens in all its inputs cannot fire unless the colors of the tokens comprise an element of the domain of its color function; we assume this firing rule.

One gets a coarser description by use of a surjective map that is not an isomorphism by dropping the color distinction and dropping the color functions from the transitions; this coarsening, however, preserves a one-to-one correspondence between the number of firings in one net and the number in another. All these maps are continuous in the net topology [14], and, as emphasized by Petri [15], nets form a category in which the morphisms are continuous maps, an idea that extends to nets with colored tokens [9].

3.2.5. *Disciplined coarsening of time.* Some other kinds of continuous coarsening maps bundle up multiple event firings into a single firing; as when one describes *e.g.* "running a program" as a single event. This brings us to the first of several areas open to future work, for, more than other computing, process control benefits from well defined timing, and in particular from machine and software design that

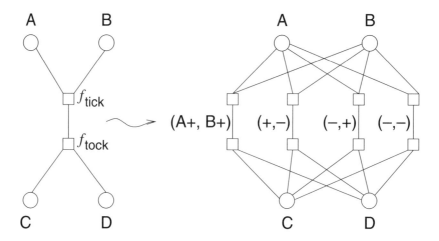

FIGURE 4. From color detail to net detail.

allows systematic, well controlled mappings that take a certain number of firings in an FSM to a single firing, so that one can think at a coarser level while still maintaining discipline in timing.

A striking example of the need to design programs that run in the same time for all inputs from some set I occurs in quantum computing. For example, suppose that $\mathbf{U}$ is the universal unitary operator defined by Deutsch to operate on basis states of the form $|s;\mathbf{n};\mathbf{m}\rangle$ where s is the location of the scanned square, $\mathbf{n}$ is the state of the FSM-processor ($\mathbf{n}=0$ is the starting state and $\mathbf{n}=1$ the halt state) and $\mathbf{m}$ is the tape [16]. For this to work in a computation that takes advantage of quantum superposition, one needs $\exists r[(\forall x \in I)\mathbf{U}^r|0;0;x,0\rangle = |0;1;x,f(x)\rangle]$; however, this is by no means implied for a program π_f for which (as is usual in borrowed classical programs) one can assure only $(\forall x \in I)(\exists r(x))[\mathbf{U}^{r(x)}|0;0;x,0\rangle = |0;1;x,f(x)\rangle]$ [17]. An interesting topic for future study is the complexity of converting various classes of programs with variable running time to programs running in a time independent of the input for some set of inputs.

3.2.6. *Cartoon of UTMP.* Ignoring the laboratory instruments for the moment, by connecting input- and output-signals from a suitable FSM to a scientist and coupling the FSM to an unlimited memory, one gets a Universal Turing Machine (UTM) that provides for continual communication with a scientist, as shown in figure 5(a), in which boxes connected by a horizontal line are read as a single event. We cartoon the UTM in the condensed form of figure 5(b). By adding input- and output-signals from the FSM to laboratory instruments and to other FSM's, one gets a Universal Turing Machine adapted for Process control (UTMP), as shown in figure 5(c); again almost all of the burden of description is in the color functions, here called T_1 and T_2 (for Turing) that define a finite state machine that operates a UTMP. We assume that at some level of description, the ticks and tocks of the UTMP slice time into moments not only for the UTMP but also for the scientist at a keyboard and the instruments on the laboratory bench; we assume that input tokens from the scientist and from the instruments arrive at the UTMP synchronized with the UTMP pendulum. If the scientist enters nothing at a given clock tick, then the token taken by the UTMP from the input buffer for the scientist

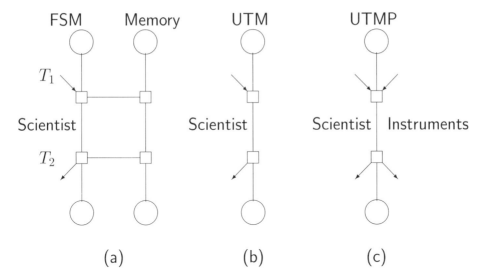

FIGURE 5. From FSM to UTMP.

carries the color "empty," and if the instruments enter nothing, the token from the input buffer for the instruments carries the color "empty"; similarly the UTMP marks output tokens with the color "empty" if it writes nothing else on them.

3.2.7. *A scientist controls a UTMP.* To see the structure imposed on physics by the UTMP, one must think as if the UTMP were delivered to a scientist in a bare condition: no installed software,[6] the FSM in a starting state, and the memory all blank. We assume that the function T_1 operating on empty input tokens, the starting state of the FSM, and a blank memory produces empty output tokens and makes no change in the FSM state or the memory or the memory location scanned. Finally, we invoke the universality of a UTM to assume that the functions T_1 and T_2 are fixed (by a manufacturer, so to speak) independent of whatever laboratory instruments need to be considered and independent of all action by the scientist. These assumptions imply

PROPOSITION 3.1. *Whatever a UTMP does besides staying in its starting state and taking in and putting out empty tokens is in response to input tokens.*

We invoke this proposition to view the scientist as precluded from defending questionable management of equations or instruments by saying "the computer did it." If a CPC does something, it executes a program; we view the scientist as responsible for any program entered (as a colored token) into the UTMP and for running the program on any particular occasion.[7]

3.2.8. *Reprogramming always an option.* We assume the UTMP is isomorphic to the net shown in figure 6, so that the scientist has a recurring choice of letting the UTMP run as programmed or of interrupting it to reprogram it. By programming a UTMP, a scientist can simulate an arbitrary special Turing machine. At will, the scientist can interrupt a program in execution to change to a program that

---

[6]The scientist can borrow software and install it, but is responsible for it.

[7]This rules out taking for granted the operating system, instrument-managing programs, a simulator, and whatever other programs come pre-installed in a commercially available CPC.

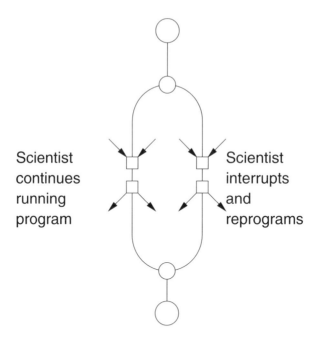

FIGURE 6. Alternative modes controlled by scientist.

simulates a different special Turing machine, corresponding to a different FSM and a different net. One can glimpse this in figure 4, where it is apparent that if the colors are limited to the sets A^+ and B^+, then six of the eight events are precluded from firing, and the net is in effect reduced to the fragment defined by the selected colors. In this way the part of the net that actually fires, corresponding to the event "Use existing program" of figure 6, is variable in how it acts and in the net by which one portrays it in more detail, according to the scientist's actions in providing and running programs.

3.2.9. *Plug and play*. To see how UTMP's can be connected (as well as the detail of how the FSM of a TM or TMP is connected to the memory), we introduce a signal that is phased just opposite to an FSM: the signal takes an input at a tock event and issues an output at a tick event. Then FSM A can send a signal (which can convey a message as a token color) to B (which can be either another FSM or a memory), as shown in figure 7, provided the signal path is short enough compared to the clock rate of the machines. This use of a signal synchronizes A with B. For two-way communication, one adds a signal going the other way. If communication over a distance long compared to the clock period is called for, then a chain of communication over intermediating UTMP's, is necessary, with the result that more firings of an event of A are required before a consequence of one firing can propagate to B and return as a property of a color on a token at a later firing of the A-event. The use of colored tokens sets up an area for future investigation of replacing the awkward definition of *synchronic distance* [**18**] with a measure of synchronization that counts firings in circuits of color effects, without having to add artificial elements to a net.

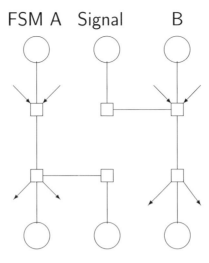

FIGURE 7. Signal from A to B.

3.3. Net fragments formalized. Portraying logic operating in CPC's calls for fragments of Petri nets, not complete nets, to allow for guesses as token colors definable neither by results of experiments nor by calculations. From among the standard definitions of a Petri net, the one we use is (S, E, F), where S is a set of states, E is a set of events, and $F \subseteq S \times E \cup E \times S$ is the flow relation. In order to make room for guesses from a scientist and results of instruments inexpressible in the logic defined by a net but essential to setting it up, the nets used are all *net fragments*, which we define as follows. A net fragment is a structure (S, S_I, S_O, E, F) where S is a set of states of CPC's, and S_I is a set of states of input signals (*e.g.* from A/D converters to a CPC input buffer), disjoint from S, allowing for input to the CPC from a scientist and laboratory instruments. S_O is a set of states of output signals disjoint from both S and S_I, allowing for output from the CPC; the flow relation is expanded so $F \subset [(S \cup S_I) \times E] \cup [E \times (S \cup S_O)]$. States of S_I are assumed to have tokens placed in them by some means beyond the net, and states of S_O are assumed to have tokens removed from them by means beyond the net. Our pictures show stubs of arcs from states of S_I to events and from events to states of S_O while omitting the circles for these states. Associated with a net fragment is a "reduced net" obtained by omitting the states of S_I and S_O (and dropping the arc stubs); using this reduced net, one can explore issues of liveness and safety [**19**]. The events of E express computer logic and nothing else. As an example of a guess used in designing contact between equations and instruments, a mathematical model entered by a scientist as a colored token in an S_I state can assert whatever rules the scientist chooses to relate tokens received from instruments in S_I states to commands sent to them as colored tokens in S_O states. In this way the net fragment expresses the difference between such a model, with its guesswork, as a color on a token and how the instruments actually behave by producing colored tokens on their own.

4. Net-based portrait of guesswork needed to test and calibrate a quantum computer

In section 2 choices of equations to link to instruments were shown inescapably open to guesswork, bidding to make guesswork part and parcel of physics. The availability of net fragments described in section 3 brings within physics the study of contacts between equations and instruments by making available to analysis relations of sequence, concurrency and choice expressed in these contacts and in the guess-dependent actions that set the contacts up. Here we turn from nets themselves to attention to an example problem in which a net illustrates an important structure needed to link equations to instruments. Besides the net explicitly shown in figure 8, the availability of nets provides a framework in which to view the main topic of this section, the problem of resolving a choice of commands by which a CPC manages a quantum computer. That framework can be used in the future to ask other questions, to do with: how do the necessities of quantum-mechanical models, classical process control, and guesswork interact; how are FSM's as program structures affected by use of models that are quantum mechanical; how does the need for CPC's to mediate between quantum-mechanical equations and instruments change our understanding of quantum mechanics?

Turning to the case at hand, some telling illustrations of guesswork needed to link models to instruments arise in quantum computing. To build a quantum computer, say to solve problems of factorizing [20] and searching [21], a scientist must choose quantum-mechanical equations and laboratory instruments to work in harmony. Quantum computational models call for quantum gates that are unitary transformations, each a tensor product of an operator on a 1-bit or 2-bit subspace of the Hilbert space $\mathcal{H}$ and identity operators for the other factors of the tensor product. Note that each permutation of a non-identity factor with an identity factor is a distinct gate, calling for a distinct command to the instruments that implement it. For this reason, the number of quantum gates for an n-bit quantum computer grows faster than n. Call this number $G(n)$ and let the set of gates be $U_1, \ldots, U_G$. The most commonly used models of quantum computers can be put in the form [22]:

- Prepare a starting state independent of the input (*e.g.* the integer to be factorized).
- Transform the state by a product of quantum gates that depends on the input.
- Make a measurement independent of the input.

For an example, suppose the scientist assumes properties of models 1 through 4 and looks for the model that gives the least mean-square deviation between relative frequencies of outcomes and probabilities calculated by (2.13). To factorize an integer I, a classical computer program is converted to a product of $K(I)$ quantum gates, a number that rises faster than linearly with $\log I$. To obtain the effect of multiplying the gate transformations, the scientist must first have solved the model to determine the command $b_{U,j}$ for each gate U_j occurring in the product. As in the portrait in section 3 of putting tokens into a net fragment, the scientist programs a CPC to transmit a command b_v to prepare an initial state $|v\rangle$, commands $b_{U,j}$ for the gates needed, and a command b_M for a measurement. This endeavor is known to exhibit the following four features:

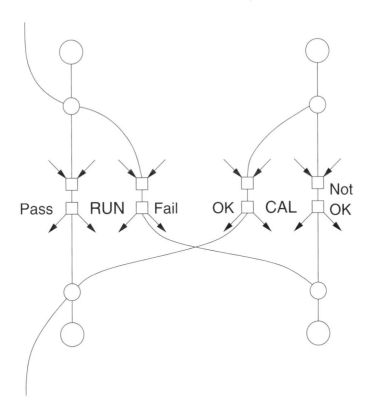

FIGURE 8. Alternating between running and testing a QC.

(1) The instruments are valuable as a quantum computer insofar as their results substitute for a more costly classical calculation defined by the model.
(2) An inexpensive classical computation (*e.g.* with the CPC) tests whether outcomes interpreted from results correctly solve the problem.
(3) Quantum indeterminacy imposes a positive probability that a result fails to provide a correct answer, so multiple tries with the instruments are the rule, and a wrong answer does not by itself imply a fault in the instruments.
(4) The tolerable imprecision of instruments implementing the chosen model of a quantum gate diminishes as the inverse of the number of gates $K(I)$ in the sequence [23].

Because the number of gates required in the product rises with the size of the integer to be factorized, feature 4 implies that passing the test for smaller integers is no guarantee against failure of the instruments to factorize larger integers, unless the model or the instruments or both are refined. This requires, in turn, that a CPC intended for use on progressively larger integers be organized to switch between a mode of using the quantum computer and a mode of inquiring into its performance, *e.g.* so as to determine commands that make it behave more precisely in accord with the desired quantum gates. This calls for a program for the CPC that expands the events "Use existing program" of figure 6 to that of figure 8.

4.1. Navigating the lattice of models to get better commands.

As an example of what goes on within the coarsely portrayed event "Calibrate," suppose a scientist who uses a model α of the form $(|v\rangle, U, M)_B$ finds it works for small integers, but fails for bigger ones, which require more precise gates, which in turn requires calibrating (*i.e.* adjusting) the commands used to generate gates. This means giving up model α and choosing some alternative model β. A scientist does not choose a model all at once, but starts with some set of models and then narrows down on a smaller set, sometimes to a single model, a process open to guesswork at various stages. At one stage, the scientist may need to relax a constraint on models, leading to a bigger set of models from which to choose; at another stage the scientist may guess a new constraint, narrowing the set of models under consideration. By such a back and forth procedure, the scientist gives up U_α and arrives at a new function U_β (and hence a new model) with the hope that solving this function for a command $b_{U,j,\beta}$ for gates U_j, $j = 1, 2 \ldots$, that will succeed for factorizing larger integers than did the commands obtained from U_α. (This makes a need for models adapted to homing in on results, with some metric on B, so that a small change in the command b_U results in a small change in *e.g.* $U(b_U)$; while properties 3 and 4 are a start, going beyond them is left to the future.)

To get a better model, the scientist guesses a set of models and hopes to find within it a model that better fits measured results interpreted as outcomes. If no model of the set adequately fits these outcomes, the scientist can first broaden the set of models and next try to guess a property that will narrow the set, not to the original model, but to one that fits better. The recognition of guesswork assures us that so long as progressively more ambitious goals of precision keep being introduced, there is no end to the need for adjusting both models and the laboratory instruments.

4.2. Sample sizes needed to choose between models of gates.

As discussed in section 2.5, the number of trials needed to statistically distinguish one model from another is bounded from below by the inverse square of a weighted statistical distance between the two models. Small numbers of experimental results can sometimes decide between distant models, but never between models that are close. In particular, distinguishing experimentally between two models for quantum gates can demand large samples:

PROPOSITION 4.1. *Models α and β that differ only in U, with spectral norm $\| U_\alpha(b_U) - U_\beta(b_U) \| = \epsilon > 0$, are statistically indistinguishable for a command b unless*

$$(4.1) \qquad N(b) \geq \epsilon^{-2}.$$

PROOF. The models α and β under the stated condition are unitarily equivalent to a pair of models that differ only in $|v\rangle$ with $\cos|\langle v_\alpha | v_\beta \rangle| \leq \epsilon$. The proposition then follows from (2.10) and (2.11). □

We argue elsewhere that this is a serious and heretofore unappreciated challenge to bringing instruments into working order as quantum computers, made visible by attention to the need for guesswork in linking of laboratory instruments to equations of quantum mechanics [24].

5. Concluding remarks

Gödel proved that no one true structure could be generated by sitting in a room with blinds drawn, writing down axioms. Quantum mechanics tells us that with the blinds up and the world of physical measurement available, the situation remains much the same. Just as the openings for new axioms are uncloseable in mathematical logic, so in physics guesswork is part of the foundation.

The net formalism can be put to use both to address improving the contacts between equations and instruments, fostering advances in theory and in instrumentation, and, at a more abstract level, to pose problems pertaining to universal Turing machines adapted to process control. By formalizing commands to instruments, the techniques presented here extend the reach of set-based mathematics into the area of contact between equations and instruments, and open to study within physics of some of what physicist do in the course of doing physics. This extends a parallel beachhead established already in mathematics by Gödel's study of what a mathematician does to prove a theorem and Turing's analysis of a mathematician who makes a note by which to resume an interrupted computation.

6. Acknowledgment

We acknowledge Amr Fahmy for showing us our debt to Gödel's proof of incompleteness in mathematical logic. We are indebted to Steffen Glaser, Raimund Marx, and Wolfgang Bermel for introducing us to the subtleties of laboratory work aimed at nuclear-magnetic-resonance quantum computers [25]. To David Mumford we owe our introduction to quantum computing from the standpoint of pure mathematics. We are greatly indebted to Anatol W. Holt and to C. A. Petri for conversations years ago, in which each pointed in his own way to the still mysterious expressive potential of nets.

References

[1] F. H. Madjid and John M. Myers, *Formal distinction between quantum states and outcomes of their measurement*, Meas. Sci. Technol. **8** (1997), 465–472.
[2] P. A. M. Dirac, *The principles of quantum mechanics*, 4-th ed., Oxford University Press, Oxford, 1958.
[3] E. G. Beltrametti and G. Cassinelli, *The Logic of Quantum Mechanics*, Addison-Wesley, Reading, MA, 1981.
[4] W. K. Wooters, *Statistical distance and Hilbert space*, Phys. Rev. D **23** (1981), 357–362.
[5] C. H. Papadimitriou, *Computational complexity*, Addison-Wesley, Reading, MA, 1994.
[6] P. A. Benioff, *Decision procedures in quantum mechanics*, J. Math. Phys. **13** (1972), 908–915.
[7] F. H. Madjid and J. M. Myers, *Linkages between the calculable and the incalculable in quantum theory*, Annals of Physics **221** (1993), 258–305.
[8] K. Gödel, *Über formal unentscheidbare Sätze der Principia mathematica und verwandter Systems I*, Monatshefte für Mathematik und Physik **38** (1931), 173–198.
[9] K. Jensen, *Coloured Petri nets: Basic concepts, analysis methods and practical use*, Monographs in Theoretical Computer Science, an EATCS Series, Springer-Verlag, Berlin, Vol. 1, 2nd ed., 1996, Vol. 2, 1995.
[10] A. W. Crosby, *The measure of reality: Quantification and western society, 1250-1600*, Cambridge University Press, Cambridge, 1997.
[11] A. M. Turing, *On computable numbers, with an application to the Entscheidungsproblem*, Proc. London Math. Soc. **42** (1937), 230–265.
[12] R. P. Feynman, *Feynman lectures on computation*, Addison-Wesley, Reading, MA, 1996.
[13] G. S. Boolos and R. C. Jeffrey, *Computability and logic*, 3rd ed., Cambridge University Press, Cambridge, 1989.

[14] H. J. Genrich, K. Lautenbach, and P. S. Thiagarajan, *Elements of general net theory*, in W. Brauer, ed., *Net theory and applications*, Lecture Notes in Computer Science, Springer-Verlag, Berlin, **254** (1987), 338–358.

[15] C. A. Petri, *General net theory*, in B. Shaw, ed., *Computing System Design*, Proceedings of the Joint IBM University of Newcastle upon Tyne Seminar, University of Newcastle upon Tyne, 7–10 September, 1976.

[16] D. Deutsch, *Quantum theory, the Church–Turing principle and the universal quantum computer*, Proc. R. Soc. Lond. A **400** (1985), 97–117.

[17] J. M. Myers, *Can a universal quantum computer be fully quantum?*, Phys. Rev. Letters **78** (1997), 1823–1824.

[18] U. Goltz, *Synchronic distance*, in W. Brauer, W. Reisig, and G. Rozenburg, eds., *Petri nets: central models and their properties*, Lecture Notes in Computer Science, Springer-Verlag, Berlin, **254** (1987), 338–358.

[19] G. Berthelot, *Transformations and decompositions of nets*, in W. Brauer, W. Reisig, and G. Rozenburg, eds., *Petri nets: central models and their properties*, Lecture Notes in Computer Science, Springer-Verlag, Berlin, **84** (1980), 359–376.

[20] P. Shor, *Algorithms for quantum computation: Discrete logarithms and factoring*, Proc. of the 35th annual symposium on foundations of computer science, IEEE Computer Society, Los Alamitos, CA, 1994, pp. 124–134.

[21] L. K. Grover, *A fast quantum mechanical algorithm for database search*, Proc. 28th annual ACM symposium on the theory of computing, ACM Press, New York, NY, 1996, pp. 212–219.

[22] D. Deutsch, *Quantum computational networks*, Proc. R. Soc. Lond. A **425** (1989), 73–90.

[23] E. Bernstein and U. Vazirani, *Quantum complexity theory*, SIAM Journal on Computing **26** (1997), 1411–1473.

[24] J. M. Myers and F. H. Madjid, "Contact between laboratory instruments and equations of quantum mechanics," to be published in E. Donker and A. R. Pirich, eds, *Quantum Computing*, Proc. SPIE, **4047**, 2000.

[25] R. Marx, A. F. Fahmy, J. M. Myers, W. Bermel, and S. J. Glaser, *Approaching five-bit NMR quantum computing*, Phys. Rev. A., **62** (2000), 012310.

(John M. Myers) GORDON MCKAY LABORATORY, HARVARD UNIVERSITY, CAMBRIDGE, MA 02138, USA

E-mail address, John M. Myers: myers@deas.harvard.edu

(F. Hadi Madjid) 82 POWERS ROAD, CONCORD, MA 01742, USA

Quantum Computation by Geometrical Means

Jiannis Pachos

ABSTRACT. A geometrical approach to quantum computation is presented, where a non-abelian connection is introduced in order to rewrite the evolution operator of an energy degenerate system as a holonomic unitary. For a simple geometrical model we present an explicit construction of a universal set of gates, represented by holonomies acting on degenerate states.

1. Prologue

Abelian [Sha] and non-abelian [Wil] geometrical phases in quantum theory have been considered as a deep and fascinating subject. They provide a natural connection between the evolution of a physical system with degenerate structure and differential geometry. Here we shall present a model where these concepts can be explicitly applied for quantum computation [Zan].

The physical setup consists of an energy degenerate quantum system on which we perform an adiabatic isospectral evolution described by closed paths in the parametric space of external variables. The corresponding evolution operators acting on the code-state in the degenerate eigenspace are given in terms of holonomies and we can use them as quantum logical gates. This is a generalization of the Berry phase or geometrical phase, to the non-abelian case, where a non-abelian adiabatic connection, A, is produced from the geometrical structure of the degenerate spaces. In particular, on each point of the manifold of the external parameters there is a code-state attached and a transformation between these bundles of codes is dictated by the connection A.

In order to apply this theoretical construction to a concrete example we employ a model with $\mathbf{CP}^2$ geometry, that is a complex projective manifold with two complex coordinates. This is interpreted as a qubit [Pac]. A further generalization with the tensor product of m $\mathbf{CP}^2$ models and additional interaction terms parametrized by the Grassmannian manifold, $\mathbf{G}(4,2)$, is interpreted as a model of quantum computer.

The initial code-state is written on the degenerate eigenspace of the system. The geometrical evolution operator is a unitary acting on it and it is interpreted as

---

2000 *Mathematics Subject Classification.* Primary 81P68.

The author was supported in part by TMR Network under the contract no. ERBFMRXCT96 - 0087.

a logical gate. Due to adiabaticity the geometrical part of the evolution operator has a dimensionality equal to the degree of degeneracy of the eigenspace. Specific logical gates given by holonomies are constructed for a system with a tensor product structure resulting in universality while at the end a quantum optical application is sketched.

2. Coset Space Geometry

A transformation $U(n)$ between the states $|\alpha\rangle$, $\alpha = 1, ..., n$ can be realized by all possible sub-$U(2)$ transformations between any two of those states, $|i\rangle$ and $|j\rangle$. A coset space can be produced as the factor with respect to some particular $U(2)$ symmetries of these transformations.

Examples of such constructions are given in the following:

- the $\mathbf{CP}^2$ projective space:

$$\mathbf{CP}^2 \cong \frac{U(3)}{U(2) \times U(1)}$$

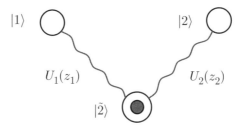

The lines denote $U(2)$ transformations between the states represented here by "holes". The $U(3)$ group could be interpreted by three lines connecting all the holes together. The distinction between "filled" and "unfilled" holes is due to the coset structure, which factors out the symmetry transformations, between $|1\rangle$ and $|2\rangle$, and denotes explicitly the non-symmetric ones between $|1\rangle$ or $|2\rangle$ and $|\tilde{2}\rangle$.

- the $\left(\mathbf{CP}^2\right)^{\times m} \times \left(\mathbf{G}(4,2)_{int}\right)^{\times (m-1)}$ product space:

$$\cdots \frac{U(3)}{U(2) \times U(1)} \times \frac{U(3)}{U(2) \times U(1)}, \left.\frac{U(4)}{U(2) \times U(2)}\right|_{int} \cdots$$

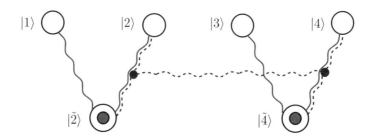

Transformations can be performed between the states $\{|1\rangle, |2\rangle\}$ and $\{|3\rangle, |4\rangle\}$ due to their connections with the states $\{|\tilde{2}\rangle, |\tilde{4}\rangle\}$, while an interaction in the tensor product space between $|24\rangle$ and $|\tilde{2}\tilde{4}\rangle$ gives transformations between those two sets. The transformations between only two states may be performed by linear operations with respect to $U(2)$ generators, while the combined transformations between the two qubits can be produced by bilinear generators which act simultaneously on the states of both of the $\mathbf{CP}^2$ models. The latter is denoted in the previous figure by the dashed lines, where the connection between the black dots indicates the simultanious action.

3. Degeneracy, Adiabaticity and Holonomies

Let us introduce the degenerate Hamiltonians H_0^1 and H_0^m as follows

$$H_0^1 = \begin{bmatrix} 0 & 0 & 0 \\ 0 & 0 & 0 \\ 0 & 0 & 1 \end{bmatrix}, \qquad H_0^m = \sum_m H_0^1 .$$

The orbit, that is the parametric manifold of the unitary transformations which preserves the degenerate spectrum of H_0^1 is given by $\mathbf{CP}^2$. A sub-manifold of the orbit of H_0^m in which we are interested in is given by the $\left(\mathbf{CP}^2\right)^{\times m} \times \left(\mathbf{G}(4,2)_{int}\right)^{\times (m-1)}$ product manifold.

A general transformation parametrized by the $\mathbf{CP}^2$ space is given by $\mathbf{U}(\mathbf{z}) := U_1(z_1)U_2(z_2)$, with $U_\alpha(z_\alpha) = \exp G_\alpha(z_\alpha) = \exp(z_\alpha|\alpha\rangle\langle\tilde{2}| - \bar{z}_\alpha|\tilde{2}\rangle\langle\alpha|)$. The complex parameter z_α may be decomposed as $z_\alpha = \theta_\alpha \exp i\phi_\alpha$. Due to the 2×2 sub-form of $G_\alpha(z_\alpha)$ we can rewrite $U_\alpha(z_\alpha)$ as

$$U_\alpha(z_\alpha) = \mathbf{1}_\alpha^\perp + \cos\theta_\alpha \mathbf{1}_\alpha + \frac{\sin\theta_\alpha}{\theta_\alpha} G_\alpha(z_\alpha) ,$$

where $\mathbf{1}_\alpha^\perp = \mathbf{1} - \mathbf{1}_\alpha$ and $\mathbf{1}_\alpha = |\alpha\rangle\langle\alpha| + |\tilde{2}\rangle\langle\tilde{2}|$. For the Grassmannian manifold $\mathbf{G}(4,2)$ we have, for example, the $U(2)$ rotation in the tensor product basis of two qubits, between the states $|24\rangle$ and $|\tilde{2}\tilde{4}\rangle$, given by $\mathbf{U}(z) = \exp(z|24\rangle\langle\tilde{2}\tilde{4}| - \bar{z}|\tilde{2}\tilde{4}\rangle\langle 24|)$, with $z = \theta \exp i\phi$. The coordinates $\{\lambda^a\} = \{\boldsymbol{\theta}, \boldsymbol{\phi}\}$ provide the parametric space which the experimenters control.

In the four dimensional manifold $\mathbf{CP}^2$ with coordinates $\{\lambda^a\}$ a closed path, C, is drawn on a two-submanifold. Consider this evolution to be adiabatic as well as isospectral which is provided by the formula $H(\lambda(t)) = \mathbf{U}(\lambda(t))H_0\mathbf{U}^\dagger(\lambda(t))$. As a result the state of the system, $|\psi(t)\rangle$, stays on the same eigenvalue of the Hamiltonian, taken in our example to be $E_0 = 0$, without level-crossing.

At the end of the loop C, spanned in time $T = N\Delta t$, when divided in N equal time intervals, we obtain

$$\begin{aligned}|\psi(T)\rangle &= \mathbf{T}e^{-i\int_0^T \mathbf{U}H_0\mathbf{U}^\dagger dt}\,|\psi(0)\rangle \\ &= \mathbf{T}\lim_{N\to\infty}\prod_{i=1}^N \mathbf{U}_i e^{-iH_0\Delta t}\mathbf{U}_i^\dagger\,|\psi(0)\rangle \\ &= \mathbf{P}\lim_{N\to\infty}\left(1 + \sum_{i=1}^N A_i\Delta\lambda_i\right)|\psi(0)\rangle\end{aligned}$$

with $\qquad A_i = \mathbf{U}_i^\dagger \dfrac{\Delta \mathbf{U}_i}{\Delta\lambda_i} \quad \text{and} \quad \mathbf{U}_i = \mathbf{U}(\lambda(t_i)) .$

Hence, the state $|\psi(0)\rangle$ acquires a geometrical unitary operator given by the holonomy of a connection A as

$$\Gamma_A(C) := \mathbf{P}\exp\oint_C A , \quad \text{where} \quad (A^{\lambda^a})_{\alpha\beta} := \langle\alpha|\mathbf{U}^\dagger(\lambda)\frac{\partial}{\partial\lambda^a}\mathbf{U}(\lambda)|\beta\rangle .$$

The states $|\alpha\rangle$ and $|\beta\rangle$ belong to the same degenerate eigenspace of H_0 and λ^a's are the real control parameters.

The produced unitary operator is an effect of the non-commutativity of the control transformations which produce effectively a curvature. In the case of the Berry phase produced for example in front of the spin states of an electron when placed in a magnetic field, the non-commutativity is between the $U(2)$ control unitaries which change the direction of the magnetic field in the three dimensional space. What is presented here is the generalization of the Berry phase to the non-abelian case.

The $\Gamma_A(C)$'s produced by $\mathbf{CP}^2$ for various loops C, generate the whole $U(2)$,

$$\{\Gamma_A(C); \ \forall \ C \in \mathbf{CP}^2\} \approx U(2) \ .$$

In the case of m qubits with their proper interactions, the produced group is $U(2^m)$,

$$\{\Gamma_A(C); \ \forall \ C \in \left(\mathbf{CP}^2\right)^{\times m} \times (\mathbf{G}(4,2)_{int})^{\times (m-1)}\} \approx U(2^m) \ .$$

4. Quantum Computation

In order to perform quantum computation by using the above constructions we consider the following identifications:

$$\text{QUANTUM CODE} \equiv \text{Degenerate States}, \ |\psi(0)\rangle$$

$$\text{LOGICAL GATES} \equiv \text{Holonomies}, \ \Gamma_A(C)$$

Let us first investigate the $\mathbf{CP}^2$ case. The basic question is how we can generate a general $U(2)$ element by moving along a closed path, C. Or in other words, for a specific $U \in U(2)$ which loop C is such that $\Gamma_A(C) = U$. In general, $\forall \ \mathbf{g} \in u(2)$ $\exists$ loop $C \in \mathbf{CP}^2$ manifold, such that $\Gamma_A(C) = \exp \mathbf{g}$, which is the statement of irreducibility of the connection A [**Zan**]. To answer the above question we perform the following analysis. The loop integral

$$\oint_C A = \oint_C A_{\lambda^1} d\lambda^1 + A_{\lambda^2} d\lambda^2 + \cdots$$

is the main ingredient of the holonomy. Due to the path ordering symbol it is not possible to just calculate it and evaluate its exponential, as in general the connection components do not commute with each other. Still it is possible to consider the following restrictions in the position of the loop. Choose C such that:

- it belongs to *one* plane $(\lambda^i, \lambda^j) = (\theta_i, \phi_j)$ or (θ_i, θ_j), hence only two components of A are involved,
- the position of the plane is such that the connection, A, restricted on it become, $A|_{(\lambda^i, \lambda^j)} = (A^{\lambda^i} = 0, A^{\lambda^j})$, that is these two components commute with each other. Still it is important that their related field strength component, $F_{ij} = \partial_i A_j - \partial_j A_i + [A_i, A_j]$, is non-vanishing in order to obtain a non-trivial holonomy. Such a requirement is possible for the $\mathbf{CP}^2$ model and for a wide class of other models.

On the planes where those conditions are satisfied the evaluation of the holonomy is trivially given by just exponentiating the loop integral of the connection without worrying about the path ordering symbol. Hence,

$$\Gamma_A(C) = \mathbf{P} \exp \oint_C A = \exp(\Sigma \mathbf{g}) = \mathbf{1}_{2\times 2} \cos \Sigma + \mathbf{g} \sin \Sigma ,$$

where Σ represents the area enclosed by the loop C projected on the sphere associated with the compactified $\mathbf{CP}^2$ manifold. This area may be varied desirably. Furthermore, we are able to obtain a complete set of generators $\mathbf{g}$ by choosing C to lie on different planes. In detail we may obtain for $\mathbf{g}$ the following forms

$$\begin{aligned} -i|\alpha\rangle\langle\alpha| &:= -i\sigma_\alpha^3 , \; \alpha = 1, \, 2 \\ -i(-i|1\rangle\langle 2| + i|2\rangle\langle 1|) &:= -i\sigma^2 , \\ -i(|1\rangle\langle 2| + |2\rangle\langle 1|) &:= -i\sigma^1 . \end{aligned}$$

The σ_α^3 generators are similar to a Berry phase and they are produced by paths C_1 on the $(\theta_\alpha, \phi_\alpha)$ planes. The corresponding holonomy is the exponential of this generator multiplied by the area, Σ_1, of the surface the path C_1 encloses when projected on a sphere $S^2(2\theta_\alpha, \phi_\alpha)$ with spherical coordinates $2\theta_\alpha$ and ϕ_α,

$$-i\sigma_\alpha^3 : C_1 \in (\theta_\alpha, \phi_\alpha) \to \Gamma_A(C_1) = \exp -i\Sigma_1 \sigma_\alpha^3 .$$

The σ^2 generator is produced by a path C_2 along the plane (θ_1, θ_2) positioned at $\phi_1 = \phi_2 = 0$, while σ^1 is produced by a path, C_3, along a parallel plane positioned at $\phi_1 = \frac{\pi}{2}$ and $\phi_2 = 0$. Their corresponding areas are Σ_2 and Σ_3. For example

$$-i\sigma^2 : C_2 \in (\theta_1, \theta_2)|_{\phi_1=0, \phi_2=0} \to \Gamma_A(C_2) = \exp -i\Sigma_2 \sigma^2 .$$

Altogether we have 2^2 independent generators spanning the Lie algebra of $U(2)$.

For the case of the two qubit interaction the corresponding connection components are given by

$$A_\theta = \text{diag}(0, 0, 0, 0) , \; A_\phi = \text{diag}(0, 0, 0, -i \sin^2 \theta)$$

which are written in the basis $\{|13\rangle, |14\rangle, |23\rangle, |24\rangle\}$. A loop C on the (θ, ϕ) plane will produce the following holonomy

$$\Gamma_A(C) = \text{diag}(1, 1, 1, e^{-i\Sigma}) , \; \Sigma = \int_{D(C)} d\theta d\phi \sin 2\theta ,$$

where Σ can also be interpreted as an area on the sphere $S^2(2\theta, \phi)$.

5. One and Two Qubit Logical Gates

By performing appropriate loops we can obtain one qubit phase rotations as well as two qubit gates such as a controlled phase rotation U_{CPH}.

Analytically, by spanning the indicated areas we may obtain

$$U_1 = \exp \begin{bmatrix} -i\Sigma_1 & 0 \\ 0 & 0 \end{bmatrix} , \; U_2 = \exp \begin{bmatrix} 0 & 0 \\ 0 & i\Sigma_1 \end{bmatrix} , \; U_3 = \exp \begin{bmatrix} 0 & -\Sigma_2 \\ \Sigma_2 & 0 \end{bmatrix} .$$

The combinations

$$U_1 U_2 = \exp(-i\sigma_3 \Sigma_1) , \; U_3 = \exp(-i\sigma_2 \Sigma_2)$$

can give any $U(2)$ transformation and hence any one qubit rotation.

For the two qubit gates we can construct easily the controlled rotation $U_{CPH} = \text{diag}(1, 1, 1, \exp -i\Sigma)$ between any pair of qubits. It is generated by a loop C on the (θ, ϕ) plane. Together with the one qubit rotations they provide a universal set of gates.

6. Epilogue

Apart from the intriguing theoretical formulation of holonomic computation there are several aspects of it, which have appealing technical advantages. Without overlooking the difficulties posed to an experimenter for performing continuous control over a system in order to span a loop, there are several unique characteristics of it, which await for exploitation. For example, robustness of the control procedure in terms of the spanned area, according to errors in the actual form of the performed loop, as well as the isolation of the degenerate states as a calculational space may prove to be advantages worth exploring.

In quantum optics displacing devices, squeezing devices and interferometers acting on laser beams can provide the control parameters for the holonomic computation. Each laser beam is placed in a non-linear Kerr medium with degenerate Hamiltonian $H_0 = n(n-1)$, where n is the photon numbering operator. The degenerate states $|0\rangle$ and $|1\rangle$ are the basis for encoding one qubit which is manipulated by displacing and squeezing devices. Any two qubit interactions can be implemented by interferometers [**Cho**].

It is challenging for the experimenters to produced the desired closed paths.

References

[Sha] *Geometric Phases in Physics*, A. Shapere and F. Wilczek, Eds. World Scientific (1989).
[Wil] F. Wilczek and A. Zee., Phys. Rev. Lett. **52**, 2111 (1984).
[Zan] P. Zanardi and M. Rasetti, Phys. Lett. A **264**, 94 (1999), quant-ph/9904011.
[Pac] J. Pachos, P. Zanardi and M. Rasetti, Phys. Rev. A **61** 010305(R), quant-ph/9907103.
[Cho] J. Pachos and S. Chountasis, quant-ph/9912093.

(Jiannis Pachos) INSTITUTE FOR SCIENTIFIC INTERCHANGE FOUNDATION, VILLA GUALINO, VIALE SETTIMIO SEVERO 65, I-10133 TORINO, ITALY
E-mail address, Jiannis Pachos: `pachos@isiosf.isi.it`

Pauli Exchange and Quantum Error Correction

Mary Beth Ruskai

ABSTRACT. In many physically realistic models of quantum computation, Pauli exchange interactions cause a special type of two-qubit errors, called exchange errors, to occur as a first order effect of couplings within the computer. We discuss the physical mechanisms behind exchange errors and codes designed to explicitly deal with them.

1. Introduction

Most discussions of quantum error correction assume, at least implicitly, that errors result from interactions with the environment[1] and that single qubit errors are much more likely than two qubit errors. Most discussions also ignore the Pauli exclusion principle and permutational symmetry of the states describing multi-qubit systems. Although this can be justified by consideration of the full wave function, including spatial as well as spin components, an analysis (given in [**27**]) of these more complete wave functions suggests that more attention should be given to the effect of exchange interactions within a quantum computer. Interactions between identical particles can cause an error in two bits simultaneously as a first order effect. Moreover, because they result from interactions within the quantum computer, exchange errors cannot be reduced by better isolating the quantum computer from its environment. The effect of a (single) exchange error is to flip two bits if, and only if, they are different. It is a non-classical type of error in the sense that it arises directly from a physical mechanism which occurs only in the case of identical particles which follow the rules of quantum theory. If classical systems were to exhibit this type of behavior, they would require unusual correlations which do not normally occur from first-order couplings.

Schemes for fault-tolerant computation [**6, 22, 28**] have been developed which treat two-bit, and even multi-bit, errors. (See, e.g., [**8, 9, 11, 29**] and references in [**7, 23**] and at the end of Chapter 10 of [**20**].) However, many of these, such as those

---

2000 *Mathematics Subject Classification.* Primary 81V70; Secondary 94B60, 81R99, 81Q05.

Supported in part by National Science Foundation Grant DMS-97-06981 and Army Research Office Grant DAAG55-98-1-0374.

[1]We consider the "environment" as external to the quantum computer in the sense that interactions among qubits, whether or not used in the implementation of quantum gates, are *not* regarded as arising from the environment. Many authors follow [**9**] in defining the environment to include all "unwanted interactions".

©2002 by author. Reproduction of this article, in its entirety, is permitted for non-commercial purposes.

arising from concatenated codes [1, 10, 23] require a large number of physical bits to represent one logical bit. Steane, in particular, has emphasized the problems associated with the size of repeatedly concatenated codes and discussed techniques [29] for coding m logical bits in n qubits. Furthermore, threshold estimates [1, 6, 12, 22, 28] are generally based on the assumption that two-bit errors are second order effects resulting from uncorrelated interactions with the environment. In those situations where exchange errors are important, shorter codes that explicitly address exchange errors can be effective.

A very different approach to fault tolerant computation is based on the assumption of highly correlated errors at low temperature, allowing the use of "'decoherence free subspaces" (DFS) [14, 15]. Shortly after [27] was posted, Lidar, et al observed [16] that the existing schemes for concatenating a DFS code with a standard 5-qubit code for correcting single-bit errors [15] could also correct exchange errors. This is because exchange errors on physical qubits appear as single Pauli errors on the logical bits used in DFS codes, i.e., it appears as if Pauli matrices act on the 4-qubit units which form DFS codes. Subsequently, they [17] turned this idea around to show how exchange interactions could be used to construct universal gates within the DFS scheme for quantum computation.

In section 2 we first review some of the basic principles underlying permutational symmetry of multi-particle quantum wave functions and then show how this leads to exchange errors. In Section 3 we discuss the issues associated with correction of exchange errors and present an explicit (non-additive) 9-bit code which can correct both exchange errors and all one-qubit errors. Section 3.4 contains an ambitious proposal for constructing powerful new codes using irreducible representations of the symmetric group.

2. The full wave function

2.1. Permutational symmetry. A (pure) state of a quantum mechanical particle with spin q corresponds to a one-dimensional subspace of the Hilbert space $= \mathbf{C}^{2q+1} \otimes L^2(\mathbf{R}^3)$ and is typically represented by a vector in that subspace. The state of a system of N such particles is then represented by a vector $\Psi(x_1, x_2, \ldots, x_N)$ in $\mathcal{H}^N$. However, when dealing with identical particles Ψ must also satisfy the Pauli principle, i.e., it must be symmetric or anti-symmetric under exchange of the coordinates $x_j \leftrightarrow x_k$ so that, e.g.,

$$(2.1) \qquad \Psi(x_2, x_1, \ldots, x_N) = \pm \Psi(x_1, x_2, \ldots, x_N).$$

depending on whether the particles in question are bosons (e.g. photons) or fermions (e.g., electrons). In either case, we can write the full wave function in the form

$$(2.2) \qquad \Psi(x_1, x_2, \ldots, x_N) = \sum_k \chi_k(s_1, s_2, \ldots, s_N) \Phi_k(\mathbf{r}_1, \mathbf{r}_2, \ldots, \mathbf{r}_N)$$

where the "space functions" Φ_k are elements of $L^2(\mathbf{R}^{3N})$, the "spin functions" χ_k are[2] in $[\mathbf{C}^{2q+1}]^N$ and $x_k = (\mathbf{r}_k, s_k)$ with $\mathbf{r}$ a vector in $\mathbf{R}^3$ and s_k (called the spin coordinate) an element of $\{0, 1, \ldots 2q\}$ corresponding to spin values going from $-\frac{1}{2}q$ to $+\frac{1}{2}q$ in integer steps. It is not necessary that χ and Φ each satisfy the Pauli principle; indeed, when $q = \frac{1}{2}$ so that $2q + 1 = 2$ and we are dealing with $\mathbf{C}^2$

---

[2] A spin state χ looks formally like a (possibly entangled) N-qubit state. However, unlike qubits which involve an implicit spatial component, we want only vectors in $[\mathbf{C}^{2q+1}]^N$ itself.

it is *not* possible for χ to be anti-symmetric when $N \geq 3$.[3] Instead, we expect that χ and Φ satisfy certain duality conditions which guarantee that Ψ has the correct permutational symmetry. In the case of anti-symmetric functions there is an extensive literature, e.g., [18, 19], about functions in which the χ_k and Φ_k are bases for irreducible representations of S_n with dual Young tableaux.

With this background, we now restrict attention to the important special case in which $q = \frac{1}{2}$ yielding two spin states labeled[4] so that $s = +\frac{1}{2}$ corresponds to $|0\rangle$ and $s = -\frac{1}{2}$ corresponds to $|1\rangle$, and the particles are electrons so that Ψ must be anti-symmetric.

To emphasize the distinction between a pure spin state as an element of $\mathbf{C}^2$ and a spin associated with a particular qubit or spatial wave function, we will replace $|0\rangle$ and $|1\rangle$ by $\uparrow$ and $\downarrow$ respectively. The notation $|01\rangle$ then describes a two-qubit state in which the particle in the first qubit has spin "up" ($\uparrow$) and that in the second has spin "down" ($\downarrow$). What does it mean for a particle to "be" in a qubit? A reasonable answer is that each qubit is identified by the spatial component of its wave function $f_A(\mathbf{r})$ where $A, B, C \ldots$ label the qubits and wave functions for different qubits are orthogonal. Thus, if the qubits did not correspond to identical particles we would have $|01\rangle = f_A(\mathbf{r_1}) \uparrow f_B(\mathbf{r_2}) \downarrow$. In the more realistic situation of identical particles

$$(2.3) \qquad |01\rangle = \frac{1}{\sqrt{2}} \big(f_A(\mathbf{r_1}) \uparrow f_B(\mathbf{r_2}) \downarrow \pm f_B(\mathbf{r_1}) \downarrow f_A(\mathbf{r_2}) \uparrow \big).$$

with the plus sign (+) for bosons and the minus sign (−) for fermions. We will henceforth consider the special case of electrons, which are fermions, in which case the antisymmetric function given by (2.3) is called a Slater determinant. Note that a function of the form (2.3) has the important property that the electron whose spatial function is f_A always has spin "up" regardless of whether its coordinates are labeled by 1 or 2. Although (2.3) is not a simple product, but a special type of superposition which is the anti-symmetrization of a product, it behaves in some ways like a product state. It should be contrasted with the a true entangled Bell state such as

$$(2.4) \qquad \frac{1}{\sqrt{2}}\big[|01\rangle - |10\rangle\big] = \tfrac{1}{2}\big(f_A(\mathbf{r_1}) \uparrow f_B(\mathbf{r_2}) \downarrow - f_B(\mathbf{r_1}) \downarrow f_A(\mathbf{r_2}) \uparrow$$
$$- f_A(\mathbf{r_1}) \downarrow f_B(\mathbf{r_2}) \uparrow + f_B(\mathbf{r_1}) \uparrow f_A(\mathbf{r_2}) \downarrow \big)$$

which is a superposition of two Slater determinants or four products.

It may be useful to observe that (2.4) has the form of a wave function associated with an entangled state shared by "Alice" and "Bob" when f_A describes a particle localized near Alice and f_B a particle localized near Bob, and discuss its interpretation in that situation. When Alice uses a detector in her lab to measure the spin, she also implicitly makes a measurement of the spatial function, i.e., a measurement which projects onto spatial functions localized in her lab. She may get electron #1 with spin "up" with probability $\frac{1}{4}$ or electron #2 with spin "up" with probability $\frac{1}{4}$. However, there is no physical way to distinguish these two possibilities. The net result is a measurement of spin "up" in Alice's lab with total probability $\frac{1}{2}$. The

---

[3]This is because an antisymmetric N-particle wave function requires at least N linearly independent one-particle functions.

[4]These labels are the reverse of the usual physicists's convention; in essence, the convention in quantum computation is to label the eigenvectors of σ_z so that the eigenvalue $= e^{i\,\text{label}}$.

other two states in the superposition would correspond to measuring some electron in her lab with spin "down", also with net probability $\frac{1}{2}$. Once Alice has made a measurement, a corresponding measurement by Bob always yields the opposite spin.

Returning to (2.3), we note that it can be rewritten in the form (2.5) as

(2.5) $\qquad |01\rangle = \frac{1}{\sqrt{2}}[\chi^+(s_1,s_2)\phi^-(\mathbf{r_1},\mathbf{r_2}) + \chi^-(s_1,s_2)\phi^+(\mathbf{r_1},\mathbf{r_2})]$

where $\chi^\pm = \frac{1}{\sqrt{2}}[\uparrow\downarrow \pm \downarrow\uparrow]$ denote the indicated Bell-like spin states and

$$\phi^\pm = \frac{1}{\sqrt{2}}[f_A(\mathbf{r_1})f_B(\mathbf{r_2}) \pm f_B(\mathbf{r_1})f_A(\mathbf{r_2})].$$

We emphasize again that the reduction to a simple expression of the form (2.5), in which each term in the product is either an *antisymmetric* spin function times a *symmetric* spatial functions or vice-versa, is possible only[3] when $N = 2$. For more than two electrons, more complex expressions, of the form (2.2) are needed.

2.2. The origin of Pauli exchange errors. We now describe the origin of Pauli exchange errors by analyzing the two-qubit case in detail, under the additional simplifying assumption that the Hamiltonian is spin-free. This is certainly not realistic; quantum computers based upon spin will involve magnetic fields and hence, not be spin-free. However, the assumption of a spin-free Hamiltonian H, merely implies that the time development of (2.3) is determined by $e^{-iHt}\phi^\pm$, and this suffices to illustrate the principles involved. With a spin-dependent Hamiltonian the time development $e^{-iHt}\chi^\pm$ would also be non-trivial.

We will also assume that the qubits are formed using charged particles, such as electrons or protons, so that H includes a term corresponding to the $\frac{1}{r_{12}} \equiv \frac{1}{|\mathbf{r_1}-\mathbf{r_2}|}$ long-range Coulomb interaction. The Hamiltonian will be symmetric so that the states $\phi^\pm$ retain their permutational symmetry; however, the interaction term implies that they will not retain the simple form of symmetrized (or antisymmetrized) product states. Hence, after some time the states $\phi^\pm$ evolve into

(2.6a) $\qquad \Phi^-(\mathbf{r}_1,\mathbf{r}_2) = \sum_{m<n} c_{mn} \frac{1}{\sqrt{2}}[f_m(\mathbf{r_1})f_n(\mathbf{r_2}) - f_n(\mathbf{r_1})f_m(\mathbf{r_2})]$

(2.6b) $\qquad \Phi^+(\mathbf{r}_1,\mathbf{r}_2) = \sum_{m\leq n} d_{mn} \frac{1}{\sqrt{2}}[f_m(\mathbf{r_1})f_n(\mathbf{r_2}) + f_n(\mathbf{r_1})f_m(\mathbf{r_2})].$

where f_m denotes any orthonormal basis whose first two elements are f_A and f_B respectively. There is no reason to expect that $c_{mn} = d_{mn}$ in general. On the contrary, only the symmetric sum includes pairs with $m = n$. Hence if one $d_{mm} \neq 0$, then one must have some $c_{mn} \neq d_{mn}$. Inserting (2.6a) in (2.5) yields

$$e^{-iHt}|01\rangle = \frac{c_{AB}+d_{AB}}{2}\left(f_A(\mathbf{r_1})\uparrow f_B(\mathbf{r_2})\downarrow - f_B(\mathbf{r_1})\downarrow f_A(\mathbf{r_2})\uparrow\right)$$

$$+ \frac{c_{AB}-d_{AB}}{2}\left(f_B(\mathbf{r_1})\uparrow f_A(\mathbf{r_2})\downarrow - f_A(\mathbf{r_1})\downarrow f_B(\mathbf{r_2})\uparrow\right) + \Psi^{\text{Remain}}$$

(2.7) $\qquad\qquad = \frac{c_{AB}+d_{AB}}{2}|01\rangle + \frac{c_{AB}-d_{AB}}{2}|10\rangle + \Psi^{\text{Remain}}$

where Ψ^{Remain} is orthogonal to $\phi^\pm$.

A measurement of qubit-A corresponds to projecting onto f_A. Hence a measurement of qubit-A on the state (2.5) yields spin "up" with probability $\frac{1}{4}|c_{AB}+d_{AB}|^2$

and spin "down" with probability $\frac{1}{4}|c_{AB} - d_{AB}|^2$. Note that the *full* wave function is *necessarily* an *entangled* state and that the measurement process leaves the system in state $|10\rangle$ or $|01\rangle$ with probabilities $\frac{1}{4}|c_{AB} \pm d_{AB}|^2$ respectively, i.e., a subsequent measurement of qubit-B always yields the opposite spin. With probability $\frac{1}{4}|c_{AB} - d_{AB}|^2$ the initial state $|10\rangle$ has been converted to $|01\rangle$.

Although the probability of this may be small, it is *not* zero. Moreover, it would seem that any implementation which provides a mechanism for two-qubit gates would not allow the qubits to be so isolated as to preclude interactions between particles in different qubits[5]. In general, one would expect qubits to be less isolated from each other than from the external environment so that the interaction between a single pair of qubits would be greater than between a qubit and a particle in the environment. However, the environment consists of a huge number of particles (in theory, the rest of the world) and it may well happen that the number of environmental particles which interact with a given qubit is several orders of magnitude greater than the number of qubits, giving a net qubit-environment interaction which is greater than a typical qubit-qubit interaction. On the other hand, the number of qubit-qubit interactions grows quadratically with the size of the computer. Thus, prototype quantum computers, using only a few qubits, may not undergo exchange errors at the same level as the larger computers needed for real computations.

It is worth emphasizing that when the implementation involves charged particles, whether electrons or nuclei, the interaction *always* includes a contribution from the $\frac{1}{r_{12}}$ Coulomb potential, which is known to have long-range[6] effects. This is true even when the interaction used to implement the gates is entirely different, and does not involve electrostatics. Screening may reduce the effective charge, but it will not, in general, remove the basic long-range behavior of the Coulomb interaction.

Precise estimates of exchange errors require more detailed models of the specific experimental implementations. The role of long-range Coulomb effects (for which exchange errors grow quadratically with the size of the computer) suggests that implementations involving neutral particles may be advantageous for minimizing exchange errors. This would include both computers based on polarized photons (rather than charged particles) and more innovative schemes, such as Briegel, et al's proposal [3] using optical lattices. On the other hand, the ease with which exchange errors can be corrected using appropriate 9-qubit codes, suggests that dealing with exchange interactions need not be a serious obstacle.

3. Correcting Exchange Errors

A Pauli exchange error is a special type of "two-qubit" error which has the same effect as "bit flips" if (and *only* if) they are different. Exchange of bits j and

---

[5]Although the gates themselves require interactions, we expect these to be short-lived and well-controlled, i.e., in a well-designed quantum computer the gates themselves should not be a significant source of error. However, the process of turning gates on and off could induce errors in other qubits. We do not consider this error mechanism.

[6]This is because even when f and g have non-overlapping compact support $[a,b]$ and $[c,d]$ respectively, such expectations as $\int \int |f(\mathbf{r_1})|^2 \frac{1}{|\mathbf{r_1}-\mathbf{r_2}|} |f(\mathbf{r_2})|^2$ will be non-zero because the integrand is non-zero on $[a,b] \times [c,d]$. Non-overlapping initial states will not prevent the system from evolving in time to one whose states are *not* simple products or (in the case of fermions) Slater determinants!

k is equivalent to acting on a state with the operator

$$E_{jk} = \tfrac{1}{2}\Big(I_j \otimes I_k + Z_j \otimes Z_k + X_j \otimes X_k + Y_j \otimes Y_k\Big) \tag{3.1}$$

where X_j, Y_j, Z_j denote the action of the Pauli matrices $\sigma_x, \sigma_y, \sigma_z$ respectively on the bit j.

3.1. Example: the 9-bit Shor code.
As an example of potential difficulties with existing codes, consider the simple 9-bit code of Shor [28]

$$|c_0\rangle = |\mathbf{000}\rangle + |\mathbf{011}\rangle + |\mathbf{101}\rangle + |\mathbf{110}\rangle \tag{3.2a}$$

$$|c_1\rangle = |\mathbf{111}\rangle + |\mathbf{100}\rangle + |\mathbf{010}\rangle + |\mathbf{001}\rangle \tag{3.2b}$$

where boldface denotes a triplet of 0's or 1's. It is clear that these code words are invariant under exchange of electrons within the 3-qubit triples (1,2,3), (4,5,6), or (7,8,9). To see what happens when electrons in different triplets are exchanged, consider the exchange E_{34} acting on $|c_0\rangle$. This yields $|000\,000\,000\rangle + |001\,011\,111\rangle + |110\,100\,111\rangle + |111\,111\,000\rangle$ so that

$$E_{34}|c_0\rangle = |c_0\rangle + Z_8|c_0\rangle + |001\,011\,111\rangle + |110\,100\,111\rangle \tag{3.3a}$$

$$E_{34}|c_1\rangle = |c_1\rangle - Z_8|c_1\rangle + |110\,100\,000\rangle + |001\,011\,000\rangle \tag{3.3b}$$

If $|\psi\rangle = a|c_0\rangle + b|c_1\rangle$ is a superposition of code words,

$$E_{34}|\psi\rangle = \tfrac{1}{2}\Big(|\psi\rangle + Z_8|\tilde{\psi}\rangle\Big) + \frac{1}{\sqrt{2}}|\gamma\rangle$$

where $|\tilde{\psi}\rangle = a|c_0\rangle - b|c_1\rangle$ differs from ψ by a "phase error" on the code words and $|\gamma\rangle$ is orthogonal to the space of codewords and single bit errors. Thus, this code cannot reliably distinguish between an exchange error E_{34} and a phase error on any of the last 3 bits. This problem occurs because if one tries to write $E_{34}|c_0\rangle = \alpha|c_0\rangle + \beta|d_0\rangle$ with $|d_0\rangle$ orthogonal to $|c_0\rangle$, then one can not also require that $|d_0\rangle$ be orthogonal to $|c_1\rangle$.

3.2. Conditions for error correction.
Before discussing specific codes for correcting exchange errors, we first review some of the basic principles of error correction. In order to be able to correct a given class of errors, we identify a set of basic errors $\{e_p\}$ in terms of which all other errors can be written as linear combinations. In the case of unitary transformations on single bit, or one-qubit errors, this set usually consists of X_k, Y_k, Z_k ($k = 1\ldots n$) where n is the number of qubits in the code and X_k, Y_k, Z_k now denote $I \otimes I \otimes I \ldots \otimes \sigma_p \otimes \ldots \otimes I$ where σ_p denotes one of the three Pauli matrices acting on qubit-k. If we let $e_0 = I$ denote the identity and $\{C_j\}$ the set of code words, then a sufficient condition for error correction is

$$\langle e_p C_i | e_q C_j\rangle = \delta_{ij}\delta_{pq} \tag{3.4}$$

However, (3.4) can be replaced [2, 4, 9] by the weaker condition

$$\langle e_p C_i | e_q C_j\rangle = \delta_{ij} d_{pq}. \tag{3.5}$$

where the matrix D with elements d_{pq} is independent of i, j. When considering Pauli exchange errors, it is natural to seek codes which are invariant under some subset of permutations. This is clearly incompatible with (3.4) since some of the exchange errors will then satisfy $E_{k\ell}|C_i\rangle = |C_i\rangle$. Hence we will need to use (3.5).

The most common code words have the property that $|C_1\rangle$ can be obtained from $|C_0\rangle$ by exchanging all 0's and 1's. For such codes, it is not hard to see that $\langle C_1|Z_k C_1\rangle = -\langle C_0|Z_k C_0\rangle$ which is consistent with (3.5) if and only if it is identically zero. Hence even when using (3.5) rather than (3.4) it is necessary to require

$$\text{(3.6)} \qquad \langle C_1|Z_k C_1\rangle = -\langle C_0|Z_k C_0\rangle = 0$$

when the code words have this type of $0 \leftrightarrow 1$ duality.

If the basic error set has size N (i.e., $p = 0, 1 \ldots N-1$), then a two-word code requires codes which lie in a space of dimension at least $2N$. For the familiar case of single-bit errors $N = 3n + 1$ and, since an n-bit code word lies in a space of dimension 2^n, any code must satisfy $3n + 1 < 2^{n-1}$ or $n \geq 5$. There are $n(n-1)/2$ possible single exchange errors compared to $9n(n-1)/2$ two-bit errors of all types. Thus, similar dimension arguments would imply that codes which can correct all one- and two-bit errors must satisfy $2N = 9n(n-1) + 2(3n+1) \leq 2^n$ or $n \geq 10$. The shortest code known [4] which can do this has n = 11. We will see that, not surprisingly, correcting both one-bit and Pauli exchange errors, can be done with shorter codes than required to correct all two-bit errors.

However, the dimensional analysis above need not yield the best bounds when exchange errors are involved. Consider the simple code $|C_0\rangle = |000\rangle, |C_1\rangle = |111\rangle$ which is optimal for single bit flips (but can not correct phase errors). In this case $N = n + 1$ and $n = 3$ yields equality in $2(n+1) \leq 2^n$. But, since this code is invariant under permutations, the basic error set can be expanded to include all 6 exchange errors E_{jk} for a total of $N = 10$ without increasing the length of the code words.

3.3. Permutationally invariant codes.

We now present a 9-bit code code which can handle both Pauli exchange errors and all one-bit errors. It is based on the realization that codes which are invariant under permutations are impervious to Pauli exchange errors. Let

$$\text{(3.7a)} \qquad |C_0\rangle = |000\,000\,000\rangle + \frac{1}{\sqrt{28}} \sum |111\,111\,000\rangle$$

$$\text{(3.7b)} \qquad |C_1\rangle = |111\,111\,111\rangle + \frac{1}{\sqrt{28}} \sum |000\,000\,111\rangle$$

where $\sum$ denotes the sum over all permutations of the indicated sequence of 0's and 1's and it is understood that we count permutations which result in identical vectors only once. This differs from the 9-bit Shor code in that *all* permutations of $|111\,111\,000\rangle$ are included, rather than only three. The normalization of the code words is

$$\langle C_i|C_i\rangle = 1 + \frac{1}{28}\binom{9}{3} = 4.$$

The coefficient $1/\sqrt{28}$ is needed to satisfy (3.6). Simple combinatorics implies

$$\langle C_i|Z_k C_i\rangle = (-1)^i \left[1 - \frac{1}{3}\binom{9}{3}\frac{1}{28}\right] = 0.$$

Moreover,

$$\langle Z_k C_i | Z_\ell C_i \rangle = 1 + \delta_{k\ell} \binom{9}{3} \frac{1}{28} = 1 + 3\delta_{k\ell}. \tag{3.8}$$

The second term in (3.8) is zero when $k \neq \ell$ because of the fortuitous fact that there are exactly the same number of positive and negative terms. If, instead, we had used all permutations of κ 1's in n qubits, this term would be $\dfrac{(n-2\kappa)^2 - n}{n(n-1)} \binom{n}{\kappa}$ when $k \neq \ell$.

Since all components of $|C_0\rangle$ have 0 or 6 bits equal to 1, any single bit flip acting on $|C_0\rangle$, will yield a vector whose components have $1, 5,$ or 7 bits equal to 1 and is thus orthogonal to $|C_0\rangle$, to $|C_1\rangle$, to a bit flip acting on $|C_1\rangle$ and to a phase error on either $|C_0\rangle$ or $|C_1\rangle$. Similarly, a single bit flip on $|C_1\rangle$ will yield a vector orthogonal to $|C_0\rangle$, to $|C_1\rangle$, to a bit flip acting on $|C_0\rangle$ and to a phase error on $|C_0\rangle$ or $|C_1\rangle$. This suffices to ensure that (3.4), and hence (3.5), holds if e_p is I or some Z_k and e_q is one of the X_ℓ or Y_ℓ.

However, single bit flips on a given code word need not be mutually orthogonal. To find $\langle X_k C_i | X_\ell C_i \rangle$ when $k \neq \ell$, consider

$$\langle X_k(\nu_1 \nu_2 \ldots \nu_9) | X_\ell(\mu_1 \mu_2 \ldots \mu_9) \rangle. \tag{3.9}$$

where ν_i, μ_i are in $0, 1$. This will be nonzero only when $\nu_k = \mu_\ell = 0$, $\nu_\ell = \mu_k = 1$ or $\nu_k = \mu_\ell = 1$, $\nu_\ell = \mu_k = 0$ and the other $n-2$ bits are equal. From $\sum$ with κ of n bits equal to 1, there are $2\binom{n-2}{\kappa-1}$ such terms. Thus, for the code (3.7), there are 42 such terms which yields an inner product of $\frac{42}{28} = \frac{3}{2}$ when $k \neq \ell$. We similarly find that

$$\langle Y_k C_i | X_\ell C_i \rangle = -i \langle X_k Z_k C_i | X_\ell C_i \rangle = 0 \text{ for all } k \neq \ell$$

because exactly half of the terms analogous to (3.9) will occur with a positive sign and half with a negative sign, yielding a net inner product of zero. We also find

$$\langle Y_k C_i | X_k C_i \rangle = -i \langle X_k Z_k C_i | X_k C_i \rangle = -i \langle Z_k C_i | C_i \rangle = 0$$

so that

$$\langle Y_k C_i | X_\ell C_i \rangle = 0 \text{ for all } k, \ell.$$

These results imply that (3.5) holds and that the matrix D is block diagonal with the form

$$D = \begin{pmatrix} D_0 & 0 & 0 & 0 \\ 0 & D_X & 0 & 0 \\ 0 & 0 & D_Y & 0 \\ 0 & 0 & 0 & D_Z \end{pmatrix} \tag{3.10}$$

where D_0 is the 37×37 matrix corresponding to the identity and the 36 exchange errors, and D_X, D_Y, D_Z are 9×9 matrices corresponding respectively to the X_k, Y_k, Z_k single bit errors. One easily finds that $d^0_{pq} = 4$ for all p, q so that D_0 is is a multiple of a one-dimensional projection. The 9×9 matrices D_X, D_Y, D_Z all have $d_{kk} = 4$ while for $k \neq \ell$, $d_{k\ell} = 3/2$ in D_X and D_Y but $d_{k\ell} = 1$ in D_Z. Orthogonalization of this matrix is straightforward. Since D has rank $28 = 3 \cdot 9 + 1$, we are using only a $54 < 2^6$ dimensional subspace of our 2^9 dimension space.

The simplicity of codes which are invariant under permutations makes them attractive. However, there are few such codes. All code words must have the form

$$\sum_{\kappa=0}^{n} a_\kappa \sum |\underbrace{1\ldots 1}_{\kappa}\underbrace{0\ldots 0}_{n-\kappa}\rangle. \qquad (3.11)$$

Condition (3.5) places some severe restrictions on the coefficient a_κ. For example, in (3.7) only a_0 and a_6 are non-zero in $|C_0\rangle$ and only a_3 and a_9 in $|C_1\rangle$. If we try to change this so that a_0 and a_3 are non-zero in $|C_0\rangle$, i.e.,

$$|C_0\rangle = a_0|000\,000\,000\rangle + a_3 \sum |111\,000\,000\rangle \qquad (3.12a)$$

$$|C_1\rangle = a_9|111\,111\,111\rangle + a_6 \sum |000\,111\,111\rangle \qquad (3.12b)$$

then it is *not* possible to satisfy (3.6).

The 5-bit error correction code in [2, 13] does not have the permutationally invariant form (3.11) because the code words include components of the form $\sum \pm|11000\rangle$, i.e., not all terms in the sum have the same sign. The non-additive 5-bit error *detection* code in [24] also requires changes in the $\sum \pm|10000\rangle$ term. Since such sign changes seem needed to satisfy (3.6), one would not expect that 5-bit codes can handle Pauli exchange errors. In fact, Rains [25] has shown that the 5-bit error correction code is essentially unique, which implies that no 5-bit code can correct both all 1-bit errors and exchange errors. In [27] the possibility of 7-bit codes of the form (3.11) was raised. However, Wallach [30] has obtained convincing evidence that no permutationally invariant 7-bit code can correct all one-qubit errors.

3.4. Proposal for a new class of codes. Permutational invariance, which is based on a one-dimensional representation of the symmetric group, is not the only approach to exchange errors. Our analysis of (3.2) suggests a construction which we first describe in over-simplified form. Let $|c_0\rangle, |d_0\rangle, |c_1\rangle, |d_1\rangle$ be four mutually orthogonal n-bit vectors such that $|c_0\rangle, |c_1\rangle$ form a code for one-bit errors and $|c_0\rangle, |d_0\rangle$ and $|c_1\rangle, |d_1\rangle$ are each bases of a two-dimensional representation of the symmetric group S_n. If $|d_0\rangle$ and $|d_1\rangle$ are also orthogonal to one-bit errors on the code words, then the code $|c_0\rangle, |c_1\rangle$ can correct Pauli exchange errors as well as one-bit errors. If, in addition, the vectors $|d_0\rangle, |d_1\rangle$ also form a code isomorphic to $|c_0\rangle, |c_1\rangle$ in the sense that the matrix D in (3.5) is identical for both codes, then the code should also be able to correct products of one-bit and Pauli exchange errors.

However, applying this scheme to an n-bit code requires a non-trivial irreducible representation of S_n of which the smallest has dimension $n-1$. Thus we will seek a set of $2(n-1)$ mutually orthogonal vectors denoted $|C_0^m\rangle, |C_1^m\rangle$ ($m = 1 \ldots n-1$) such that $|C_0^1\rangle, |C_1^1\rangle$ form a code for one bit errors and $|C_0^m\rangle$ ($m = 1 \ldots n-1$) and $|C_1^m\rangle$ ($m = 1 \ldots n-1$) each form basis of the same irreducible representation of S_n. Such code will be able to correct *all* errors which permute qubits; not just single exchanges. If, in addition, (3.5) is extended to

$$\langle e_p C_i^m | e_q C_j^{m'}\rangle = \delta_{ij}\delta_{mm'}d_{pq} \qquad (3.13)$$

with the matrix $D = \{D_{pq}\}$ independent of both i and m, then this code will also be able to correct products of one bit errors and permutation errors.

In the construction proposed above, correction of exchange and one-bit errors would require a space of dimension $2(n-1)(3n+1) \leq 2^n$ or $n \geq 9$. If codes satisfying (3.13) exist, they could correct *all* permutation errors as well as products of permutations and one-bit errors (which includes a very special subclass of 3-bit errors and even a few higher ones). Thus exploiting permutational symmetry may yield powerful new codes.

In some sense, the strategy proposed here is the opposite of that of Section 3.3 (despite the fact that both are based on representations of S_n). In Section 3.3 we sought code words with the maximum symmetry of being invariant under all permutatations. Now, we seek instead, a pair of dual code words $|C_0^1\rangle, |C_1^1\rangle$ with "minimal" symmetry in the sense that a set of generators of S_n acting on each of these code words yields an orthogonal basis for a non-trivial irreducible representation of S_n. If the code words $|C_0^m\rangle, |C_1^m\rangle$ $n = 2 \ldots n-1$ can be obtained in this way, then each pair should also be a single-bit error correction code, as desired.

3.5. Non-additive codes. Most existing codes used for quantum error correction are obtained by a process [**4, 5, 7**] through which the codes can be described in terms of a subgroup, called the *stabilizer*, of the error group. Such codes are called "stabilizer codes" or "additive codes". In [**24**] an example of a non-additive 5-bit code was given, establishing the existence of non-additive codes. However, this was only an error detection code and, hence, less powerful than the 5-bit error correction code [**2, 13**] obtained using the stabilizer formalism. Subsequently, V. P. Roychowdhury and F. Vatan [**26**] showed that many non-additive codes exist; however, it was not clear how useful such codes might be.

H. Pollatsek [**21**] has pointed out that the 9-qubit code (3.7) is a non-additive code. This establishes that non-additive codes may well have an important role to play in quantum error correction, particularly in situations in which exchange errors and permutational symmetry are important. The non-additivity is immediate if one accepts that for additive codes all non-zero blocks of D consist entirely of $d_{pq} = 1$. In (3.10), this is true only for D_0 (after suitable normalization); but not for D_X, D_Y, D_Z.

Nevertheless, it may be instructive to present an argument is based on the observation that the set of vectors which occur in $\sum$ in (3.7b) spans the vector space of binary 9-tuples $\mathbf{Z}_2^9$. More generally let $\Gamma_{\kappa,n}$ denote the set of all vectors $\mathbf{a} = (a_1, a_2, \ldots a_n)$ in $\mathbf{Z}_2^n$ with precisely κ of the a_j taking the value 1 and $n - \kappa$ the value 0 as in (3.11). Then span $\{\Gamma_{\kappa,n}\} = \mathbf{Z}_2^n$ if κ is odd and $\kappa \neq 0, n$. (If $\kappa \neq 0, n$ is even, $\{\Gamma_{\kappa,n}\}$ spans the even subspace of $\mathbf{Z}_2^n$.) By definition, an additive (or stabilizer) code forms an eigenspace for an abelian subgroup S of the error group $E = \{i^\ell X(\mathbf{a})Z(\mathbf{b}) : \mathbf{a}, \mathbf{b} \in \mathbf{Z}_2^9\}$. When the stabilizer S consists of only the scalar multiples of the identity I, then the corresponding eigenspace is all of $\mathbf{C}^{2^9}$. Consequently, to show that (3.7) is *not* a stabilizer code, it suffices to show that no vector of the subspace spanned by the codewords $|C_0\rangle$ and $|C_1\rangle$ can be an eigenvector for an element of E other than I.

It suffices to consider the image of $|C_1\rangle$ under $X(a)Z(b)$, i.e.,

$$X(\mathbf{a})Z(\mathbf{b})|C_1\rangle = (-1)^{\mathbf{b} \cdot \mathbf{e}}|\mathbf{a} + \mathbf{e}\rangle + \frac{1}{\sqrt{28}} \sum_{\mathbf{v} \in \Gamma_{3,9}} (-1)^{\mathbf{b} \cdot \mathbf{v}} |\mathbf{v} + \mathbf{a}\rangle$$

where $\mathbf{e} = (111\,111\,111)$. If $X(\mathbf{a})Z(\mathbf{b})|C_1\rangle = \lambda|C_1\rangle$, we must have, in particular, $X(\mathbf{a})Z(\mathbf{b})|\mathbf{e}\rangle = \lambda|\mathbf{e}\rangle$ which implies $\mathbf{a} = (000\,000\,000)$ or, equivalently, $X(\mathbf{a}) = I$. Then $\lambda = (-1)^{\mathbf{b}\cdot\mathbf{e}}$ and the eigenvalue equation reduces to

$$Z(\mathbf{b})|C_1\rangle = \lambda \left(|\mathbf{e}\rangle + + \frac{1}{\sqrt{28}} \sum_{\mathbf{v}\in\Gamma_{3,9}} (-1)^{\mathbf{b}\cdot(\mathbf{v}+\mathbf{e})}|\mathbf{v}\rangle\right)$$

which implies $\mathbf{b}\cdot(\mathbf{v}+\mathbf{e}) = 0$ for every $\mathbf{v} \in \Gamma_{3,9}$. But since (as noted above) $\Gamma_{3,9}$ spans $\mathbf{Z}_2^9$ this implies that $\mathbf{b}$ is orthogonal to all of $\mathbf{Z}_2^9$, which implies $\mathbf{b} = (000\,000\,000)$ so that $Z(\mathbf{b}) = I$ as well. Thus, since any element of E can be written as a multiple of $X(\mathbf{a})Z(\mathbf{b})$, the stabilizer S contains only multiples of the identity.

4. Conclusion

Although codes which can correct Pauli exchange errors will be larger than the minimal 5-qubit codes obtained for single-bit error correction, this may not be a serious drawback. For implementations of quantum computers which have a grid structure (e.g., solid state or optical lattices) it may be natural and advantageous to use 9-qubit codes which can be implemented in 3×3 blocks. (See, e.g., [3].) However, codes larger than 9-bits may be impractical for a variety of reasons. Hence it is encouraging that both the code in section 3.3 and the construction proposed in section 3.4 do *not* require $n > 9$.

It may be worth investigating whether or not the codes proposed here can be used advantageously in combination with other schemes, particularly those [10] based on hierarchical nesting. Since the code in sections 3.3 and 3.4 can already handle some types of multiple errors, concatenation of one of these 9-bit codes with itself will contain some redundancy and concatenation with a 5-bit code may be worth exploring. Indeed, when exchange correlations are the prime mechanism for multi-bit errors, the need for repeated concatenation may be significantly reduced.

Construction of codes of the type proposed in Section 3.4 remains a significant challenge. However, development of such new methods of may be precisely what is needed to obtain codes powerful enough to correct multi-qubit errors efficiently, without the large size drawback of codes based on repeated concatenation.

Acknowledgment It is a pleasure to thank Professor Eric Carlen for a useful comment which started my interest in exchange interactions, Dr. Daniel Gottesman for helpful information on existing codes and error correction procedures, Professor Chris King for several helpful discussions and comments on earlier drafts, Professor Harriet Pollatsek for additional comments, discussions and permission to include her observations about the non-additivity of the 9-bit code presented here, Professor Nolan Wallach for communications about 7-bit codes, and the five anonymous referees of *Physical Review Letters* for their extensive commentary on [27].

Note added in proof: Recent work with H. Pollatsek suggests that condition (3.13) is too strong and the simultaneous orthogonality conditions cannot be satisfied. Other methods for using codes which involve $(n-1)$-dimensional representations of S_n are under investigation.

References

[1] D. Aharanov, M. Ben-Or, "Fault-Tolerant Quantum Computation With Constant Error Rate" lanl preprints quant-ph/9611025 and quant-ph/9906129.

[2] C.H. Bennett, D.P. DiVincenzo, J.A. Smolin and W.K. Wooters, "Mixed State Entanglement and Quantum Error Correction" *Phys. Rev. A* **54**, 3824–3851 (1996) [lanl preprint quant-ph/9604024].

[3] H.J. Briegel, T. Calarco, D. Jaksch, J.I. Cirac, and P. Zoller "Quantum computing with neutral atoms" lanl preprint quant-ph/9904010.

[4] R. Calderbank, E.M. Rains, P.W. Shor and N.J.A. Sloane, "Quantum Error Correction and Orthogonal Geometry" *Phys. Rev. Lett.* **78**, 405–408 (1997) [lanl preprint quant-ph/9605005]; and "Quantum Error Correction via Codes over GF(4)" *IEEE Trans. Info. Theory* **44**, 1369–1387 (1998) [lanl preprint quant-ph/9608006].

[5] D. Gottesman "Stabilizer Codes and Quantum Error Correction" PhD thesis, Caltech (1997). [lanl preprint quant-ph/9705052].

[6] D. Gottesman "A Theory of Fault-Tolerant Quantum Computation" *Phys.Rev. A* **57**, 127– (1998) [lanl preprint quant-ph/9702029].

[7] D. Gottesman, "An Introduction to Quantum Error Correction," in "Quantum Computation: A Grand Mathematical Challenge for the Twenty-First Century and the Millennium," edited by Samuel J. Lomonaco, Jr, Proceedings of the Symposia of Applied Mathematics, Volume 58, American Mathematical Society, Providence, RI, (2002).

[8] A. Y. Kitaev "Fault-tolerant Quantum Computation by Anyons" lanl preprint quant-ph/9707021.

[9] E. Knill and R Laflamme, "A Theory of Quantum Error-Correcting Codes" *Phys. Rev. A* **55**, 900-911 (1997).

[10] E. Knill and R Laflamme, "Concatenated Quantum Codes" lanl preprint quant-ph/9608012.

[11] E. Knill, R Laflamme, and L. Viola "Theory of Quantum Error Correction for General Noise" *Phys. Rev. Lett.* **84**, 25254–28 (2000) [lanl preprint quant-ph/9908066].

[12] E. Knill, R Laflamme, W. H. Zurek "Resilient Quantum Computation: Error Models and Thresholds" *Proc. Roy. Soc. A* **454**, 365–384 (1998). [lanl preprint quant-ph/9702058]

[13] R. Laflamme, C. Miquel, J.P. Paz, W.H. Zurek, "Perfect Quantum Error Correction Code" *Phys. Rev. Lett.* **77**, 198–201 (1996).

[14] D.A. Lidar, I.L. Chuang, and K.B. Whaley "Decoherence Free Subspaces for Quantum Computation" *Phys. Rev. Lett.* **81**, 2594–97 (1998) [lanl preprint quant-ph/9807004].

[15] D.A. Lidar, D. Bacon, and K.B. Whaley "Concatenating Decoherence Free Subspaces with Quantum Error Correcting Codes" *Phys. Rev. Lett.* **82**, 4556–59 (1999) [lanl preprint quant-ph/9809081].

[16] D.A. Lidar, J. Kempe, D. Bacon, and K.B. Whaley "Protecting Quantum Information Encoded in Decoherence Free States Against Exchange Errors" *Phys. Rev. A* **61**, 052307 (2000) [lanl preprint quant-ph/0004064].

[17] D. Bacon, J. Kempe, D.A. Lidar, and K.B. Whaley, "Universal Fault-Tolerant Computation on Decoherence-Free Subspaces" *Phys. Rev. Lett.* **85**, 1758-61 (2000) [lanl preprint quant-ph/9909058]; and J. Kempe, D. Bacon, D.A. Lidar, and K.B. Whaley, " Theory of Decoherence-Free Fault-Tolerant Universal Quantum Computation" lanl preprint quant-ph/0004064.

[18] M. Hamermesh, *Group Theory* (Addison-Wesley Publishing, 1962).

[19] L. Landau and L. Lifshitz, *Quantum Mechanics* (Second edition of English translation, Pergamon Press, 1965).

[20] M.A. Nielsen and I.L. Chuang, *Quantum Computation and Quantum Information* (Cambridge University Press, 2000).

[21] H. Pollatsek, private communication

[22] J. Preskill, "Reliable Quantum Computers" *Proc. Roy. Soc. A* **454**, 385–410 (1998) [lanl preprint quant-ph/9705031], and "Fault-Tolerant Quantum Computation " lanl preprint quant-ph/9712048

[23] J. Preskill, "Battling Decoherence: The Fault Tolerant Quantum Computer" *Physics Today* (6)**52**, 24–30 (June, 1999).

[24] E.M. Rains, R. H. Hardin, P.W. Shor and N.J.A. Sloane, "A nonadditive quantum code" *Phys. Rev. Lett.* **79**, 953–954 (1997).

[25] E.M. Rains, "Quantum Codes of Minimum Distance Two" lanl preprint quant-ph/9704043

[26] V. P. Roychowdhury and F. Vatan, "On the Structure of Additive Quantum Codes and the Existence of Nonadditive Codes" lanl preprint quant-ph/9710031
[27] M.B. Ruskai, "Pauli Exchange Errors in Quantum Computation" *Phys. Rev. Lett.* **85**, 194–197 (2000); [lanl preprint quant-ph/9906114]
[28] P. Shor, "Scheme for Reducing Decoherence in Quantum Computer Memory" *Phys. Rev. A* **52**, 2493-2496 (1995).
[29] A.M. Steane, "Efficient Fault-tolerant Quantum Computing" *Nature* **399**, 124–126 (May 1999).
[30] N. Wallach, private communication.

(Mary Beth Ruskai) DEPARTMENT OF MATHEMATICS, UNIVERSITY OF MASSACHUSETTS LOWELL, LOWELL, MA 01854 USA

E-mail address, Mary Beth Ruskai: `bruskai@cs.uml.edu`

Relative Entropy in Quantum Information Theory

Benjamin Schumacher and Michael D. Westmoreland

ABSTRACT. We review the properties of the quantum relative entropy function and discuss its application to problems of classical and quantum information transfer and to quantum data compression. We then outline further uses of relative entropy to quantify quantum entanglement and analyze its manipulation.

1. Quantum relative entropy

In this paper we discuss several uses of the quantum relative entropy function in quantum information theory. Relative entropy methods have a number of advantages. First of all, the relative entropy functional satisfies some strong identities and inequalities, providing a basis for good theorems. Secondly, the relative entropy has a natural interpretation in terms of the statistical distinguishability of quantum states; closely related to this is the picture of relative entropy as a "distance" measure between density operators. These interpretations of the relative entropy give insight about the meaning of the mathematical constructions that use it. Finally, relative entropy has found a wide variety of applications in quantum information theory.

The usefulness of relative entropy in quantum information theory should come as no surprise, since the classical relative entropy has shown its power as a unifying concept in classical information theory [1]. Indeed, some of the results we will describe have close analogues in the classical domain. Nevertheless, the quantum relative entropy can provide insights in contexts (such as the quantification of quantum entanglement) that have no parallel in classical ideas of information.

Let Q be a quantum system described by a Hilbert space $\mathcal{H}$. (Throughout this paper, we will restrict our attention to systems with Hilbert spaces having a finite number of dimensions.) A pure state of Q can be described by a normalized vector $|\psi\rangle$ in $\mathcal{H}$, but a general (mixed) state requires a density operator ρ, which is a positive semi-definite operator on $\mathcal{H}$ with unit trace. For the pure state $|\psi\rangle$, the density operator ρ is simply the projection operator $|\psi\rangle\langle\psi|$; otherwise, ρ is a convex combination of projections. The entropy $S(\rho)$ is defined to be

(1) $$S(\rho) = -\operatorname{Tr} \rho \log \rho.$$

2000 *Mathematics Subject Classification.* Primary 81P68.

©2002 American Mathematical Society

The entropy is non-negative and equals zero if and only if ρ is a pure state. (By "log" we will mean a logarithm with base 2.)

Closely related to the entropy of a state is the relative entropy of a pair of states. Let ρ and σ be density operators, and define the quantum relative entropy $\mathcal{S}(\rho||\sigma)$ to be

$$\mathcal{S}(\rho||\sigma) = \text{Tr}\,\rho\log\rho - \text{Tr}\,\rho\log\sigma. \tag{2}$$

(We read this as "the relative entropy of ρ with respect to σ".) This function has a number of useful properties: [**2**]

1. $\mathcal{S}(\rho||\sigma) \geq 0$, with equality if and only if $\rho = \sigma$.
2. $\mathcal{S}(\rho||\sigma) < \infty$ if and only if $\text{supp}\,\rho \subseteq \text{supp}\,\sigma$. (Here "supp ρ" is the subspace spanned by eigenvectors of ρ with non-zero eigenvalues.)
3. The relative entropy is continuous where it is not infinite.
4. The relative entropy is jointly convex in its arguments [**3**]. That is, if ρ_1, ρ_2, σ_1 and σ_2 are density operators, and p_1 and p_2 are non-negative numbers that sum to unity (i.e., probabilities), then

$$\mathcal{S}(\rho||\sigma) \leq p_1\mathcal{S}(\rho_1||\sigma_1) + p_2\mathcal{S}(\rho_2||\sigma_2) \tag{3}$$

where $\rho = p_1\rho_1 + p_2\rho_2$ and $\sigma = p_1\sigma_1 + p_2\sigma_2$. Joint convexity automatically implies convexity in each argument, so that (for example)

$$\mathcal{S}(\rho||\sigma) \leq p_1\mathcal{S}(\rho_1||\sigma) + p_2\mathcal{S}(\rho_2||\sigma). \tag{4}$$

The properties, especially property (1), motivate us to think of the relative entropy as a kind of "distance" between density operators. The relative entropy, which is not symmetric and which lacks a triangle inequality, is not technically a metric; but it is a positive definite directed measure of the separation of two density operators.

Suppose the density operator ρ_k occurs with probability p_k, yielding an average state $\rho = \sum_k p_k \rho_k$, and suppose σ is some other density operator. Then

$$\sum_k p_k \mathcal{S}(\rho_k||\sigma) = \sum_k p_k \left(\text{Tr}\,\rho_k \log\rho_k - \text{Tr}\,\rho_k \log\sigma \right)$$

$$= \sum_k p_k \left(\text{Tr}\,\rho_k \log\rho_k - \text{Tr}\,\rho_k \log\rho + \text{Tr}\,\rho_k \log\rho - \text{Tr}\,\rho_k \log\sigma \right)$$

$$= \sum_k p_k \left(\text{Tr}\,\rho_k \log\rho_k - \text{Tr}\,\rho_k \log\rho \right) + \text{Tr}\,\rho\log\rho - \text{Tr}\,\rho\log\sigma$$

$$\sum_k p_k \mathcal{S}(\rho_k||\sigma) = \sum_k p_k \mathcal{S}(\rho_k||\rho) + \mathcal{S}(\rho||\sigma). \tag{5}$$

Equation 5 is known as Donald's identity. [**4**]

The classical relative entropy of two probability distributions is related to the probability of distinguishing the two distributions after a large but finite number of independent samples. This is called Sanov's theorem [**1**], and this result has quantum analogue [**5**]. Suppose ρ and σ are two possible states of the quantum system Q, and suppose we are provided with N identically prepared copies of Q. A measurement is made to determine whether the prepared state is ρ, and the probability P_N that the state σ passes this test—in other words, is confused with ρ—is

$$P_N \approx 2^{-N\mathcal{S}(\rho||\sigma)} \tag{6}$$

as $N \to \infty$. (We have assumed that the measurement made is an optimal one for the purpose, and it is possible to show that an asymptotically optimal measurement strategy can be found that depends on ρ but not σ.)

The quantum version of Sanov's theorem tells us that the quantum relative entropy governs the asymptotic distinguishability of one quantum state from another by means of measurements. This further supports the view of $\mathcal{S}(\cdot||\cdot)$ as a measure of "distance"; two states are "close" if they are difficult to distinguish, but "far apart" if the probability of confusing them is small.

The remainder of this paper is organized as follows. Sections 2–5 apply relative entropy methods to the problem of sending classical information by means of a (possibly noisy) quantum channel. Sections 6–7 consider the transmission and compression of quantum information. Sections 8–9 then apply relative entropy methods to the discussion of quantum entanglement and its manipulation by local operations and classical communication. We conclude with a few remarks in Section 10.

2. Classical communication via quantum channels

One of the oldest problems in quantum information theory is that of sending classical information via quantum channels. A sender ("Alice") wishes to transmit classical information to a receiver ("Bob") using a quantum system as a communication channel. Alice will represent the message a, which occurs with probability p_a, by preparing the channel in the "signal state" represented by the density operator ρ_a. The average state of the channel will thus be $\rho = \sum_a p_a \rho_a$. Bob will attempt to recover the message by making a measurement of some "decoding observable" on the channel system.

The states ρ_a should be understood here as the "output" states of the channel, the states that Bob will attempt to distinguish in his measurement. In other words, the states ρ_a already include the effects of the dynamical evolution of the channel (including noise) on its way from sender to receiver. The dynamics of the channel will be described by a trace-preserving, completely positive map $\mathcal{E}$ on density operators [6]. The effect of $\mathcal{E}$ is simply to restrict the set of output channel states that Alice can arrange for Bob to receive. If $\mathcal{D}$ is the set of all density operators, then Alice's efforts can only produce output states in the set $\mathcal{A} = \mathcal{E}(\mathcal{D})$, a convex, compact set of density operators.

Bob's decoding observable is represented by a set of positive operators E_b such that $\sum_b E_b = 1$. If Bob makes his measurement on the state ρ_a, then the conditional probability of measurement outcome b is

(7) $$P(b|a) = \mathrm{Tr}\, \rho_a E_b.$$

This yields a joint distribution over Alice's input messages a and Bob's decoded messages b:

(8) $$P(a,b) = p_a P(b|a).$$

Once a joint probability distribution exists between the input and output messages (random variables A and B, respectively), the information transfer can be analyzed by classical information theory. The information obtained by Bob is given by the *mutual information* $I(A:B)$:

(9) $$I(A:B) = H(A) + H(B) - H(A,B)$$

where H is the Shannon entropy function

(10) $$H(X) = -\sum_x p(x) \log p(x).$$

Shannon showed that, if the channel is used many times with suitable error-correcting codes, then any amount of information up to $I(A:B)$ bits (per use of the channel) can be sent from Alice to Bob with arbitrarily low probability of error [1]. The classical capacity of the channel is $C = \max I(A:B)$, where the maximum is taken over all input probability distributions. C is thus the maximum amount of information that may be reliably conveyed per use of the channel.

In the quantum mechanical situation, for a given ensemble of signal states ρ_a, Bob has many different choices for his decoding observable. Unless the signal states happen to be orthogonal, no choice of observable will allow Bob to distinguish perfectly between them. A theorem stated by Gordon[7] and Levitin[8] and first proved by Holevo[9] states that the amount of information accessible to Bob is limited by $I(A:B) \leq \chi$, where

(11) $$\chi = S(\rho) - \sum_a p_a S(\rho_a).$$

The quantity χ is non-negative, since the entropy S is concave.

More recently, Holevo [10] and Schumacher and Westmoreland [11] have shown that this upper bound on $I(A:B)$ is asymptotically achievable. If Alice uses the same channel many times and prepares long codewords of signal states, and Bob uses an entangled decoding observable to distinguish these codewords, then Alice can convey to Bob up to χ bits of information per use of the channel, with arbitrarily low probability of error. (This fact was established for pure state signals $\rho_a = |\psi_a\rangle\langle\psi_a|$ in [12]. In this case, $\chi = S(\rho)$.)

The Holevo bound χ can be expressed in terms of the relative entropy:

$$\chi = -\text{Tr}\,\rho \log \rho + \sum_a p_a \text{Tr}\,\rho_a \log \rho_a$$
$$= \sum_a p_a \left(\text{Tr}\,\rho_a \log \rho_a - \text{Tr}\,\rho_a \log \rho \right)$$
(12) $$\chi = \sum_a p_a \mathcal{S}(\rho_a \| \rho).$$

In geometric terms, χ is the average relative entropy "directed distance" from the average state ρ to the members of the signal ensemble.

Donald's identity (Equation 5) has a particularly simple form in terms of χ. Given an ensemble and an additional state σ,

(13) $$\sum_a p_a \mathcal{S}(\rho_a \| \sigma) = \chi + \mathcal{S}(\rho \| \sigma).$$

This implies, among other things, that

(14) $$\chi \leq \sum_a p_a \mathcal{S}(\rho_a \| \sigma)$$

with equality if and only if $\sigma = \rho$, the ensemble average state.

3. Thermodynamic cost of communication

In this section and the next, we focus on the transfer of classical information by means of a quantum channel.

Imagine a student who attends college far from home [13]. Naturally, the student's family wants to know that the student is passing his classes, and so they want the student to report to them frequently over the telephone. But the student is poor and cannot afford very many long-distance telephone calls. So they make the following arrangement: each evening at the same time, the poor student will call home only if he is failing one or more of this classes. Otherwise, he will save the phone charges by *not* calling home.

Every evening that the poor student does not call, therefore, the family is receiving a message *via the telephone* that his grades are good. (That the telephone is being used for this message can be seen from the fact that, if the phone lines are knocked out for some reason, the family can no longer make any inference from the absence of a phone call.)

For simplicity, imagine that the student's grades on successive days are independent and that the probability that the student will be failing on a given evening is p. Then the information conveyed each evening by the presence or absence of a phone call is

$$(15) \qquad H(p) = -p \log p - (1-p) \log(1-p).$$

The cost of making a phone call is c, while not making a phone call is free. Thus, the student's average phone charge is cp per evening. The number of bits of information per unit cost is thus

$$(16) \qquad \frac{H(p)}{cp} = \frac{1}{c}\left(-\log p - \left(\frac{1}{p}-1\right)\log(1-p)\right).$$

If the poor student is very successful in his studies, so that $p \to 0$, then this ratio becomes unboundedly large, even though both $H(p) \to 0$ and $cp \to 0$. That is, the student is able to send an arbitrarily large number of bits per unit cost. There is no irreducible cost for sending one bit of information over the telephone.

The key idea in the story of the poor student is that one possible signal—no phone call at all—has no cost to the student. The student can exploit this fact to use the channel in a cost-effective way, by using the zero cost signal almost all of the time.

Instead of a poor student using a telephone, we can consider an analogous quantum mechanical problem. Suppose that a sender can manipulate a quantum channel to produce (for the receiver) one of two possible states, ρ_0 or ρ_1. The state ρ_0 can be produced at "zero cost", while the state ρ_1 costs a finite amount $c_1 > 0$ to produce. In the signal ensemble, the signal state ρ_1 is used with probability η and ρ_0 with probability $1 - \eta$, leading to an average state

$$(17) \qquad \rho = (1-\eta)\rho_0 + \eta \rho_1.$$

The average cost of creating a signal is thus $c = \eta c_1$. For this ensemble,

$$(18) \qquad \chi = (1-\eta)\mathcal{S}(\rho_0 || \rho) + \eta \mathcal{S}(\rho_1 || \rho).$$

As discussed in the previous section, χ is an asymptotically achievable upper bound for the information transfered by the channel.

An upper bound for χ can be obtained from Donald's identity. Letting ρ_0 be the "additional" state,

$$(19) \qquad \chi \leq (1-\eta)\mathcal{S}(\rho_0||\rho_0) + \eta\mathcal{S}(\rho_1||\rho_0) = \eta\mathcal{S}(\rho_1||\rho_0).$$

Combining this with a simple lower bound, we obtain

$$(20) \qquad \eta\mathcal{S}(\rho_1||\rho) \leq \chi \leq \eta\mathcal{S}(\rho_1||\rho_0).$$

If we divide χ by the average cost, we find an asymptotically achievable upper bound for the number of bits sent through the channel per unit cost. That is,

$$(21) \qquad \frac{\chi}{c} \leq \frac{1}{c_1}\mathcal{S}(\rho_1||\rho_0).$$

Furthermore, equality holds in the limit that $\eta \to 0$. Thus,

$$(22) \qquad \sup \frac{\chi}{c} = \frac{1}{c_1}\mathcal{S}(\rho_1||\rho_0).$$

In short, the relative entropy "distance" between the signal state ρ_1 and the "zero cost" signal ρ_0 gives the largest possible number of bits per unit cost that may be sent through the channel—the "cost effectiveness" of the channel. If the state ρ_0 is a pure state, or if we can find a usable signal state ρ_1 whose support is not contained in the support of ρ_0, then $\mathcal{S}(\rho_1||\rho_0) = \infty$ and the cost effectiveness of the channel goes to infinity as $\eta \to 0$. (This is parallel to the situation of the poor student, who can make the ratio of "bits transmitted" to "average cost" arbitrarily large.)

What if there are many possible signal states ρ_1, ρ_2, etc., with positive costs c_1, c_2, and so on? If we assign the probability ηq_k to ρ_k for $k = 1, 2, \ldots$ (where $\sum_k q_k = 1$), and use ρ_0 with probability $1 - \eta$, then we obtain

$$(23) \qquad \eta \sum_k q_k \mathcal{S}(\rho_k||\rho) \leq \chi \leq \eta \sum_k q_k \mathcal{S}(\rho_k||\rho_0).$$

The average cost of the channel is $c = \eta \sum_k q_k c_k$. This means that

$$(24) \qquad \frac{\chi}{c} \leq \frac{\sum_k q_k \mathcal{S}(\rho_k||\rho_0)}{\sum_k q_k c_k}.$$

We now note the following fact about real numbers. Suppose $a_n, b_n > 0$ for all n. Then

$$(25) \qquad \frac{\sum_n a_n}{\sum_n b_n} \leq \max_n \frac{a_n}{b_n}.$$

This can be proven by letting $R = \max(a_n/b_n)$ and pointing out that $a_n \leq R b_n$ for all n. Then

$$\sum_n a_n \leq R \sum_n b_n$$

$$\frac{\sum_n a_n}{\sum_n b_n} \leq R.$$

In our context, this implies that

$$(26) \qquad \frac{\sum_k q_k \mathcal{S}(\rho_k||\rho_0)}{\sum_k q_k c_k} \leq \max_k \frac{q_k \mathcal{S}(\rho_k||\rho_0)}{q_k c_k}$$

and thus
$$\frac{\chi}{c} \leq \max_k \frac{S(\rho_k\|\rho_0)}{c_k}. \tag{27}$$

By using only the "most efficient state" (for which the maximum on the right-hand side is achieved) and adopting the "poor student" strategy of $\eta \to 0$, we can show that
$$\sup \frac{\chi}{c} = \max_k \frac{S(\rho_k\|\rho_0)}{c_k}. \tag{28}$$

These general considerations of an abstract "cost" of creating various signals have an especially elegant development if we consider the thermodynamic cost of using the channel. The thermodynamic entropy S_θ is related to the information-theoretic entropy $S(\rho)$ of the state ρ of the system by
$$S_\theta = k \ln 2 \, S(\rho). \tag{29}$$

The constant k is Boltzmann's constant. If our system has a Hamiltonian operator H, then the thermodynamic energy E of the state is the expectation of the Hamiltonian:
$$E = \langle H \rangle = \operatorname{Tr} \rho H. \tag{30}$$

Let us suppose that we have access to a thermal reservoir at temperature T. Then the "zero cost" state ρ_0 is the thermal equilibrium state
$$\rho_0 = \frac{1}{Z} e^{-\beta H}, \tag{31}$$

where $\beta = 1/kT$ and $Z = \operatorname{Tr} e^{-\beta H}$. ($Z$ is the partition function.)

The free energy of the system in the presence of a thermal reservoir at temperature T is $F = E - TS_\theta$. For the equilibrium state ρ_0,
$$\begin{aligned} F_0 &= \operatorname{Tr} \rho_0 H + kT \ln 2 \left(-\log Z - \frac{\beta}{\ln 2} \operatorname{Tr} \rho_0 H \right) \\ &= -kT \ln 2 \log Z \end{aligned} \tag{32}$$

The thermodynamic cost of the state ρ_1 is just the difference $F_1 - F_0$ between the free energies of ρ_1 and the equilibrium state ρ_0. But this difference has a simple relation to the relative entropy. First, we note
$$\operatorname{Tr} \rho_1 \log \rho_0 = -\log Z - \beta \operatorname{Tr} \rho_1 H, \tag{33}$$

from which it follows that [14]
$$\begin{aligned} F_1 - F_0 &= \operatorname{Tr} \rho_1 H + kT \ln 2 \operatorname{Tr} \rho_1 \log \rho_1 + kT \ln 2 \log Z \\ &= kT \ln 2 \left(\operatorname{Tr} \rho_1 \log \rho_1 - \operatorname{Tr} \rho_1 \log \rho_0 \right) \\ F_1 - F_0 &= kT \ln 2 \, S(\rho_1\|\rho_0). \end{aligned} \tag{34}$$

If we use the signal state ρ_1 with probability η, then the average thermodynamic cost is $f = \eta(F_1 - F_0)$. The number of bits sent per unit free energy is therefore
$$\frac{\chi}{f} \leq \eta \frac{S(\rho_1\|\rho_0)}{f} = \frac{1}{kT \ln 2}. \tag{35}$$

The same bound holds for all choices of the state ρ_1, and therefore for all ensembles of signal states.

We can approach this upper bound if we make η small, so that

$$\sup \frac{\chi}{f} = \frac{1}{kT \ln 2} \tag{36}$$

In short, for *any* coding and decoding scheme that makes use of the quantum channel, the maximum number of bits that can be sent per unit free energy is just $(kT \ln 2)^{-1}$. Phrased another way, the minimum free energy cost per bit is $kT \ln 2$.

This analysis can shed some light on Landauer's principle [15], which states that the minimum thermodynamic cost of information erasure is $kT \ln 2$ per bit. From this point of view, information erasure is simply information transmission into the environment, which requires the expenditure of an irreducible amount of free energy.

4. Optimal signal ensembles

Now we consider χ-maximizing ensembles of states from a given set $\mathcal{A}$ of available (output) states, without regard to the "cost" of each state. Our discussion in Section 2 tells us that the χ-maximizing ensemble is the one to use if we wish to maximize the classical information transfer from Alice to Bob via the quantum channel. Call an ensemble that maximizes χ an "optimal" signal ensemble, and denote the maximum value of χ by χ^*. (The results of this section are developed in more detail in [16].)

The first question is, of course, whether an optimal ensemble exists. It is conceivable that, though there is a least upper bound χ^* to the possible values of χ, no particular ensemble in $\mathcal{A}$ achieves it. (This would be similar to the results in the last section, in which the optimal cost effectiveness of the channel is only achieved in a limit.) However, an optimal ensemble does exist. Uhlmann [17] has proven a result that goes most of the way. Suppose our underlying Hilbert space $\mathcal{H}$ has dimension d and the set $\mathcal{A}$ of available states is convex and compact. Then given a fixed average state ρ, there exists an ensemble of at most d^2 signal states ρ_a that achieves the maximum value of χ for that particular ρ. The problem we are considering is to maximize χ over all choices of ρ in $\mathcal{A}$. Since Uhlmann has shown that each ρ-fixed optimal ensemble need involve no more than d^2 elements, we only need to maximize χ over ensembles that contain d^2 or fewer members. The set of such ensembles is compact and χ is a continuous function on this set, so χ achieves its maximum value χ^* for some ensemble with at most d^2 elements.

Suppose that the state ρ_a occurs with probability p_a in some ensemble, leading to the average state ρ and a Holevo quantity χ. We will now consider how χ changes if we modify the ensemble slightly. In the modified ensemble, a new state ω occurs with probability η and the state ρ_a occurs with probability $(1-\eta)p_a$. For the modified ensemble,

$$\rho' = \eta\omega + (1-\eta)\rho \tag{37}$$

$$\chi' = \eta \mathcal{S}(\omega\|\rho') + (1-\eta)\sum_a p_a \mathcal{S}(\rho_a\|\rho'). \tag{38}$$

We can apply Donald's identity to these ensembles in two different ways. First, we can take the original optimal ensemble and treat ρ' as the other state (σ in Eq. 5), obtaining:

$$\sum_a p_a \mathcal{S}(\rho_a\|\rho') = \chi + \mathcal{S}(\rho\|\rho'). \tag{39}$$

Substituting this expression into the expression for χ' yields:
$$\chi' = \eta S(\omega||\rho') + (1-\eta)(\chi + S(\rho||\rho'))$$
$$\Delta\chi = \chi' - \chi$$
(40)
$$= \eta(S(\omega||\rho') - \chi) + \eta S(\rho||\rho')$$

Our second application of Donald's identity is to the modified ensemble, taking the original average state ρ to play the role of the other state:

(41) $$\eta S(\omega||\rho) + (1-\eta)\chi = \chi' + S(\rho'||\rho)$$
(42) $$\Delta\chi = \eta(S(\omega||\rho) - \chi) - S(\rho'||\rho).$$

Since the relative entropy is never negative, we can conclude that

(43) $$\eta(S(\omega||\rho') - \chi) \leq \Delta\chi \leq \eta(S(\omega||\rho) - \chi).$$

This gives upper and lower bounds for the change in χ if we mix in an additional state ω to our original ensemble. The bounds are "tight", since as $\eta \to 0$, $S(\omega||\rho') \to S(\omega||\rho)$.

Very similar bounds for $\Delta\chi$ apply if we make more elaborate modifications of our original ensemble, involving more than one additional signal state. This is described in [**16**].

We say that an ensemble has the *maximal distance property* if and only if, for any ω in $\mathcal{A}$,

(44) $$S(\omega||\rho) \leq \chi,$$

where ρ is the average state and χ is the Holevo quantity for the ensemble. This property gives an interesting characterization of optimal ensembles:

Theorem: *An ensemble is optimal if and only if it has the maximum distance property.*

We give the essential ideas of the proof here; further details can be found in [**16**].

Suppose our ensemble has the maximum distance property. Then, if we add the state ω with probability η, the change $\Delta\chi$ satisfies

(45) $$\Delta\chi \leq \eta(S(\omega||\rho) - \chi) \leq 0.$$

In other words, we cannot increase χ by mixing in an additional state. Consideration of more general changes to the ensemble leads to the same conclusion that $\Delta\chi \leq 0$. Thus, the ensemble must be optimal, and $\chi = \chi^*$.

Conversely, suppose that the ensemble is optimal (with $\chi = \chi^*$). Could there be a state ω in $\mathcal{A}$ such that $S(\omega||\rho) > \chi^*$? If there were such an ω, then by choosing η small enough we could make $S(\omega||\rho') > \chi^*$, and so

(46) $$\Delta\chi \geq \eta(S(\omega||\rho') - \chi^*) > 0.$$

But this contradicts the fact that, if the original ensemble is optimal, $\Delta\chi \leq 0$ for any change in the ensemble. Thus, no such ω exists and the optimal ensemble satisfies the maximal distance property.

Two corollaries follow immediately from this theorem. First, we note that the support of the average state ρ of an optimal ensemble must contain the support of every state ω in $\mathcal{A}$. Otherwise, the relative entropy $S(\omega||\rho) = \infty$, contradicting the maximal distance property. The fact that ρ has the largest support possible could be called the *maximal support property* of an optimal ensemble.

Second, we recall that χ^* is just the average relative entropy distance of the members of the optimal ensemble from the average state ρ:
$$\chi^* = \sum_a p_a \mathcal{S}(\rho_a || \rho).$$
Since $\mathcal{S}(\rho_a || \rho) \leq \chi^*$ for each a, it follows that whenever $p_a > 0$ we must have
$$(47) \qquad \mathcal{S}(\rho_a || \rho) = \chi^*.$$
We might call this the *equal distance property* of an optimal ensemble.

We can now give an explicit formula for χ^* that does not optimize over ensembles, but only over states in $\mathcal{A}$. ¿From Equation 14, for any state σ,
$$(48) \qquad \chi \leq \sum_a p_a \mathcal{S}(\rho_a || \sigma)$$
and thus
$$(49) \qquad \chi \leq \max_\omega \mathcal{S}(\omega || \sigma)$$
where the maximum is taken over all ω in $\mathcal{A}$. We apply this inequality to the optimal ensemble, finding the lowest such upper bound for χ^*:
$$(50) \qquad \chi^* \leq \min_\sigma \left(\max_\omega \mathcal{S}(\omega || \sigma) \right).$$
But since the optimal ensemble has the maximal distance property, we know that
$$(51) \qquad \chi^* = \max_\omega \mathcal{S}(\omega || \rho)$$
for the optimal average state ρ. Therefore,
$$(52) \qquad \chi^* = \min_\sigma \left(\max_\omega \mathcal{S}(\omega || \sigma) \right).$$

5. Additivity for quantum channels

The quantity χ^* is an asymptotically achievable upper bound to the amount of classical information that can be sent using available states of the channel system Q. It is therefore tempting to identify χ^* as the classical capacity of the quantum channel. But there is a subtlety here, which involves an important unsolved problem of quantum information theory.

Specifically, suppose that two quantum systems A and B are available for use as communication channels. The two systems evolve independently according the product map $\mathcal{E}^A \otimes \mathcal{E}^B$. Each system can be considered as a separate channel, or the joint system AB can be analyzed as a single channel. It is not known whether the following holds in general:
$$(53) \qquad \chi^{AB*} \stackrel{?}{=} \chi^{A*} + \chi^{B*}.$$
Since separate signal ensembles for A and B can be combined into a product ensemble for AB, it is clear that $\chi^{AB*} \geq \chi^{A*} + \chi^{B*}$. However, the joint system AB also has other possible signal ensembles that use entangled input states and that might perhaps have a Holevo bound for the output states greater than $\chi^{A*} + \chi^{B*}$.

Equation 53 is the "additivity conjecture" for the classical capacity of a quantum channel. If the conjecture is false, then the use of entangled input states would sometimes increase the amount of classical information that can be sent over two or more independent channels. The classical capacity of a channel (which is defined asymptotically, using many instances of the same channel) would thus be greater than χ^* for a single instance of a channel. On the other hand, if the conjecture holds, then χ^* is the classical capacity of the quantum channel.

Numerical calculations to date [18] support the additivity conjecture for a variety of channels. Recent work [19, 20] gives strong evidence that Equation 53 holds for various special cases, including channels described by unital maps. We present here another partial result: χ^* is additive for any "half-noisy" channel, that is, a dual channel that is represented by an map of the form $\mathcal{I}^A \otimes \mathcal{E}^B$, where $\mathcal{I}^A$ is the identity map on A.

Suppose the joint system AB evolves according to the map $\mathcal{I}^A \otimes \mathcal{E}^B$, and let ρ^A and ρ^B be the average output states of optimal signal ensembles for A and B individually. We will show that the product ensemble (with average state $\rho^A \otimes \rho^B$) is optimal by showing that this ensemble has the maximal distance property. That is, suppose we have another, possibly entangled input state of AB that leads to the output state ω^{AB}. Our aim is to prove that $\mathcal{S}\left(\omega^{AB} \| \rho^A \otimes \rho^B\right) \leq \chi^{A*} + \chi^{B*}$. From the definition of $\mathcal{S}(\cdot \| \cdot)$ we can show that

$$
\begin{aligned}
\mathcal{S}\left(\omega^{AB} \| \rho^A \otimes \rho^B\right) &= -S\left(\omega^{AB}\right) - \mathrm{Tr}\, \omega^A \log \rho^A - \mathrm{Tr}\, \omega^B \log \rho^B \\
&= S\left(\omega^A\right) + S\left(\omega^B\right) - S\left(\omega^{AB}\right) \\
&\quad + \mathcal{S}\left(\omega^A \| \rho^A\right) + \mathcal{S}\left(\omega^B \| \rho^B\right).
\end{aligned}
\tag{54}
$$

(The right-hand expression has an interesting structure; $S(\omega^A) + S(\omega^B) - S(\omega^{AB})$ is clearly analogous to the mutual information defined in Equation 9.)

Since A evolves according to the identity map $\mathcal{I}^A$, it is easy to see that $\chi^{A*} = d = \dim \mathcal{H}^A$ and

$$
\rho^A = \left(\frac{1}{d}\right) 1^A.
\tag{55}
$$

¿From this it follows that

$$
S\left(\omega^A\right) + \mathcal{S}\left(\omega^A \| \rho^A\right) = \log d = \chi^{A*}
\tag{56}
$$

for any ω^A. This accounts for two of the terms on the right-hand side of Equation 54. The remaining three terms require a more involved analysis.

The final joint state ω^{AB} is a mixed state, but we can always introduce a third system C that "purifies" the state. That is, we can find $\left|\Omega^{ABC}\right\rangle$ such that

$$
\omega^{AB} = \mathrm{Tr}_C \left|\Omega^{ABC}\right\rangle\!\left\langle\Omega^{ABC}\right|.
\tag{57}
$$

Since the overall state of ABC is a pure state, $S(\omega^{AB}) = S(\omega^C)$, where ω^C is the state obtained by partial trace over A and B. Furthermore, imagine that a complete measurement is made on A, with the outcome k occuring with probability p_k. For a given measurement outcome k, the subsequent state of the remaining system BC will be $\left|\Omega_k^{BC}\right\rangle$. Letting

$$
\begin{aligned}
\omega_k^B &= \mathrm{Tr}_C \left|\Omega_k^{BC}\right\rangle\!\left\langle\Omega_k^{BC}\right| \\
\omega_k^C &= \mathrm{Tr}_B \left|\Omega_k^{BC}\right\rangle\!\left\langle\Omega_k^{BC}\right|,
\end{aligned}
\tag{58}
$$

we have that $S(\omega_k^B) = S(\omega_k^C)$ for all k. Furthermore, by locality,

$$
\begin{aligned}
\omega^B &= \sum_k p_k \omega_k^B \\
\omega^C &= \sum_k p_k \omega_k^C.
\end{aligned}
\tag{59}
$$

In other words, we have written both ω^B and ω^C as ensembles of states.

We can apply this to get an upper bound on the remaining terms in Equation 54

$$S\left(\omega^B\right) - S\left(\omega^{AB}\right) + \mathcal{S}\left(\omega^B\|\rho^B\right)$$
$$= S\left(\omega^B\right) - \sum_k p_k S\left(\omega_k^B\right)$$
$$-S\left(\omega^C\right) + \sum_k p_k S\left(\omega_k^C\right) + \mathcal{S}\left(\omega^B\|\rho^B\right)$$
(60) $$\leq \chi_\omega^B + \mathcal{S}\left(\omega^B\|\rho^B\right),$$

where χ_ω^B is the Holevo quantity for the ensemble of ω_k^B states. Donald's identity permits us to write

(61) $$S\left(\omega^B\right) - S\left(\omega^{AB}\right) + \mathcal{S}\left(\omega^B\|\rho^B\right) = \sum_k p_k \mathcal{S}\left(\omega_k^B\|\rho^B\right).$$

The B states ω_k^B are all available output states of the B channel. These states are obtained by making a complete measurement on system A when the joint system AB is in the state ω^{AB}. But this state was obtained from some initial AB state and a dynamical map $\mathcal{I}^A \otimes \mathcal{E}^B$. This map commutes with the measurement operation on A alone, so we could equally well make the measurement *before* the action of $\mathcal{I}^A \otimes \mathcal{E}^B$. The A-measurement outcome k would then determine the input state of B, which would evolve into ω_k^B. Thus, for each k, ω_k^B is a possible output of the $\mathcal{E}^B$ map.

Since ρ^B has the maximum distance property and the states ω_k^B are available outputs of the channel, $\mathcal{S}\left(\omega_k^B\|\rho^B\right) \leq \chi^{B*}$ for every k. Combining Equations 54, 56 and 61, we find the desired inequality:

(62) $$\mathcal{S}\left(\omega^{AB}\|\rho^A \otimes \rho^B\right) \leq \chi^{A*} + \chi^{B*}.$$

This demonstrates that the product of optimal ensembles for A and B also has the maximum distance property for the possible outputs of the joint channel, and so this product ensemble must be optimal. It follows that $\chi^{AB*} = \chi^{A*} + \chi^{B*}$ in this case.

Our result has been phrased for the case in which A undergoes "trivial" dynamics $\mathcal{I}^A$, but the proof also works without modification if the time evolution of A is unitary—that is, A experiences "distortion" but not "noise". If only one of the two systems is noisy, then χ^* is additive.

The additivity conjecture for χ^* is closely related to another additivity conjecture, the "minimum output entropy" conjecture [19, 20]. Suppose A and B are systems with independent evolution described by $\mathcal{E}^A \otimes \mathcal{E}^B$, and let ρ^{AB} be an output state of the channel with minimal entropy $S(\rho^A B)$. Is ρ^{AB} a product state $\rho^A \otimes \rho^B$? The answer is not known in general; but it is quite easy to show this in the half-noisy case that we consider here.

6. Maximizing coherent information

When we turn from the transmission of classical information to the transmission of quantum information, it will be helpful to adopt an explicit description of the channel dynamics, instead of merely specifying the set of available output states $\mathcal{A}$. Suppose the quantum system Q undergoes a dynamical evolution described by the map $\mathcal{E}$. Since $\mathcal{E}$ is a trace-preserving, completely positive map, we can always find a representation of $\mathcal{E}$ as a unitary evolution of a larger system [6]. In this representation, we imagine that an additonal "environment" system E is present, initially in a pure state $|\breve{0}^E\rangle$, and that Q and E interact via the unitary evolution operator U^{QE}. That is,

$$\rho^Q = \mathcal{E}(\breve{\rho}^Q) = \mathrm{Tr}\,_E U^{QE}\left(\breve{\rho}^Q \otimes |\breve{0}^E\rangle\langle\breve{0}^E|\right) U^{QE\dagger}. \tag{63}$$

For convenience, we denote an initial state of a system by the breve accent (as in $\breve{\rho}^Q$), and omit this symbol for final states.

The problem of sending quantum information through our channel can be viewed in one of two ways:

(1) An unknown pure quantum state of Q is to be transmitted. In this case, our criterion of success is the *average fidelity* $\bar{F}$, defined as follows. Suppose the input state $\left|\breve{\phi}_k\right\rangle$ occurs with probability p_k and leads to the output state ρ_k. Then

$$\bar{F} = \sum_k p_k \left\langle \breve{\phi}_k \right| \rho_k \left| \breve{\phi}_k \right\rangle. \tag{64}$$

In general, $\bar{F}$ depends not only on the average input state $\breve{\rho}^Q$ but also on the particular pure state input ensemble. [21]

(2) A second "bystander" system R is present, and the joint system RQ is initially in a pure entangled state $\left|\breve{\Psi}^{RQ}\right\rangle$. The system R has "trivial" dynamics described by the identity map $\mathcal{I}$, so that the joint system evolves according to $\mathcal{I} \otimes \mathcal{E}$, yielding a final state ρ^{RQ}. Success is determined in this case by the *entanglement fidelity* F_e, defined by

$$F_e = \left\langle \breve{\Psi}^{RQ} \right| \rho^{RQ} \left| \breve{\Psi}^{RQ} \right\rangle. \tag{65}$$

It turns out, surprisingly, that F_e is only dependent on $\mathcal{E}$ and the input state $\breve{\rho}^Q$ of Q alone. That is, F_e is an "intrinsic" property of Q and its dynamics. [22]

These two pictures of quantum information transfer are essentially equivalent, since F_e approaches unity if and only if $\bar{F}$ approaches unity for every ensemble with the same average input state $\breve{\rho}^Q$. For now we adopt the second point of view, in which the transfer of quantum information is essentially the transfer of quantum entanglement (with the bystander system R) through the channel.

The quantum capacity of a channel should be defined as the amount of entanglement that can be transmitted through the channel with $F_e \to 1$, if we allow ourselves to use the channel many times and employ quantum error correction schemes [23]. At present it is not known how to calculate this *asymptotic* capacity of the channel in terms of the properties of a single instance of the channel.

Nevertheless, we can identify some quantities that are useful in describing the quantum information conveyed by the channel [24]. A key quantity is the *coherent information* I^Q, defined by

$$I^Q = S\left(\rho^Q\right) - S\left(\rho^{RQ}\right). \tag{66}$$

This quantity is a measure of the final entanglement between R and Q. (The initial entanglement is measured by the entropy $S(\breve{\rho}^Q)$ of the initial state of Q, which of course equals $S(\breve{\rho}^R)$. See Section 7 below.) If we adopt a unitary representation for $\mathcal{E}$, then the overall system RQE including the environment remains in a pure state from beginning to end, and so $S(\rho^{RQ}) = S(\rho^E)$. Thus,

$$I^Q = S(\rho^Q) - S(\rho^E). \tag{67}$$

Despite the apparent dependence of I^Q on the systems R and E, it is in fact a function only of the map $\mathcal{E}$ and the initial state $\breve{\rho}^Q$ of Q. Like the entanglement fidelity F_e, it is an "intrinsic" characteristic of the channel system Q and its dynamics.

It can be shown that the coherent information I^Q does not increase if the map $\mathcal{E}$ is followed by a second independent map $\mathcal{E}'$, giving an overall dynamics described by $\mathcal{E}' \circ \mathcal{E}$. That is, the coherent information cannot be increased by any "quantum data processing" on the channel outputs. The coherent information is also closely related to quantum error correction. Perfect quantum error correction—resulting in $F_e = 1$ for the final state—is possible if and only if the channel loses no coherent information, so that $I^Q = S(\breve{\rho}^Q)$. These and other properties lead us to consider I^Q as a good measure of the quantum information that is transmitted through the channel [24].

The coherent information has an intriguing relation to the Holevo quantity χ, and thus to classical information transfer (and to relative entropy) [25]. Suppose we describe that the input state $\breve{\rho}^Q$ by an ensemble of pure states $\left|\breve{\phi}_k^Q\right\rangle$:

$$\breve{\rho}^Q = \sum_k p_k \left|\breve{\phi}_k^Q\right\rangle\left\langle\breve{\phi}_k^Q\right|. \tag{68}$$

We adopt a unitary representation for the evolution and note that the initial pure state $\left|\breve{\phi}_k^Q\right\rangle \otimes \left|\breve{0}^E\right\rangle$ evolves into a pure, possibly entangled state $\left|\phi_k^{QE}\right\rangle$. Thus, for each k the entropies of the final states of Q and E are equal:

$$S(\rho_k^Q) = S(\rho_k^E). \tag{69}$$

It follows that

$$\begin{aligned} I^Q &= S(\rho^Q) - S(\rho^E) \\ &= S(\rho^Q) - \sum_k p_k S(\rho_k^Q) - S(\rho^E) + \sum_k p_k S(\rho_k^E) \\ I^Q &= \chi^Q - \chi^E. \end{aligned} \tag{70}$$

Remarkably, the difference $\chi^Q - \chi^E$ depends only on $\mathcal{E}$ and the average input state $\breve{\rho}^Q$, not the details of the environment E or the exact choice of pure state input ensemble.

The quantities χ^Q and χ^E are related to the classical information transfer to the output system Q and to the environment E, respectively. Thus, Equation 70 relates the classical and quantum information properties of the channel. This relation has been used to analyze the privacy of quantum cryptographic channels [25]. We will use it here to give a relative entropy characterization of the the input state $\breve{\rho}^Q$ that maximizes the coherent information of the channel.

Let us suppose that $\breve{\rho}^Q$ is an input state that maximizes the coherent information I^Q. If we change the input state to

$$\breve{\rho}^{Q\prime} = (1-\eta)\breve{\rho}^Q + \eta\breve{\omega}^Q, \tag{71}$$

for some pure state $\breve{\omega}^Q$, we produces some change ΔI^Q in the coherent information. Viewing $\breve{\rho}^Q$ as an ensemble of pure states, this change amounts to a modification of that ensemble; and such a modification leads to changes in the output ensembles for both system Q and system E. Thus,

$$\Delta I^Q = \Delta \chi^Q - \Delta \chi^E. \tag{72}$$

We can apply Equation 43 to bound both $\Delta \chi^Q$ and $\Delta \chi^E$ and obtain a lower bound for ΔI^Q:

$$\Delta I^Q \geq \eta \left(\mathcal{S} \left(\omega^Q || \rho^{Q'} \right) - \chi^Q \right) - \eta \left(\mathcal{S} \left(\omega^E || \rho^E \right) - \chi^E \right)$$
$$\Delta I^Q \geq \eta \left(\mathcal{S} \left(\omega^Q || \rho^{Q'} \right) - \mathcal{S} \left(\omega^E || \rho^E \right) - I^Q \right). \tag{73}$$

Since we assume that I^Q is maximized for the input $\breve{\rho}^Q$, then $\Delta I^Q \leq 0$ when we modify the input state. This must be true for every value of η in the relation above. Whenever $\mathcal{S}\left(\omega^Q || \rho^Q\right)$ is finite, we can conclude that

$$\mathcal{S}\left(\omega^Q || \rho^Q\right) - \mathcal{S}\left(\omega^E || \rho^E\right) \leq I^Q. \tag{74}$$

This is analogous to the maximum distance property for optimal signal ensembles, except that it is the difference of two relative entropy distances that is bounded above by the maximum of I^Q.

Let us write Equation 70 in terms of relative entropy, imagining that the input state $\breve{\rho}^Q$ is written in terms of an ensemble of pure states $\left| \breve{\phi}_k^Q \right\rangle$:

$$I^Q = \sum_k p_k \left(\mathcal{S}\left(\rho_k^Q || \rho^Q\right) - \mathcal{S}\left(\rho_k^E || \rho^E\right) \right). \tag{75}$$

Every input pure state $\left| \breve{\phi}_k^Q \right\rangle$ in the input ensemble with $p_k > 0$ will be in the support of $\breve{\rho}^Q$, and so Equation 74 holds. Therefore, we can conclude that

$$I^Q = \mathcal{S}\left(\rho_k^Q || \rho^Q\right) - \mathcal{S}\left(\rho_k^E || \rho^E\right) \tag{76}$$

for every such state in the ensemble. Furthermore, *any* pure state in the support of $\breve{\rho}^Q$ is a member of some pure state ensemble for $\breve{\rho}^Q$.

This permits us to draw a remarkable conclusion. If $\breve{\rho}^Q$ is the input state that maximizes the coherent information I^Q of the channel, then for any pure state $\bar{\omega}^Q$ in the support of $\breve{\rho}^Q$,

$$I^Q = \mathcal{S}\left(\omega^Q || \rho^Q\right) - \mathcal{S}\left(\omega^E || \rho^E\right). \tag{77}$$

This result is roughly analogous to the equal distance property for optimal signal ensembles. Together with Equation 74, it provides a strong characterization of the state that maximizes coherent information.

The additivity problem for χ^* leads us to ask whether the maximum of the coherent information is additive when independent channels are combined. In fact, there are examples known where $\max I^{AB} > \max I^A + \max I^B$; in other words, entanglement between independent channels can increase the amount of coherent information that can be sent through them [**26**]. The asymptotic behavior of coherent information and its precise connection to quantum channel capacities are questions yet to be resolved.

7. Indeterminate length quantum coding

In the previous section we saw that the relative entropy can be used to analyze the coherent information "capacity" of a quantum channel. Another issue in quantum information theory is *quantum data compression* [21], which seeks to represent quantum information using the fewest number of qubits. In this section we will see that the relative entropy describes the cost of suboptimal quantum data compression.

One approach to classical data compression is to use variable length codes, in which the codewords are finite binary strings of various lengths [1]. The best-known examples are the Huffman codes. The Shannon entropy $H(X)$ of a random variable X is a lower bound to the average codeword length in such codes, and for Huffman codes this average codeword length can be made arbitrarily close to $H(X)$. Thus, a Huffman code optimizes the use of a communication resources (number of bits required) in classical communication without noise.

There are analogous codes for the compression of quantum information. Since coherent superpositions of codewords must be allowed as codewords, these are called *indeterminate length* quantum codes [27]. A quantum analogue to Huffman coding was recently described by Braunstein et al. [28] An account of the theory of indeterminate length quantum codes, including the quantum Kraft inequality and the condensability condition (see below), will be presented in a forthcoming paper [29]. Here we will outline a few results and demonstrate a connection to the relative entropy.

The key idea in constructing an indeterminate length code is that the codewords themselves must carry their own length information. For a classical variable length code, this requirement can be phrased in two ways. A *uniquely decipherable* code is one in which any string of N codewords can be correctly separated into its individual codewords, while a *prefix-free* code is one in which no codeword is an initial segment of another codeword. The lengths of the codewords in each case satisfy the Kraft-McMillan inequality:

$$(78) \qquad \sum_k 2^{-l_k} \leq 1,$$

where is the sum is over the codewords and l_k is the length of the kth codeword. Every prefix-free code is uniquely decipherable, so the prefix-free property is a more restrictive property. Nevertheless, it turns out that any uniquely decipherable code can be replaced by a prefix-free code with the same codeword lengths.

There are analogous conditions for indeterminate length quantum codes, but these properties must be phrased carefully because we allow coherent superpositions of codewords. For example, a classical prefix-free code is sometimes called an "instantaneous" code, since as soon as a complete codeword arrives we can recognize it at once and decipher it immediately. However, if an "instantaneous" decoding procedure were to be attempted for a quantum prefix-free code, it would destroy coherences between codewords of different lengths. Quantum codes require that the entire string of codewords be deciphered together.

The property of an indeterminate length quantum code that is analogous to unique decipherability is called *condensability*. We digress briefly to describe the condensability condition. We focus on *zero-extended forms* (zef) of our codewords. That is, we cosider that our codewords occupy an initial segment of a qubit register of fixed length n, with $|0\rangle$'s following. (Clearly n must be chosen large enough to contain the longest codeword.) The set of all zef codewords spans a subspace of the Hilbert space of register states. We imagine that the output of a quantum information source has been mapped unitarily to the zef codeword space of the register. Our challenge is to take N such registers and "pack" them together in a way that can exploit the fact that some of the codewords are shorter than others.

If codeword states must carry their own length information, there must be a *length observable* Λ on the zef codeword space with the following two properties:

- The eigenvalues of Λ are integers $1, \ldots, n$, where n is the length of the register.
- If $|\psi_{\sf zef}\rangle$ is an eigenstate of Λ with eigenvalue l, then it has the form

(79) $$|\psi_{\sf zef}\rangle = |\psi^{1\cdots l} 0^{l+1\cdots n}\rangle.$$

That is, the last $n - l$ qubits in the register are in the state $|0\rangle$ for a zef codeword of length l.

For register states not in the zef subspace, we can take $\Lambda = \infty$.

A code is *condensable* if the following condition holds: For any N, there is a unitary operator U (depending on N) that maps

$$\underbrace{|\psi_{1,{\sf zef}}\rangle \otimes \cdots \otimes |\psi_{N,{\sf zef}}\rangle}_{Nn\,\text{qubits}} \to \underbrace{|\Psi_{1\cdots N}\rangle}_{Nn\,\text{qubits}}$$

with the property that, if the individual codewords are all length eigenstates, then U maps the codewords to a zef string of the Nn qubits—that is, one with $|0\rangle$'s after the first $L = l_1 + \cdots + l_N$ qubits:

$$\left|\psi_1^{1\cdots l_1} 0^{l_1+1\cdots n}\right\rangle \otimes \cdots \otimes \left|\psi_N^{1\cdots l_N} 0^{l_N+1\cdots n}\right\rangle \to \left|\Psi^{1\cdots L} 0^{L+1\cdots Nn}\right\rangle.$$

The unitary operator U thus "packs" N codewords, given in their zef forms, into a "condensed" string that has all of the trailing $|0\rangle$'s at the end. The unitary character of the packing protocol automatically yields an "unpacking" procedure given by U^{-1}. Thus, if the quantum code is condensable, a packed string of N codewords can be coherently sorted out into separated zef codewords.

The quantum analogue of the Kraft-McMillan inequality states that, for any indeterminate length quantum code that is condensable, the length observable Λ on the subspace of zef codewords must satisfy

(80) $$\text{Tr}\, 2^{-\Lambda} \leq 1,$$

where we have restricted our trace to the zef subspace. We can construct a density operator ω (a positive operator of unit trace) on the zef subspace by letting $K = \text{Tr}\, 2^{-\Lambda} \leq 1$ and

(81) $$\omega = \frac{1}{K} 2^{-\Lambda}.$$

The density operator ω is generally not the same as the actual density operator ρ of the zef codewords produced by the quantum information source. The average codeword length is

$$\begin{aligned}
\bar{l} &= \operatorname{Tr} \rho \Lambda \\
&= -\operatorname{Tr} \rho \log\left(2^{-\Lambda}\right) \\
&= -\operatorname{Tr} \rho \log \omega - \log K
\end{aligned}$$

(82) $$\bar{l} = S(\rho) + \mathcal{S}(\rho\|\omega) - \log K.$$

Since $\log K \leq 0$ and the relative entropy is positive definite,

(83) $$\bar{l} \geq S(\rho).$$

The average codeword length must always be at least as great as the von Neuman entropy of the information source.

Equality for Equation 83 can be approached asymptotically using block coding and a quantum analogue of Huffman (or Shannon-Fano) coding. For special cases in which the eigenvalues of ρ are of the form 2^{-m}, then a code exists for which $\bar{l} = S(\rho)$, without the asymptotic limit. In either case, we say that a code satisfying $\bar{l} = S(\rho)$ is a length optimizing quantum code. Equation 82 tells us that, if we have a length optimizing code, $K = 1$ and

(84) $$\rho = \omega = 2^{-\Lambda}.$$

The condensed string of N codewords has Nn qubits, but we can discard all but about $N\bar{l}$ of them and still retain high fidelity. That is, $\bar{l}$ is the asymptotic number of qubits that must be used per codeword to represent the quantum information faithfully.

Suppose that we have an indeterminate length quantum code that is designed for the wrong density operator. That is, our code is length optimizing for some other density operator ω, but $\rho \neq \omega$. Then (recalling that $K = 1$ for a length optimizing code, even if it is optimizing for the wrong density operator),

(85) $$\bar{l} = S(\rho) + \mathcal{S}(\rho\|\omega).$$

$S(\rho)$ tells us the number of qubits necessary to represent the quantum information if we used a length optimizing code for ρ. (As we have mentioned, such codes always exist in an asymptotic sense.) However, to achieve high fidelity in the situation where we have used a code designed for ω, we have to use at least $\bar{l}$ qubits per codeword, an additional cost of $\mathcal{S}(\rho\|\omega)$ qubits per codeword.

This result gives us an interpretation of the relative entropy function $\mathcal{S}(\rho\|\omega)$ in terms of the physical resources necessary to accomplish some task—in this case, the additional cost (in qubits) of representing the quantum information described by ρ using a coding scheme optimized for ω. This is entirely analogous to the situation for classical codes and classical relative entropy [1]. A fuller development of this analysis will appear in [29].

8. Relative entropy of entanglement

One recent application of relative entropy has been to quantify the entanglement of a mixed quantum state of two systems [30]. Suppose Alice and Bob share a joint quantum system AB in the state ρ^{AB}. This state is said to be *separable* if it is a product state or else a probabilistic combination of product states:

(86) $$\rho^{AB} = \sum_k p_k \rho_k^A \otimes \rho_k^B.$$

Without loss of generality, we can if we wish take the elements in this ensemble of product states to be pure product states. Systems in separable states display statistical correlations having perfectly ordinary "classical" properties—that is, they do not violate any sort of Bell inequality. A separable state of A and B could also be created from scratch by Alice and Bob using only local quantum operations (on A and B separately) and the exchange of classical information.

States which are not separable are said to be *entangled*. These states cannot be made by local operations and classical communication; in other words, their creation requires the exchange of *quantum* information between Alice and Bob. The characterization of entangled states and their possible transformations has been a central issue in much recent work on quantum information theory.

A key question is the quantification of entanglement, that is, finding numerical measures of the entanglement of a quantum state ρ^{AB} that have useful properties. If the joint system AB is in a pure state $|\Psi^{AB}\rangle$, so that the subsystem states are

$$(87) \qquad \begin{aligned} \rho^A &= \operatorname{Tr}_B |\Psi^{AB}\rangle\langle\Psi^{AB}| \\ \rho^B &= \operatorname{Tr}_A |\Psi^{AB}\rangle\langle\Psi^{AB}| \end{aligned}$$

then the entropy $S(\rho^A) = S(\rho^B)$ can be used to measure the entanglement of A and B. This measure has many appealing properties. It is zero if and only if $|\Psi^{AB}\rangle$ is separable (and thus a product state). For an "EPR pair" of qubits—that is, a state of the general form

$$(88) \qquad |\phi^{AB}\rangle = \frac{1}{\sqrt{2}} \left(|0^A 0^B\rangle + |1^A 1^B\rangle \right),$$

the susbsystem entropy $S(\rho^A) = 1$ bit.

The subsystem entropy is also an asymptotic measure, both of the resources necessary to create the particular entangled pure state, and of the value of the state as a resource [31]. That is, for sufficiently large N,

- approximately $NS(\rho^A)$ EPR pairs are required to create N copies of $|\Psi^{AB}\rangle$ by local operations and classical communication; and
- approximately $NS(\rho^A)$ EPR pairs can be created from N copies of $|\Psi^{AB}\rangle$ by local operations and classical communication.

For mixed entangled states ρ^{AB} of the joint system AB, things are not so well established. Several different measures of entanglement are known, including [32]

- the *entanglement of formation* $E(\rho^{AB})$, which is the minimum asymptotic number of EPR pairs required to create ρ^{AB} by local operations and classical communication; and
- the *distillable entanglement* $D(\rho^{AB})$, the maximum asymptotic number of EPR pairs that can be created from ρ^{AB} by entanglement purification protocols involving local operations and classical communication.

Bennett et al. [32] further distinguish D_1 and D_2, the distillable entanglements with respect to purification protocols that allow one-way and two-way classical communication, respectively. All of these measures reduce to the subsystem entropy $S(\rho^A)$ if ρ^{AB} is a pure entangled state.

These entanglement measures are not all equal; furthermore, explicit formulas for their calculation are not known in most cases. This motivates us to consider alternate measures of entanglement with more tractable properties and which have useful relations to the asymptotic measures E, D_1 and D_2.

A state ρ^{AB} is entangled inasmuch as it is not a separable state, so it makes sense to adopt as a measure of entanglement a measure of the distance of ρ^{AB} from the set Σ^{AB} of separable states of AB. Using relative entropy as our "distance", we define the *relative entropy of entanglement* E_r to be [30]

$$E_r\left(\rho^{AB}\right) = \min_{\sigma^{AB} \in \Sigma^{AB}} S\left(\rho^{AB} || \sigma^{AB}\right). \tag{89}$$

The relative entropy of entanglement has several handy properties. First of all, it reduces to the subsystem entropy $S(\rho^A)$ whenever ρ^{AB} is a pure state. Second, suppose we write ρ^{AB} as an ensemble of pure states $\left|\psi_k^{AB}\right\rangle$. Then

$$E_r\left(\rho^{AB}\right) \leq \sum_k p_k S\left(\rho_k^A\right) \tag{90}$$

where $\rho_k^A = \mathrm{Tr}_B \left|\psi_k^{AB}\right\rangle\left\langle\psi_k^{AB}\right|$. It follows from this that $E_r \leq E$ for any state ρ^{AB}.

Even more importantly, the relative entropy of entanglement E_r can be shown to be non-increasing on average under local operations by Alice and Bob together with classical communication between them.

The quantum version of Sanov's theorem gives the relative entropy of entanglement an interpretation in terms of the statistical distinguishability of ρ^{AB} and the "least distinguishable" separable state σ^{AB}. The relative entropy of entanglement is thus a useful and well-motivated measure of the entanglement of a state ρ^{AB} of a joint system, both on its own terms and as a surrogate for less tractable asymptotic measures.

9. Manipulating multiparticle entanglement

The analysis in this section closely follows that of Linden et al. [33], who provides a more detailed discussion of the main result here and its applications.

Suppose Alice, Bob and Claire initially share three qubits in a "GHZ state"

$$\left|\Psi^{ABC}\right\rangle = \frac{1}{\sqrt{2}}\left(\left|0^A 0^B 0^C\right\rangle + \left|1^A 1^B 0^C\right\rangle\right). \tag{91}$$

The mixed state ρ^{BC} shared by Bob and Claire is, in fact, not entangled at all:

$$\rho^{BC} = \frac{1}{2}\left(\left|0^B 0^C\right\rangle\left\langle 0^B 0^C\right| + \left|1^B 1^C\right\rangle\left\langle 1^B 1^C\right|\right). \tag{92}$$

No local operations performed by Bob and Claire can produce an entangled state from this starting point. However, Alice can create entanglement for Bob and Claire. Alice measures her qubit in the basis $\{\left|+^A\right\rangle, \left|-^A\right\rangle\}$, where

$$\left|\pm^A\right\rangle = \frac{1}{\sqrt{2}}\left(\left|0^A\right\rangle \pm \left|1^A\right\rangle\right). \tag{93}$$

It is easy to verify that the state of Bob and Claire's qubits after this measurement, depending on the measurement outcome, must be one of the two states

$$\left|\phi_\pm^{BC}\right\rangle = \frac{1}{\sqrt{2}}\left(\left|0^A 0^B\right\rangle \pm \left|1^A 1^B\right\rangle\right), \tag{94}$$

both of which are equivalent (up to a local unitary transformation by either Bob or Claire) to an EPR pair. In other words, if Alice makes a local measurement on her qubit and then announces the result by classical communication, the GHZ triple can be converted into an EPR pair for Bob and Claire.

When considering the manipulation of quantum entanglement shared among several parties, we must therefore bear in mind that the entanglement between subsystems can both increase and decrease, depending on the situation. This raises several questions: Under what circumstances can Alice increase Bob and Claire's entanglement? How much can she do so? Are there any costs involved in the process?

To study these questions, we must give a more detailed account of "local operations and classical communication". It turns out that Alice, Bob and Claire can realize any local operation on their joint system ABC by a combination of the following:

- Local unitary transformations on the subsystems A, B and C;
- Adjoining to a subsystem additional local "ancilla" qubits in a standard state $|0\rangle$;
- Local ideal measurements on the (augmented) subsystems A, B and C; and
- Discarding local ancilla qubits.

Strictly speaking, though, we do not need to include the last item. That is, any protocol that involves discarding ancilla qubits can be replaced by one in which the ancillas are simply "set aside"—not used in future steps, but not actually gotten rid of. In a similar vein, we can imagine that the ancilla qubits required are already present in the subsystems A, B and C, so the second item in our list is redundant. We therefore need to consider only local unitary transformations and local ideal measurements.

What does classical communication add to this? It is sufficient to suppose that Alice, Bob and Claire have complete information—that is, they are aware of all operations and the outcomes of all measurements performed by each of them, and thus know the global state of ABC at every stage. Any protocol that involved an incomplete sharing of information could be replaced by one with complete sharing, simply by ignoring some of the messages that are exchanged.

Our local operations (local unitary transformations and local ideal measurements) always take an initial pure state to a final pure state. That is, if ABC starts in the joint state $|\Psi^{ABC}\rangle$, then the final state will be a pure state $|\Psi_k^{ABC}\rangle$ that depends on the joint outcome k of all the measurements performed. Thus, ABC is always in a pure state known to all parties.

It is instructive to consider the effect of local operations on the entropies of the various subsystems of ABC. Local unitary transformations leave $S(\rho^A)$, $S(\rho^B)$ and $S(\rho^C)$ unchanged. But suppose that Alice makes an ideal measurement on her subsystem, obtaining outcome k with probability p_k. The initial global state is $|\Psi^{ABC}\rangle$ and the final global state is $|\Psi_k^{ABC}\rangle$, depending on k. For the initial subsystem states, we have that

$$S\left(\rho^A\right) = S\left(\rho^{BC}\right) \tag{95}$$

since the overall state is a pure state. Similarly, the various final subsystem states satisfy

$$S\left(\rho_k^A\right) = S\left(\rho_k^{BC}\right). \tag{96}$$

But an operation on A cannot change the average state of BC:

$$\rho^{BC} = \sum_k p_k \rho_k^{BC}. \tag{97}$$

Concavity of the entropy gives

$$S\left(\rho^{BC}\right) \geq \sum_k p_k S\left(\rho_k^{BC}\right) \tag{98}$$

and therefore

$$S\left(\rho^A\right) \geq \sum_k p_k S\left(\rho_k^A\right). \tag{99}$$

Concavity also tells us that $S(\rho^B) \geq \sum_k p_k S(\rho_k^B)$, etc., and similar results hold for local measurements performed by Bob or Claire.

We now return to the question of how much Alice can increase the entanglement shared by Bob and Claire. Let us measure the bipartite entanglement of the system BC (which may be in a mixed state) by the relative entropy of entanglement $E_r(\rho^{BC})$, and let σ^{BC} be the separable state of BC for which

$$E_r(\rho^{BC}) = \mathcal{S}\left(\rho^{BC} \| \sigma^{BC}\right). \tag{100}$$

No local unitary operation can change $E_r(\rho^{BC})$; furthermore, no local measurement by Bob or Claire can increase $E_r(\rho^{BC})$ on average. We need only consider an ideal measurement performed by Alice on system A. Once again we suppose that outcome k of this measurement occurs with probability p_k, and once again Equation 97 holds. Donald's identity tells us that

$$\sum_k p_k \mathcal{S}\left(\rho_k^{BC} \| \sigma^{BC}\right) = \sum_k p_k \mathcal{S}\left(\rho_k^{BC} \| \rho^{BC}\right) + \mathcal{S}\left(\rho^{BC} \| \sigma^{BC}\right). \tag{101}$$

But $E_r(\rho_k^{BC}) \leq \mathcal{S}\left(\rho_k^{BC} \| \sigma^{BC}\right)$ for every k, leading to the following inequality:

$$\sum_k p_k E_r(\rho_k^{BC}) - E_r(\rho^{BC}) \leq \sum_k p_k \mathcal{S}\left(\rho_k^{BC} \| \rho^{BC}\right). \tag{102}$$

We recognize the left-hand side of this inequality χ for the ensemble of post-measurement states of BC, which we can rewrite using the definition of χ in Equation 11. This yields:

$$\begin{aligned}\sum_k p_k E_r(\rho_k^{BC}) - E_r(\rho^{BC}) &\leq S\left(\rho^{BC}\right) - \sum_k p_k S\left(\rho_k^{BC}\right) \\ &= S\left(\rho^A\right) - \sum_k p_k S\left(\rho_k^A\right),\end{aligned} \tag{103}$$

since the overall state of ABC is pure at every stage.

To summarize, in our model (in which all measurements are ideal, all classical information is shared, and no classical or quantum information is ever discarded), the following principles hold:

- The entropy of any subsystem A cannot be increased on average by any local operations.
- The relative entropy of entanglement of two subsystems B and C cannot be increased on average by local operations on those two subsystems.

- The relative entropy of entanglement of B and C can be increased by a measurement performed on a third subsystem A, but the average increase in E_r^{BC} is no larger than the average decrease in the entropy of A.

We say that a joint state $\left|\Psi_1^{ABC}\right\rangle$ can be transformed *reversibly* into $\left|\Psi_2^{ABC}\right\rangle$ if, for sufficiently large N, N copies of $\left|\Psi_1^{ABC}\right\rangle$ can be transformed with high probability (via local operations and classical communication) to approximately N copies of $\left|\Psi_2^{ABC}\right\rangle$, and *vice versa*. The qualifiers in this description are worth a comment or two. "High probability" reflects the fact that, since the local operations may involve measurements, the actual final state may depend on the exact measurement outcomes. "Approximately N copies" means more than $(1-\epsilon)N$ copies, for some suitably small ϵ determined in advance. We denote this reversibility relation by

$$\left|\Psi_1^{ABC}\right\rangle \leftrightarrow \left|\Psi_2^{ABC}\right\rangle.$$

Two states that are related in this way are essentially equivalent as "entanglement resources". In the large N limit, they may be interconverted with arbitrarily little loss.

Our results for entropy and relative entropy of entanglement allow us to place necessary conditions on the reversible manipulation of multiparticle entanglement. For example, if $\left|\Psi_1^{ABC}\right\rangle \leftrightarrow \left|\Psi_2^{ABC}\right\rangle$, then the two states must have exactly the same subsystem entropies. Suppose instead that $S(\rho_1^A) < S(\rho_2^A)$. Then the transformation of N copies of $\left|\Psi_1^{ABC}\right\rangle$ into about N copies of $\left|\Psi_2^{ABC}\right\rangle$ would involve an increase in the entropy of subsystem A, which cannot happen on average.

In a similar way, we can see that $\left|\Psi_1^{ABC}\right\rangle$ and $\left|\Psi_2^{ABC}\right\rangle$ must have the same relative entropies of entanglement for every pair of subsystems. Suppose instead that $E_{r,1}^{BC} < E_{r,2}^{BC}$. Then the transformation of N copies of $\left|\Psi_1^{ABC}\right\rangle$ into about N copies of $\left|\Psi_2^{ABC}\right\rangle$ would require an increase in E_r^{BC}. This can take place if a measurement is performed on A, but as we have seen this would necessarily involve a decrease in $S(\rho^A)$. Therefore, reversible transformations of multiparticle entanglement must preserve both subsystem entropies and the entanglement (measured by E_r) of pairs of subsystems.

As a simple example of this, suppose Alice, Bob and Claire share two GHZ states. Each subsystem has an entropy of 2.0 bits. This would also be the case if Alice, Bob and Claire shared three EPR pairs, one between each pair of participants. Does it follow that two GHZs can be transformed reversibly (in the sense described above) into three EPRs?

No. If the three parties share two GHZ triples, then Bob and Claire are in a completely unentangled state, with $E_r^{BC} = 0$. But in the "three EPR" situation, the relative entropy of entanglement E_r^{BC} is 1.0 bits, since they share an EPR pair. Thus, two GHZs cannot be reversibly transformed into three EPRs; indeed, $2N$ GHZs are inequivalent to $3N$ EPRs.

Though we have phrased our results for three parties, they are obviously applicable to situations with four or more separated subsystems. In reversible manipulations of multiparticle entanglement, all subsystem entropies (including the entropies of clusters of subsystems) must remain constant, as well as the relative entropies of entanglement of all pairs of subsystems (or clusters of subsystems).

10. Remarks

The applications discussed here show the power and the versatility of relative entropy methods in attacking problems of quantum information theory. We have derived useful fundamental results in classical and quantum information transfer, quantum data compression, and the manipulation of quantum entanglement. In particular, Donald's identity proves to be an extremely useful tool for deriving important inequalities.

One of the insights provided by quantum information theory is that the von Neumann entropy $S(\rho)$ has an interpretation (actually several interpretations) as a measure of the resources necessary to perform an information task. We have seen that the relative entropy also supports such interpretations. We would especially like to draw attention to the results in Sections 3 on the cost of communication and Section 7 on quantum data compression, which are presented here for the first time.

We expect that relative entropy techniques will be central to further work in quantum information theory. In particular, we think that they show promise in resolving the many perplexing additivity problems that face the theory at present. Section 5, though not a very strong result in itself, may point the way along this road.

The authors wish to acknowledge the invaluable help of many colleagues. T. Cover, M. Donald, M. Neilsen, M. Ruskai, A. Uhlmann and V. Vedral have given us indispensible guidance about the properties and meaning of the relative entropy function. Our work on optimal signal ensembles and the additivity problem was greatly assisted by conversations with C. Fuchs, A. Holevo, J. Smolin, and W. Wootters. Results described here on reversibility for transformations of multiparticle entanglement were obtained in the course of joint work with N. Linden and S. Popescu. We would like to thank the organizers of the AMS special session on "Quantum Information and Computation" for a stimulating meeting and an opportunity to pull together several related ideas into the present paper. We hope it will serve as a spur for the further application of relative entropy methods to problems of quantum information theory.

References

[1] T. M. Cover and J. A. Thomas, *Elements of Information Theory* (Wiley, New York, 1991).
[2] A. Wehrl, *Rev. Mod. Phys.* **50**, 221 (1978).
[3] E. Leib and M. B. Ruskai, *Phys. Rev. Lett.* **30**, 434 (1973); E. Leib and M. B. Ruskai, *J. Math. Phys.* **14**, 1938 (1973).
[4] M. J. Donald, Math. Proc. Cam. Phil. Soc. **101**, 363 (1987).
[5] F. Hiai and D. Petz, Comm. Math. Phys. **143**, 99 (1991). V. Vedral, M. B. Plenio, K. Jacobs and P. L. Knight, Phys. Rev. A **56**, 4452 (1997).
[6] W. F. Stinespring, *Proc. of Am. Math. Soc.* **6**, 211 (1955); K. Kraus, *Annals of Phys.* **64**, 311 (1971); K. Hellwig and K. Kraus, *Comm. Math. Phys.* **16**, 142 (1970); M.-D. Choi, *Lin. Alg. and Its Applications* **10**, 285 (1975); K. Kraus, *States, Effects and Operations: Fundamental Notions of Quantum Theory* (Springer-Verlag, Berlin, 1983).
[7] J. P. Gordon, "Noise at optical frequencies; information theory," in *Quantum Electronics and Coherent Light; Proceedings of the International School of Physics "Enrico Fermi," Course XXXI*, P. A. Miles, ed., (Academic Press, New York, 1964), pp. 156–181.
[8] L. B. Levitin, "On the quantum measure of the amount of information," in *Proceedings of the IV National Conference on Information Theory*, Tashkent, 1969, pp. 111–115 (in Russian); "Information Theory for Quantum Systems," in *Information, Complexity, and Control in Quantum Physics*, edited by A. Blaquière, S. Diner, and G. Lochak (Springer, Vienna, 1987).

[9] A. S. Holevo, *Probl. Inform. Transmission* **9**, 177 (1973) (translated from *Problemy Peredachi Informatsii*).
[10] A. S. Holevo, IEEE Trans. Inform. Theory **44**, 269 (1998).
[11] B. Schumacher and M. Westmoreland, *Phys. Rev. A* **51**, 2738 (1997).
[12] P. Hausladen, R. Josza, B. Schumacher, M. Westmoreland, and W. K. Wootters, *Phys. Rev. A* **54**, 1869 (1996).
[13] B. Schumacher, *Communication, Correlation and Complementarity*, Ph.D. thesis, the University of Texas at Austin (1990).
[14] G. Lindblad, *Non-Equilibrium Entropy and Irreversibility*, (Reidel, Dordrecht, 1983); M. Donald, *J. Stat. Phys.* **49**, 81 (1987); H. M. Partovi, *Phys. Lett. A* **137**, 440 (1989).
[15] R. Landauer, *IBM J. Res. Develop.* **5**, 183 (1961); V. Vedral, *Proc. Royal Soc.* (to appear, 2000). LANL e-print quant-ph/9903049.
[16] B. Schumacher and M. Westmoreland, "Optimal signal ensembles", submitted *Phys. Rev. A*. LANL e-print quant-ph/9912122.
[17] A. Uhlmann, Open Sys. and Inf. Dynamics **5**, 209 (1998).
[18] C. H. Bennett, C. Fuchs and J. A. Smolin, "Entanglement enhanced classical communication on a noisy quantum channel", in: Proc. 3d Int. Conf. on Quantum Communication and Measurement, ed. by C. M. Caves, O. Hirota, A. S. Holevo, Plenum, NY 1997. LANL e-print quant-ph/9611006.
[19] C. King and M. B. Ruskai, "Minimal entropy of states emerging from noisy quantum channels". LANL e-print quant-ph/9911079.
[20] G. G. Amosov, A. S. Holveo and R. F. Werner, "On some additivity problems in quantum information theory", LANL e-print quant-ph/0003002.
[21] B. Schumacher, *Phys. Rev. A* **51**, 2738 (1995); R. Jozsa and B. Schumacher, *J. Mod. Opt.* **41**, 2343 (1994); H. Barnum, C. A. Fuchs, R. Jozsa and B. Schumacher, *Phys. Rev. A* **54**, 4707 (1996).
[22] B. Schumacher, *Phys. Rev. A* **54**, 2614 (1996).
[23] H. Barnum, M. A. Nielsen and B. Schumacher, *Phys. Rev. A* **57**, 4153 (1998).
[24] B. Schumacher and M. A. Nielsen, *Phys. Rev. A* **54**, 2629 (1996).
[25] B. Schumacher and M. Westmoreland, *Physical Review Letters*, **80** (June, 1998), 5695 - 5697.
[26] D. P. DiVincenzo, P. Shor and J. Smolin, *Phys. Rev. A* **57**, 830 (1998).
[27] B. Schumacher, presentation at the Santa Fe Institute workshop on Complexity, Entropy and the Physics of Information (1994).
[28] S. L. Braunstein, C. A. Fuchs, D. Gottesman and H.-K. Lo, *IEEE Trans. Inf. Theory* (to appear, 2000).
[29] B. Schumacher and M. Westmoreland, "Indeterminate Length Quantum Coding" (in preparation).
[30] V. Vedral, M. B. Plenio, M. A. Rippin and P. L. Knight, *Phys. Rev. Lett.* **78**, 2275.
[31] C. H. Bennett, H. Bernstein, S. Popescu and B. W. Schumacher, *Phys. Rev. A* **53**, 3824 (1996).
[32] C. H. Bennett, D. P. DiVincenzo, J. A. Smolin and W. K. Wootters, *Phys. Rev. A* **54**, 3824 (1996).
[33] N. Linden, S. Popescu, B. Schumacher and M. Westmoreland, "Reversibility of local transformations of multiparticle entanglement", submitted to *Phys. Rev. Lett.* LANL e-print quant-ph/9912039.

(Benjamin Schumacher) DEPARTMENT OF PHYSICS, KENYON COLLEGE, GAMBIER, OH 43022 USA

(Michael D. Westmoreland) DEPARTMENT OF MATHEMATICAL SCIENCES, DENISON UNIVERSITY, GRANVILLE, OH 43023 USA

An Unentangled Gleason's Theorem

Nolan R. Wallach

ABSTRACT. The purpose of this note is to give a generalization of Gleason's theorem inspired by recent work in quantum information theory. For multipartite quantum systems, each of dimension three or greater, the only nonnegative frame functions over the set of unentangled states are those given by the standard Born probability rule. However, if one system is of dimension 2 this is not necessarily the case.

1. Introduction.

Let H be a Hilbert space with unit sphere $S(H)$. Following Gleason ([Gleason]) we will call a function $f : S(H) \to \mathbb{C}$ a frame function of weight w if for every orthonormal basis $\{v_i\}$ of $S(H)$

$$\sum_i f(v_i) = w. \tag{1}$$

In [Gleason] the following theorem was proved

THEOREM 1. *If* $\dim H \geq 3$ *and* f *is a frame function that takes non-negative real values then there exists a self adjoint trace class operator* $T : H \to H$ *such that*

$$f(v) = \langle v|T|v\rangle, v \in S(H).$$

This theorem is of importance to quantum mechanics because it allows a significant weakening of the axioms, showing that the Born probability rule [Born] provides the unique class of probability assignments for measurement outcomes so long as those probabilities are specified by frame functions [Pitowsky]. The theorem also rules out a large class of hidden-variable explanations for quantum statistics, the so-called noncontextual hidden variables, in dimension 3 or greater. The interested reader should consult [Bell] for a discussion of this point. If the Hilbert space is of dimension 2, then the statement in the theorem is easily seen to be false.

The purpose of this note is to give a generalization of Gleason's theorem inspired by recent work in quantum information theory. In that context the issue of *local* measurements and operations on multipartite quantum systems (as opposed to the full set of operations) is of the utmost importance [BDFMRSSW]. For instance,

---

2000 *Mathematics Subject Classification.* Primary 81P68.

it has been pointed out that probabilities for the outcomes of local measurements are enough to uniquely specify the quantum state from which they arise if the field of the Hilbert space is complex, though this fails for real and quaternionic Hilbert spaces [Araki,Wootters].

Chris Fuchs has asked to what extent local and semi-local measurements not only uniquely specify the quantum state, but also a Born-like rule as in Gleason's result [Fuchs]. In this regard, the following formalization appears natural. We confine our attention to finite dimensional Hilbert spaces for the sake of simplicity. Let $H_1, ..., H_n$ be Hilbert spaces. Set $H = H_1 \otimes H_2 \otimes \cdots \otimes H_n$. Let $\Sigma = \Sigma(H_1, ..., H_n)$ denote the subset of $S(H)$ consisting of those elements of the form $a_1 \otimes \cdots \otimes a_n$ with $a_i \in S(H_i)$ for $i = 1, ..., n$. In the jargon of quantum information theory such states are called *unentangled* or *product states*. The ones that are not of this form are said to be *entangled*. An orthonormal basis $\{v_i\}$ of H is said to be unentangled if $v_i \in \Sigma$ for all i. We say that $f : \Sigma \to \mathbb{C}$ is an unentangled frame function of weight w if whenever $\{v_i\}$ is an unentangled orthonormal basis of H then f satisfies (1) above. We establish the following result.

THEOREM 2. *If* $\dim H_i \geq 3$ *for all* i *and if* $f : \Sigma \to \mathbb{R}$ *is a non-negative unentangled frame function then there exists* $T : H \to H$ *a self adjoint trace class operator such that* $f(v) = \langle v|T|v \rangle$ *for all* $v \in \Sigma$.

This theorem is an almost direct consequence of Gleason's original theorem. We will give a proof of it in the next section. The second result in this paper shows that the dimensional condition is necessary.

It should be noted however, that despite the absence of entangled or "nonlocal" states in Σ, in [BDFMRSSW] it is asserted that not all unentangled bases correspond to quantum measurements that can be carried out by local means alone (even with iterative procedures based on weak local measurements and unlimited amounts of classical communication between the measurers at each site). The simplest kind of purely local measurement is given by an alternative type of basis adapted to the tensor product structure. This is a *product basis* and is defined as to be a basis of the form $\{u_{i_1 1} \otimes u_{i_2 2} \otimes \cdots \otimes u_{i_n n}\}$ where $u_{1j}, ..., u_{n_j j}$ is an orthonormal basis of H_j. We could define a product frame function in the same way as we did for an unentangled frame function except that we only assume that there exists a weight w such that $\sum_{i_1, i_2, ..., i_n} f(u_{i_1 1} \otimes u_{i_2 2} \otimes \cdots \otimes u_{i_n n}) = w$ for every product basis. One can ask whether this is all that us necessary for the conclusion of the theorem above. The answer is no and a method of "finding" a large class of examples will be given at the end of the next section (see the proposition at the end of the section). This result amasses some evidence that the structure of local measurements alone is not enough to establish the Born rule for multipartite systems, but a full answer would require consideration of the largest class of local measurements in [BDFMRSSW].

These issues also spawn another theorem.

THEOREM 3. *Let* $\dim H_1 = 2$ *and let* $f : S(H_1) \to \mathbb{C}$ *be a frame function of weight* w_1 *and* $g : \Sigma(H_2, ..., H_n) \to \mathbb{C}$ *be an unentangled frame function of weight* w_2. *We set* $h(v_1 \otimes u) = f(v_1)g(u)$ *for* $u \in \Sigma(H_2, ..., H_n)$. *Then* h *is an unentangled frame function of weight* $w_1 w_2$.

This result is a bit harder and the proof involves a method (see Theorem 5) that describes a combinatorial scheme for finding all unentangled orthonormal bases where all of the spaces, H_i, have dimension 2. This analysis in turn leads to a natural question. Given and unentangled orthonormal set can it be extended to an unentangled orthonormal basis? Or even stronger: Can it be a proper subset of an unentangled orthonormal set? This question was studied in [BDMSST]. We conclude the paper by giving a proof based on simple algebraic geometry of the following theorem which is related to the bound that occurs in [BDMSST].

THEOREM 4. *Let V be a subspace of $H_1 \otimes \cdots \otimes H_n$ such that if $v \in V$ and $v \neq 0$ then v is entangled. Then $\dim V \leq \dim(H_1) \cdots \dim(H_n) - \sum(\dim H_i - 1) - 1$. Furthermore, the upper bound is attained.*

2. The unentangled Gleason theorem.

In this section we will give a proof of Theorem 2. If $n = 1$ the statement is just Gleason's theorem. We consider the situation of $H = H_0 \otimes V$ with $V = H_1 \otimes H_2 \otimes \cdots \otimes H_n$ and $\dim H_i \geq 3$ for all i. We prove Theorem 1 by induction (i.e. assume the result for n). We note that if $\{v_i\}$ is an orthonormal basis of H_0 and if for each i, $\{u_{ij}\}$ is an unentangled orthonormal basis of V then the set $\{v_i \otimes u_{ij}\}$ is an unentangled orthonormal basis of H. Thus if w is the weight of f then we have

$$\sum_j f(v_1 \otimes u_{1j}) = w - \sum_{i \geq 2, j} f(v_i \otimes u_{ij}).$$

Thus for each $v \in S(H_0)$ the function $f_v(u) = f(v \otimes u)$ is an unentangled frame function. The inductive hypothesis implies that for each $v \in S(H_0)$ there exists a self adjoint (due to the reality of f) linear operator $T(v)$ such that $f(v \otimes u) = \langle u|T(v)|u\rangle$ for $u \in \Sigma(H_1, ..., H_n)$. Similarly, if $\{u_i\}$ is an unentangled orthonormal basis of V and for each i, $\{v_{ij}\}$ is an orthonormal basis of H_0 then $\{v_{ij} \otimes u_i\}$ is an unentangled orthonormal basis of H. We therefore conclude as above that if $u \in \Sigma(H_1, ..., H_n)$ then there exists $S(u)$ a self adjoint linear operator on H_0 so that $f(v \otimes u) = \langle v|S(u)|v\rangle$ for all $v \in H_0$.

Let $\{u_i\}$ be an unentangled orthonormal basis of V and let $\{v_j\}$ be an orthonormal basis of H_0. Set

$$a_{ij}(v) = \langle u_i|T(v)|u_j\rangle$$

and

$$b_{ij}(u) = \langle v_i|S(u)|v_j\rangle, u \in \Sigma(H_1, ..., H_n).$$

We now observe that if $v = \sum_i x_i v_i$ and if $u = \sum_j y_j u_j$ then we have

$$\sum_{p,q} a_{p,q}(v) \bar{y}_p y_q = \sum_{r,s} b_{r,s}(u) \bar{x}_r x_s.$$

If we substitute $v = v_r$ then we have

$$b_{rr}(u) = \sum_{p,q} a_{p,q}(v_r) \bar{y}_p y_q.$$

Now assuming that $r \neq s$ and taking $v = \frac{1}{\sqrt{2}}(v_r + v_s)$ we have

$$\operatorname{Re} b_{rs}(u) = \sum_{p,q} a_{p,q}(\frac{1}{\sqrt{2}}(v_r + v_s))\bar{y}_p y_q -$$
$$\frac{1}{2}\left(\sum_{p,q} a_{p,q}(v_r)\bar{y}_p y_q + \sum_{p,q} a_{p,q}(v_s)\bar{y}_p y_q\right).$$

Also if we take $v = \frac{1}{\sqrt{2}}(v_r + iv_s)$ then we have

$$-\operatorname{Im} b_{rs}(u) = \sum_{p,q} a_{p,q}(\frac{1}{\sqrt{2}}(v_r + iv_s))y_p\bar{y}_q -$$
$$\frac{1}{2}\left(\sum_{p,q} a_{p,q}(v_r)y_p\bar{y}_q + \sum_{p,q} a_{p,q}(v_s)y_p\bar{y}_q\right).$$

Thus if we set
$$c_{rrpq} = a_{pq}(v_r)$$
and if $r \neq s$ then

$$c_{rspq} = a_{p,q}\left(\frac{1}{\sqrt{2}}(v_r + v_s) - \frac{1}{2}(a_{p,q}(v_r) + a_{p,q}(v_s))\right) +$$
$$a_{p,q}\left(\frac{1}{\sqrt{2}}(v_r + iv_s) - \frac{1}{2}(a_{p,q}(v_r) + a_{p,q}(v_s))\right)$$

Then
$$f(v \otimes u) = \sum_{rspq} c_{rspq} \bar{x}_r \bar{y}_p x_s y_q.$$

This is the content of the theorem.

We will now give a counterexample to the analogous assertion for product bases.

PROPOSITION 5. *Let H_1 and H_2 be finite dimensional Hilbert spaces of dimension greater than 1. Then there exists $f : \Sigma(H_1, H_2) \to [0, \infty)$ such that $\sum_{i,j} f(u_i \otimes v_j) = w$, with $w \in \mathbb{R}$ fixed, for all choices $\{u_i\}$ and $\{v_j\}$ of orthonormal bases of H_1 and H_2 respectively but there is no linear endomorphism, T, on $H_1 \otimes H_2$ such that $f(u \otimes v) = \langle u \otimes v | T | u \otimes v \rangle$ for $u \in S(H_1)$ and $v \in S(H_2)$.*

Proof. Let for $w > 0$, $\mathcal{P}_w$ denote the set of all Hermitian positive semi-definite endomorphisms, A, of H_2 such that $tr(A) = w$. Fix $w_o = \frac{w}{\dim H_1}$. Let $\varphi : S(H_1) \to \mathcal{P}_{w_o}$ be a mapping (completely arbitrary). Set $f(u \otimes v) = \langle v | \varphi(u) | v \rangle$, for $u \in S(H_1)$ and $v \in S(H_2)$. If $\{u_i\}$ is an orthonormal basis of H_1 and if $\{v_j\}$ is an orthonormal basis of H_2 then

$$\sum_{i,j} f(u_i \otimes v_j) = \sum_i \left(\sum_j \langle v_j | \varphi(u_i) | v_j \rangle\right) = \sum_i tr(\varphi(u_i)) = \dim(H_1)w_o.$$

Note: In this argument only one factor need be finite dimensional. Also note that f can be chosen to be continuous.

3. Unentangled Bases

In this section we will develop the material on "unentangled bases" that we will need to prove Theorem 3 (in fact as we shall see a generalization). Let V be a 2-dimensional Hilbert space and let H be an n-dimensional Hilbert space. Fix $\Sigma \subset S(H)$ such that $\lambda \Sigma = \Sigma$ for all $\lambda \in \mathbb{C}$ with $|\lambda| = 1$. We will use the notation $S(V) \otimes \Sigma = \{v \otimes w | v \in S(V), w \in \Sigma\}$.

If $a \in S(V)$ then up to scalar multiple there is exactly one element of $S(V)$ that is perpendicular to a. We will denote a choice of such an element by $\widehat{a}$. The main result of this section is

THEOREM 6. *If $\{u_j\}_{j=1}^{2n}$ is an orthonormal basis of $V \otimes H$ with $u_j \in S(V) \otimes \Sigma$ for $j = 1, ..., 2n$ then there exists a partition*

$$n_1 \geq n_2 \geq ... \geq n_r > 0$$

of n, an orthogonal decomposition

$$H = U_1 \oplus ... \oplus U_r,$$

elements $a_1, ..., a_r \in S(H)$, and for each $i = 1, ..., r$ orthonormal bases $\{b_{i1}, ..., b_{in_i}\}$ and $\{c_{i1}, ..., c_{in_i}\}$ of U_i such that

$$\{u_i | i = 1, ..., 2n\} = \bigcup_{i=1}^{r} \left(\{a_i \otimes b_{ij} | j = 1, ..., n_i\} \cup \{\widehat{a}_i \otimes c_{ij} | j = 1, ..., n_i\} \right).$$

Before we prove the theorem we will make several preliminary observations. Let $\{u_i\}$ be as in the statement of the theorem. Then each $u_i = a_i \otimes h_i$ with $a_i \in S(V)$ and $h_i \in \Sigma$.

1. For each i there exists j such that a_j is a multiple of $\widehat{a}_i$.

If not then we would have $\langle a_i | a_j \rangle \neq 0$ for all j. Since $\langle a_i \otimes h_i | a_j \otimes h_j \rangle = \langle a_i | a_j \rangle \langle h_i | h_j \rangle$, $\langle h_i | h_j \rangle = 0$ for all $j \neq i$. This implies that $\{u_j\}_{j \neq i} \subset V \otimes \{h_i^{\perp}\}$. This space has dimension equal to $2(n-1)$. So it could not contain $2n-1$ orthonormal elements. This contradiction implies that assertion 1 is true.

2. Assume that $i \neq j$. If $\langle a_i | a_j \rangle \neq 0$ then $\langle h_i | h_j \rangle = 0$. If $\langle h_i | h_j \rangle \neq 0$ then $\langle a_i | a_j \rangle = 0$.

This is clear (see the proof of 1.)

We will now prove the theorem by induction on n. If $n = 1$ the result is trivial. We assume the result for all H with $\dim H < n$ and all possible choices for Σ. We now prove it for n.

For each i let m_i denote the number of j such that a_j is a multiple of a_i. Let $m = \max\{m_i | i = 1, ..., 2n\}$. If we relabel we may assume that the first m of the a_i are equal to a_1 (we may have to multiply h_i by a scalar of norm 1). By 1. above we may assume that the next k of the a_i are equal to $\widehat{a}_1$ with $1 \leq k \leq m$ and if $i > m+k$ then a_i is not a multiple of either a_1 or $\widehat{a}_1$. This implies by 2. above that $\langle h_i | h_j \rangle = 0$ for $j > m+k$ and $i = 1, ..., m$. Also $\{h_1, ..., h_m\}$ is an orthonormal set. Thus $u_i \in V \otimes (\{h_1, ..., h_m\}^{\perp})$ for $i > m+k$. This implies that $V \otimes (\{h_1, ..., h_m\}^{\perp})$ contains

$2n-(m+k)$ orthonormal elements. Since $\dim V \otimes (\{h_1,...,h_m\}^\perp) = 2(n-m)$ this implies that $k=m$. We now rewrite the first $2m$ elements of the basis as

$$a_1 \otimes b_1, ..., a_1 \otimes b_m, \widehat{a}_1 \otimes c_1, ..., \widehat{a}_1 \otimes c_m.$$

If we apply observation 2. again we see that the elements h_i for $i > 2m$ must be orthogonal to $\{b_1,...,b_m\}$ and to $\{c_1,...,c_m\}$. A dimension count says that they must span the orthogonal complements of both $\{b_1,...,b_m\}$ and $\{c_1,...,c_m\}$. But then $\{b_1,...,b_m\}$ and $\{c_1,...,c_m\}$ must span the same space, $U \subset H$. We have therefore shown that $\{u_i\}_{i>2m}$ is an orthonormal basis of $V \otimes U^\perp$. We may thus apply the inductive hypothesis to $U^\perp$ and $\Sigma \cap U^\perp$. This completes the inductive step and hence the proof.

If W is a Hilbert space and if Ξ is a subset of $S(W)$ that is invariant under multiplication by scalars of absolute value 1 then a function $f : \Xi \to \mathbb{C}$ is said to be a Ξ-frame function of weight $w = w_f$ if whenever $\{u_i\}$ is an orthonormal basis of W with $u_i \in \Xi$ (i.e. $\{u_i\}$ is a Ξ-frame) we have $\sum_i f(u_i) = w$. We note

3. Let f be a Ξ-frame function. If $\{u_i\}$ is a Ξ-frame for W and if F is a subset of $\{u_i\}$ then $f_{|F^\perp \cap \Xi}$ is a $F^\perp \cap \Xi$-frame function of weight $w_f - \sum_{u_i \in F} f(u_i)$.

This is pretty obvious. Let $\{v_j\}$ be a $\Xi \cap F^\perp$-frame for $F^\perp$. Then $\{v_j\} \cup F$ is a Ξ-frame for W.

PROPOSITION 7. *Let V be a two dimensional Hilbert space and let H be an n-dimensional Hilbert space. Let $\Sigma \subset S(H)$ be as in the rest of this section and let $g : S(V) \to \mathbb{C}$ and $h : \Sigma \to \mathbb{C}$ be respectively a frame function and a Σ-frame function. Then if $f(v \otimes w) = g(v)h(w)$ for $v \in S(H)$ and $w \in \Sigma$ then f is an $S(V) \otimes \Sigma$-frame function of weight $w_g w_h$.*

Proof. Let $\{u_i\}$ be an $S(V) \otimes \Sigma$-frame. Then Theorem 5 implies that we may assume that there is partition $n_1 \geq n_2 \geq ... \geq n_r > 0$ of n and elements a_i, b_{ij} and c_{ij} as in the statement so that

$$\{u_i\} = \bigcup_{i=1}^{r} \left(\{a_i \otimes b_{ij} | j=1,...,n_i\} \cup \{\widehat{a}_i \otimes c_{ij} | j=1,...,n_i\} \right).$$

Thus

$$\sum_i f(u_i) = \sum_i g(a_i) \sum_{j=1}^{n_i} h(b_{ij}) + \sum_i g(\widehat{a}_i) \sum_{j=1}^{n_i} h(c_{ij}).$$

Observation 3. above implies that for each i we have $\sum_{j=1}^{n_i} h(b_{ij}) = \sum_{j=1}^{n_i} h(c_{ij})$. Now $g(a_i) + g(\widehat{a}_i) = w_g$. Hence since $\{b_{ij}\}$ is a Σ-frame the result follows.

Theorem 3 is an immediate consequence of the above proposition.

4. Entangled subspaces.

Let $H_1, ..., H_n$ be finite dimensional Hilbert spaces and set $H = H_1 \otimes H_2 \otimes \cdots \otimes H_n$. If $V \subset H$ is a subspace than we will say then V is *entangled* if whenever $v \in V$ and $v \neq 0$ then v is entangled (i.e. v cannot be written in the form $v = h_1 \otimes h_2 \otimes \cdots \otimes h_n$ for any choice of $h_i \in H_i$). The purpose of this section is to give a proof of Theorem 4 using basic algebraic geometry. That is, we will prove that

$$\dim V \leq \dim(H_1) \cdots \dim(H_n) - \sum(\dim H_i - 1) - 1$$

and that this estimate is best possible. The reader should consult [Hartshorne] for the algebraic geometry used in the proof of this result.

Let $L = \{\lambda \in H^* \,|\, \lambda(V) = 0\}$ (H^* the complex dual space of H). Let $X = \{h_1 \otimes \cdots \otimes h_n \,|\, h_i \in H_i\}$. We consider the map $\Phi : H_1 \times ... \times H_n \to X$ given by $\Phi(h_1, ..., h_n) = h_1 \otimes \cdots \otimes h_n$. Then Φ is a surjective polynomial mapping. If we denote by $\overline{\Phi}$ the corresponding mapping of projective spaces we have $\overline{\Phi} : P(H_1) \times ... \times P(H_n) \to P(H)$. General theory implies that the image of $\overline{\Phi}$ is Zariski closed in $P(H)$. Since X is clearly the cone on that image we see that X is Zariski closed and irreducible. Also the map $\overline{\Phi}$ is injective so the dimension over $\mathbb{C}$ of its image is $\sum(\dim H_i - 1)$. Thus the dimension over $\mathbb{C}$ of X is $d = \sum(\dim H_i - 1) + 1$.

Since V is entangled $X \cap V = \{0\}$. This implies that $\{x \in X \,||\, \lambda(x) = 0, \lambda \in L\} = \{0\}$. Thus $\dim L \geq \dim X = d$. Hence $\dim V = \dim H - \dim L \leq \dim H - d$. This is the asserted upper bound. The fact that this upper bound is best possible follows from the Noether normalization theorem which implies that there exist $\lambda_1, ..., \lambda_d \in H^*$ such that $\{x \in X \,||\, \lambda_i(x) = 0 \text{ for all } i\} = \{0\}$ (i.e. a linear system of parameters).

References

[Gleason] A. M. Gleason, Measures on the closed subspaces of a Hilbert space, J. Math. and Mech. 6 (1957), 885–893.

[Born] M. Born, Zur Quantenmechanik der Stossvorgänge, Zeits. Phys. 37 (1926), 863–867. Reprinted and translated in *Quantum Theory and Measurement*, edited by J. A. Wheeler and W. H. Zurek (Princeton U. Press, Princeton, NJ, 1983), pp. 52–55.

[Pitowsky] I. Pitowsky, Infinite and finite Gleason's theorems and the logic of indeterminacy, J. Math. Phys. 39 (1998), 218–228.

[Bell] J. S. Bell, On the problem of hidden variables in quantum mechanics, Rev. Mod. Phys. 38 (1966), 447–452.

[BDFMRSSW] C. H. Bennett, D. P. DiVincenzo, C. A. Fuchs, T. Mor, E. Rains, P. W. Shor, J. A. Smolin, and W. K. Wootters, Quantum nonlocality without entanglement, Phys. Rev. A 59 (1999), 1070–1091.

[Araki] H. Araki, On a characterization of the state space of quantum mechanics, Comm. Math. Phys 75 (1980), 1–24.

[Wootters] W. K. Wootters, Local accessibility of quantum states, in *Complexity, Entropy and the Physics of Information*, edited by W. H. Zurek (Addison-Wesley, Redwood City, CA, 1990), pp. 39–46.

[Fuchs] C. A. Fuchs, private communication.

[BDMPSST] C. H. Bennett, D. P. DiVincenzo, T.Mor, P. W. Shor, J. A. Smolin, B. M. Terhal, Unextendible product bases and bound entanglement, Phys. Rev. Lett. 82 (1999) 5385–5388.

[Hartshorne] R. Hartshorne, *Algebraic Geometry*, Graduate Texts in Mathematics, 52, Springer-Verlag, New York, 1977.

(Nolan R. Wallach) UNIVERSITY OF CALIFORNIA, SAN DIEGO
E-mail address, Nolan R. Wallach: `nwallach@ucsd.edu`

Entangled Chains

William K. Wootters

ABSTRACT. Consider an infinite collection of qubits arranged in a line, such that every pair of nearest neighbors is entangled: an "entangled chain." In this paper we consider entangled chains with translational invariance and ask how large one can make the nearest neighbor entanglement. We find that it is possible to achieve an entanglement of formation equal to 0.285 ebits between each pair of nearest neighbors, and that this is the best one can do under certain assumptions about the state of the chain.

1. Introduction: Example of an entangled chain

Quantum entanglement has been studied for decades, first because of its importance in the foundations of quantum mechanics, and more recently for its potential technological applications as exemplified by a quantum computer [1]. The new focus has led to a quantitative theory of entanglement [2, 3, 4] that, among other things, allows us to express analytically the degree of entanglement between simple systems [5]. This development makes it possible to pose new quantitative questions about entanglement that could not have been raised before and that promise fresh perspectives on this remarkable phenomenon. In this paper I would like to raise and partially answer such a question, concerning the extent to which a collection of binary quantum objects (qubits) can be linked to each other by entanglement.

Imagine an infinite string of qubits, such as two-level atoms or the spins of spin-1/2 particles. Let us label the locations of the qubits with an integer j that runs from negative infinity to positive infinity. I wish to consider special states of the string, satisfying the following two conditions: (i) each qubit is entangled with its nearest neighbors; (ii) the state is invariant under all translations, that is, under transformations that shift each qubit from its original position j to position $j + n$ for some integer n. Let us call a string of qubits satisfying the first condition an entangled chain, and if it also satisfies the second condition, a uniform entangled chain. Note that each qubit need not be entangled with any qubits other than its two nearest neighbors. In this respect an entangled chain is like an ordinary chain, whose links are directly connected only to two neighboring links. By virtue of the translational invariance, the degree of entanglement between nearest neighbors in a uniform entangled chain must be constant throughout the chain. The main

---

2000 *Mathematics Subject Classification.* Primary 81Q99; Secondary 81P68, 81V70.
Key words and phrases. Quantum mechanics, entanglement.

question I wish to pose is this: How large can the nearest-neighbor entanglement be in a uniform entangled chain?

This problem belongs to a more general line of inquiry about how entanglement can be shared among more than two objects. Some work on this subject has been done in the context of the cloning of entanglement [6–11], where one finds limits on the extent to which entanglement can be copied. In a different setting not particularly involving cloning, one finds an inequality bounding the amount of entanglement that a single qubit can have with each of two other qubits [12]. One can imagine more general "laws of entanglement sharing" that apply to a broad range of configurations of quantum objects. The present work provides further data that might be used to discover and formulate such laws. The specific problem addressed in this paper could also prove relevant for analyzing models of quantum computers in which qubits are arranged along a line, as in an ion trap [13]. The infinite chain can be thought of as an idealization of such a computer. Moreover, the analysis of our question turns out to be interesting in its own right, being related, as we will see, to a familiar problem in many-body physics.

To make the question precise we need a measure of entanglement between two qubits. We will use a reasonably simple and well-justified measure called the "concurrence," which is defined as follows [5].

Consider first the case of pure states. A general pure state of two qubits can be written as

$$(1.1) \qquad |\psi\rangle = \alpha|00\rangle + \beta|01\rangle + \gamma|10\rangle + \delta|11\rangle.$$

One can verify that such a state is factorizable into single-qubit states—that is, it is unentangled—if and only if $\alpha\delta = \beta\gamma$. The quantity $C = 2|\alpha\delta - \beta\gamma|$, which ranges from 0 to 1, is thus a plausible measure of the degree of entanglement. We take this expression as the definition of concurrence for a pure state of two qubits. For mixed states, we define the concurrence to be the greatest convex function on the set of density matrices that gives the correct values for pure states [14].

Though this statement defines concurrence, it does not tell us how to compute it for mixed states. Remarkably, there exists an explicit formula for the concurrence of an arbitrary mixed state of two qubits [5]: Let ρ be the density matrix of the mixed state, which we imagine expressed in the standard basis $\{|00\rangle, |01\rangle, |10\rangle, |11\rangle\}$. Let $\tilde{\rho}$, the "spin-flipped" density matrix, be $(\sigma_y \otimes \sigma_y)\rho^*(\sigma_y \otimes \sigma_y)$, where the asterisk denotes complex conjugation in the standard basis and σ_y is the matrix $\begin{pmatrix} 0 & -i \\ i & 0 \end{pmatrix}$. Finally, let $\lambda_1, \lambda_2, \lambda_3, \lambda_4$ be the square roots of the eigenvalues of $\rho\tilde{\rho}$ in descending order—one can show that these eigenvalues are all real and non-negative. Then the concurrence of ρ is given by the formula

$$(1.2) \qquad C(\rho) = \max\{\lambda_1 - \lambda_2 - \lambda_3 - \lambda_4, 0\}.$$

The best justification for using concurrence as a measure of entanglement comes from a theorem [5] showing that concurrence is a monotonically increasing function of the "entanglement of formation," which quantifies the non-local resources needed to create the given state [4].[1] As mentioned above, the values of C range from zero

---

[1]One can define the entanglement of formation as follows. Let ρ be a mixed state of a pair of quantum objects, to be shared between two separated observers who can communicate with each other only via classical signals. The entanglement of formation of ρ is the asymptotic number of singlet states the observers need, per pair, in order to create a large number of pairs in *pure*

to one: an unentangled state has $C = 0$, and a completely entangled state such as the singlet state $\frac{1}{\sqrt{2}}(|01\rangle - |10\rangle)$ has $C = 1$. Our problem is to find the greatest possible nearest-neighbor concurrence of a uniform entangled chain. At the end of the calculation we can easily re-express our results in terms of entanglement of formation.

Another issue that needs to be addressed in formulating our question is the meaning of the word "state" as applied to an infinite string of qubits; in particular we need to discuss how such a state is to be normalized. Formally, we can define a state of our system as follows. A state w of the infinite string is a function that assigns to every finite set S of integers a normalized (i.e., trace one) density matrix $w(S)$, which we interpret to be the density matrix of the qubits specified by the set S; moreover the function w must be such that if S_2 is a subset of S_1, then $w(S_2)$ is obtained from $w(S_1)$ by tracing over the qubits whose labels are not in S_2. This formal definition is perfectly sensible but somewhat bulky in practice. In what follows we will usually specify states of the string more informally when it is clear from the informal specification how to generate the density matrix of any finite subset of the string. We will also usually use the symbol ρ instead of $w(S)$ to denote the density matrix of a pair of nearest neighbors.

It is not immediately obvious that there exists even a single example of an entangled chain. Note, for example, that the limit of a Schrödinger cat state—an equal superposition of an infinite string of zeros with an infinite string of ones—is not an entangled chain. In the cat state, the reduced density matrix of a pair of neighboring qubits is an incoherent mixture of $|00\rangle$ and $|11\rangle$, which exhibits a classical correlation but no entanglement. (Note, by the way, that our informal statement "an equal *superposition* of an infinite string of zeros with an infinite string of ones," specifies exactly the same state as if we had taken an incoherent *mixture* of these two infinite strings: no finite set of qubits contains information about the phase of the superposition.)

We can, however, construct a simple example of an entangled chain in the following way. Let w_0 be the state such that for each *even* integer j, the qubits at sites j and $j+1$ are entangled with each other in a singlet state. We can write this state informally as[2]

(1.3)
$$\cdots \otimes \left(\frac{|0\rangle_{-2}|1\rangle_{-1} - |1\rangle_{-2}|0\rangle_{-1}}{\sqrt{2}} \right) \otimes \left(\frac{|0\rangle_0|1\rangle_1 - |1\rangle_0|0\rangle_1}{\sqrt{2}} \right) \otimes \left(\frac{|0\rangle_2|1\rangle_3 - |1\rangle_2|0\rangle_3}{\sqrt{2}} \right) \otimes \cdots$$

---

states whose average density matrix is ρ. (This is conceptually different from the *regularized* entanglement of formation, which measures the cost of creating many copies of the *mixed* state ρ [15]. However, it is conceivable that the two quantities are identical.) Entanglement of formation is conventionally measured in "ebits," and for a pair of binary quantum objects it takes values ranging from 0 to 1 ebit.

[2]Alternatively, we can characterize the state w_0 according to our formal definition by specifying the density matrix of each finite collection of qubits: Let S define such a collection. Then for each even integer j such that both j and $j+1$ are in S, the corresponding pair of qubits is in the singlet state; all other qubits (i.e., the unpaired ones) are in the completely mixed state $\begin{pmatrix} \frac{1}{2} & 0 \\ 0 & \frac{1}{2} \end{pmatrix}$, and the full density matrix $w(S)$ is obtained by taking the tensor product of the pair states and single-qubit states.

The state w_0 is not an entangled chain because the qubits are not entangled with both of their nearest neighbors: qubits at even-numbered locations are not entangled with their neighbors on the left. However, if we let w_1 be the state obtained by translating w_0 one unit to the left (or to the right—the result is the same), and let w be an equal mixture of w_0 and w_1—that is, $w = (w_0 + w_1)/2$—then w is a uniform entangled chain, as we now show.

That w is translationally invariant follows from the fact that both w_0 and w_1 are invariant under *even* displacements and that they transform into each other under odd displacements. Thus we need only show that neighboring states are entangled. For definiteness let us consider the qubits in locations $j = 1$ and $j = 2$. In the state w_0, the density matrix for these two qubits is

$$(1.4) \qquad \rho^{(0)} = \begin{pmatrix} \frac{1}{4} & 0 & 0 & 0 \\ 0 & \frac{1}{4} & 0 & 0 \\ 0 & 0 & \frac{1}{4} & 0 \\ 0 & 0 & 0 & \frac{1}{4} \end{pmatrix},$$

that is, the completely mixed state. (The two qubits are from distinct singlet pairs.) The density matrix of the same two qubits in the state w_1 is

$$(1.5) \qquad \rho^{(1)} = \begin{pmatrix} 0 & 0 & 0 & 0 \\ 0 & \frac{1}{2} & \frac{1}{2} & 0 \\ 0 & \frac{1}{2} & \frac{1}{2} & 0 \\ 0 & 0 & 0 & 0 \end{pmatrix},$$

that is, the singlet state. In the state w, the qubits are in an equal mixture of these two density matrices, which is

$$(1.6) \qquad \rho = (\rho^{(0)} + \rho^{(1)})/2 = \begin{pmatrix} \frac{1}{8} & 0 & 0 & 0 \\ 0 & \frac{3}{8} & \frac{1}{4} & 0 \\ 0 & \frac{1}{4} & \frac{3}{8} & 0 \\ 0 & 0 & 0 & \frac{1}{8} \end{pmatrix}.$$

It is easy to compute the concurrence of this density matrix, because $\tilde{\rho}$ is the same as ρ itself. The values λ_i in this case are the eigenvalues of ρ, which are $\frac{5}{8}, \frac{1}{8}, \frac{1}{8}, \frac{1}{8}$. The concurrence is therefore $C = \frac{5}{8} - \frac{1}{8} - \frac{1}{8} - \frac{1}{8} = \frac{1}{4}$. This same value of the concurrence applies to any other pair of neighboring qubits in the string because of the translational invariance. The fact that the concurrence is non-zero implies that neighboring qubits are entangled, so that the state w is indeed an entangled chain. For uniform entangled chains, we will call the common value of C for neighboring qubits the concurrence of the chain. Thus in the above example the concurrence of the chain is $\frac{1}{4}$.

As we will see, it is possible to find uniform entangled chains with greater concurrence. Let $C_{\max}$ be the least upper bound on the concurrences of all uniform entangled chains. We would like to find this number. We know that $C_{\max}$ is no larger than 1, since concurrence never exceeds 1. In fact we can quickly get a somewhat better upper bound, using the following fact: when a qubit is entangled with each of two other qubits, the sum of the squares of the two concurrences is less than or equal to one [12]. In a uniform entangled chain, each qubit must be equally entangled with its two nearest neighbors; so the concurrence with each of them cannot exceed $1/\sqrt{2}$. Thus, so far what we know about $C_{\max}$ is this:

$$(1.7) \qquad 1/4 \leq C_{\max} \leq 1/\sqrt{2}.$$

This is still a wide range. Most of the rest of this paper is devoted to getting a better fix on $C_{\max}$ by explicitly constructing entangled chains.

2. Building chains out of blocks

Using the above example as a model, we will use the following construction to generate other uniform entangled chains. (1) Break the string into blocks of n qubits, and define a state w_0 in which each block is in the same n-qubit state $|\xi\rangle$; that is, w_0 is a tensor product of an infinite number of copies of $|\xi\rangle$. (In the above example n had the value 2 and $|\xi\rangle$ was the singlet state.) (2) Define w_k, $k = 1, \ldots, n-1$, to be the state obtained by shifting w_0 to the left by k units. (3) Let the final state w be the average $(w_0 + \cdots + w_{n-1})/n$. A state generated in this way will automatically be translationally invariant. In order that the chain have a large concurrence, we will need to choose the state $|\xi\rangle$ carefully. Finding an optimal $|\xi\rangle$ and proving that it is optimal may turn out to be a difficult problem. In this paper I will choose $|\xi\rangle$ according to a strategy that makes sense and may well be optimal but is not proven to be so.

In the final state w, each pair of neighboring qubits has the same density matrix because of the translational invariance. Our basic strategy for choosing $|\xi\rangle$, described below, is designed to give this neighboring-pair density matrix the following form:

$$(2.1) \qquad \rho = \begin{pmatrix} \rho_{11} & 0 & 0 & 0 \\ 0 & \rho_{22} & \rho_{23} & 0 \\ 0 & \rho_{32} & \rho_{33} & 0 \\ 0 & 0 & 0 & 0 \end{pmatrix}.$$

(The ordering of the four basis states is the one given above: $|00\rangle, |01\rangle, |10\rangle, |11\rangle$.) One can show that the concurrence of such a density matrix is simply

$$(2.2) \qquad C = 2|\rho_{23}|.$$

Besides making the concurrence easy to compute, the form (2.1) seems a reasonable goal because it picks out a specific kind of entanglement, namely, a coherent superposition of $|01\rangle$ and $|10\rangle$, and limits the ways in which this entanglement can be contaminated or diluted by being mixed with other states. In particular, the form (2.1) does not allow contamination by an orthogonal entangled state of the form $\alpha|00\rangle + \beta|11\rangle$—orthogonal entangled states when mixed together tend to cancel each other's entanglement—or by the combination of the two unentangled states $|00\rangle$ and $|11\rangle$. If the component ρ_{44} were not equal to zero and the form were otherwise unchanged, the concurrence would be $C = \max\{2(|\rho_{23}| - \sqrt{\rho_{11}\rho_{44}}), 0\}$; so it is good to make either ρ_{11} or ρ_{44} equal to zero if this can be done without significantly reducing ρ_{23}. We have chosen to make ρ_{44} equal to zero.

As it happens, one can guarantee the form (2.1) for the density matrix of neighboring qubits by imposing the following three conditions on the n-qubit state $|\xi\rangle$: (i) $|\xi\rangle$ is an eigenstate of the operator that counts the number of qubits in the state $|1\rangle$. That is, each basis state represented in $|\xi\rangle$ must have the same number p of qubits in the state $|1\rangle$. (ii) $|\xi\rangle$ has no component in which two neighboring qubits are both in the state $|1\rangle$. (iii) The nth qubit is in the state $|0\rangle$. (This last condition effectively extends condition (ii) to the boundary between successive blocks.) Condition (i) guarantees that the density matrix ρ for a pair of nearest neighbors is block diagonal, each block corresponding to a fixed number of 1's in the pair. That is, there are two single-element blocks corresponding to $|00\rangle$ and $|11\rangle$, and a 2x2 block corresponding to $|01\rangle$ and $|10\rangle$. Conditions (ii) and (iii) guarantee that ρ_{44} is zero. The conditions thus give us the form (2.1). We impose these three conditions because they seem likely to give the best results; we do not prove that they are optimal.

To illustrate the three conditions and how they can be used, let us consider in detail the case where the block size n is 5 and the number p of 1's in each block is 2. (Our strategy does not specify the value of either n or p; these values will ultimately have to be determined by explicit maximization.) In this case, the only basis states our conditions allow in the construction of $|\xi\rangle$ are $|10100\rangle$, $|10010\rangle$, and $|01010\rangle$. Any other basis state either would have a different number of 1's or would violate one of conditions (ii) and (iii). Thus we write

$$(2.3) \qquad |\xi\rangle = a_{13}|10100\rangle + a_{14}|10010\rangle + a_{24}|01010\rangle.$$

The subscripts in a_{ij} indicate which qubits are in the state $|1\rangle$. The state w of the infinite string is derived from $|\xi\rangle$ as described above. We now want to use Eq. (2.3) to write the density matrix ρ of a pair of nearest neighbors when the infinite string is in the state w. For definiteness let us take the two qubits of interest to be in locations $j = 1$ and $j = 2$, and let us take the 5-qubit blocks in the state w_0 to be given by $j = 1, \ldots, 5$, $j = 6, \ldots, 10$, and so on. Our final density matrix ρ will be an equal mixture of five density matrices, corresponding to the five different displacements of w_0 (including the null displacement).

For w_0 itself, the qubits at $j = 1$ and $j = 2$ are the first two qubits of $|\xi\rangle$. The density matrix for these two qubits, obtained by tracing out the other three qubits of the block, is

$$(2.4) \qquad \rho^{(0)} = \begin{pmatrix} 0 & 0 & 0 & 0 \\ 0 & |a_{24}|^2 & \bar{a}_{14}a_{24} & 0 \\ 0 & a_{14}\bar{a}_{24} & |a_{13}|^2 + |a_{14}|^2 & 0 \\ 0 & 0 & 0 & 0 \end{pmatrix}.$$

For w_1, the qubits at $j = 1$ and $j = 2$ are now the second and third qubits of the block, since the block has been shifted to the left. Thus we trace over the first, fourth, and fifth qubits to obtain

$$(2.5) \qquad \rho^{(1)} = \begin{pmatrix} |a_{14}|^2 & 0 & 0 & 0 \\ 0 & |a_{13}|^2 & 0 & 0 \\ 0 & 0 & |a_{24}|^2 & 0 \\ 0 & 0 & 0 & 0 \end{pmatrix}.$$

In a similar way one can find $\rho^{(2)}$ and $\rho^{(3)}$:

$$\rho^{(2)} = \begin{pmatrix} 0 & 0 & 0 & 0 \\ 0 & |a_{14}|^2 + |a_{24}|^2 & \bar{a}_{13}a_{14} & 0 \\ 0 & a_{13}\bar{a}_{14} & |a_{13}|^2 & 0 \\ 0 & 0 & 0 & 0 \end{pmatrix}; \rho^{(3)} = \begin{pmatrix} |a_{13}|^2 & 0 & 0 & 0 \\ 0 & 0 & 0 & 0 \\ 0 & 0 & |a_{14}|^2 + |a_{24}|^2 & 0 \\ 0 & 0 & 0 & 0 \end{pmatrix}.$$

The density matrix corresponding to w_4 is different in that the two relevant qubits now come from different blocks: the qubit at $j = 1$ is the last qubit of one block and the qubit at $j = 2$ is the first qubit of the next block. The corresponding density matrix is thus the tensor product of two single-qubit states:

$$\rho^{(4)} = \begin{pmatrix} 1 & 0 \\ 0 & 0 \end{pmatrix} \otimes \begin{pmatrix} |a_{24}|^2 & 0 \\ 0 & |a_{13}|^2 + |a_{14}|^2 \end{pmatrix} = \begin{pmatrix} |a_{24}|^2 & 0 & 0 & 0 \\ 0 & |a_{13}|^2 + |a_{14}|^2 & 0 & 0 \\ 0 & 0 & 0 & 0 \\ 0 & 0 & 0 & 0 \end{pmatrix}.$$

To get the neighboring-pair density matrix corresponding to our final state w, we average the above five density matrices, with the following simple result:

$$(2.6) \qquad \rho = \frac{1}{5} \begin{pmatrix} 1 & 0 & 0 & 0 \\ 0 & 2 & x & 0 \\ 0 & \bar{x} & 2 & 0 \\ 0 & 0 & 0 & 0 \end{pmatrix},$$

where

$$(2.7) \qquad x = \bar{a}_{13} a_{14} + \bar{a}_{14} a_{24}.$$

According to Eq. (2.2), the concurrence of the pair is

$$(2.8) \qquad C = \frac{2}{5} |\bar{a}_{13} a_{14} + \bar{a}_{14} a_{24}|.$$

Continuing with this example—$n = 5$ and $p = 2$—let us find out what values we should choose for a_{13}, a_{14}, and a_{24} in order to maximize C. First, it is clear that we cannot go wrong by taking each a_{ij} to be real and non-negative—any complex phases could only reduce the absolute value in Eq. (2.8)—so let us restrict our attention to such values. To take into account the normalization condition, we use a Lagrange multiplier $\gamma/2$ and extremize the quantity

$$(2.9) \qquad a_{13} a_{14} + a_{14} a_{24} - (\gamma/2)(a_{13}^2 + a_{14}^2 + a_{24}^2).$$

Differentiating, we arrive at three linear equations expressed by the matrix equation

$$(2.10) \qquad \begin{pmatrix} 0 & 1 & 0 \\ 1 & 0 & 1 \\ 0 & 1 & 0 \end{pmatrix} \begin{pmatrix} a_{13} \\ a_{14} \\ a_{24} \end{pmatrix} = \gamma \begin{pmatrix} a_{13} \\ a_{14} \\ a_{24} \end{pmatrix}.$$

Of the three eigenvalues, only one allows an eigenvector with non-negative components, namely, $\gamma = \sqrt{2}$. The normalized eigenvector is

$$(2.11) \qquad \begin{pmatrix} a_{13} \\ a_{14} \\ a_{24} \end{pmatrix} = \begin{pmatrix} \frac{1}{2} \\ \frac{1}{\sqrt{2}} \\ \frac{1}{2} \end{pmatrix},$$

which gives $C = \sqrt{2}/5 = 0.283$. This is greater than the value 0.25 that we obtained in our earlier example.

Before generalizing this calculation to arbitrary values of n and p, we adopt some terminology that will simplify the discussion. Let us think of the qubits as "sites," and let us call the two states of each qubit "occupied" ($|1\rangle$) and "unoccupied" ($|0\rangle$). The states $|\xi\rangle$ that we are considering have a fixed number p of occupied sites in a string of n sites; so we can regard the system as a collection of p "particles" in a one-dimensional lattice of length n. Condition (ii) requires that two particles never be in adjacent sites; it is as if each particle is an extended object, taking up two lattice sites, and two particles cannot overlap. Thus the number of particles is limited by the inequality $2p \leq n$.

3. Generalization to blocks of arbitrary size

We now turn to the calculation of the optimal concurrence for general n and p assuming our conditions are satisfied. It will turn out that this calculation can be done exactly.

For any values of n and p, the most general form of $|\xi\rangle$ consistent with condition (i) is

$$(3.1) \qquad |\xi\rangle = \sum_{j_1 < \cdots < j_p} a_{j_1,\ldots,j_p} |j_1,\ldots,j_p\rangle,$$

where $|j_1,\ldots,j_p\rangle$ is the state of n sites $j = 1,\ldots,n$ in which sites $j_1,\ldots,j_p$ are occupied and the rest are unoccupied. Because of conditions (ii) and (iii), $a_{j_1,\ldots,j_p}$ must be zero if two of the indices differ by 1 or if j_p has the value n. The coefficients in Eq. (3.1) satisfy the normalization condition

$$(3.2) \qquad \sum_{j_1 < \cdots < j_p} |a_{j_1,\ldots,j_p}|^2 = 1.$$

Going through the same steps as in the above example, we find that in the state w the density matrix of any pair of neighboring sites is

$$(3.3) \qquad \rho = \frac{1}{n} \begin{pmatrix} n-2p & 0 & 0 & 0 \\ 0 & p & y & 0 \\ 0 & \bar{y} & p & 0 \\ 0 & 0 & 0 & 0 \end{pmatrix},$$

where

$$(3.4) \qquad y = \sum_{q=1}^{p} \sum_{j_1 < \cdots < j_p} \sum_{j'_1 < \cdots < j'_p} \left[\bar{a}_{j_1,\ldots,j_p} a_{j'_1,\ldots,j'_p} \delta_{j'_q, j_q+1} \prod_{r \neq q} \delta_{j'_r, j_r} \right].$$

Here δ is the Kronecker delta, and we define $a_{j_1,\ldots,j_p}$ to be zero if any two of the indices are equal. In words, y is constructed as follows: Let two coefficients $a_{j_1,\ldots,j_p}$ and $a_{j'_1,\ldots,j'_p}$ be called adjacent if they differ in only one index and if the difference in that index is exactly one; then y is the sum of all products of adjacent pairs of coefficients, the coefficient with the smaller value of the special index being complex conjugated in each case. In the above example there were only two such products, $\bar{a}_{13} a_{14}$ and $\bar{a}_{14} a_{24}$; hence the form of Eq. (2.7).

As before, the concurrence of the chain is equal to $2|\rho_{23}|$; that is, $C = (2/n)|y|$. We want to maximize the concurrence over all possible values of the coefficients that are consistent with conditions (ii) and (iii). These conditions are somewhat awkward to enforce directly: one has to make sure that certain of the coefficients $a_{j_1,\ldots,j_p}$ are zero. However, this problem is easily circumvented by defining a new set of indices. Let $k_1 = j_1$, $k_2 = j_2 - 1$, $k_3 = j_3 - 2$, and so on up to $k_p = j_p - (p-1)$, and let $b_{k_1,\ldots,k_p} = a_{j_1,\ldots,j_p}$. The constraints on the new indices k_r are simply that $0 < k_1 < k_2 < \cdots < k_p < n'$, where $n' = n - (p-1)$. Finally, in place of $|\xi\rangle$, define a new vector $|\zeta\rangle$:

$$(3.5) \qquad |\zeta\rangle = \sum_{k_1 < \cdots < k_p} b_{k_1,\ldots,k_p} |k_1,\ldots,k_p\rangle,$$

where $|k_1,\ldots,k_p\rangle$ is the state of a lattice of length $n'-1$ in which the sites $k_1,\ldots,k_p$ are occupied. In effect we have removed from the lattice the site lying to the right of each occupied site. Note that our earlier inequality $2p \leq n$ becomes, in terms of

n', simply $p \leq n' - 1$, which reflects the fact that the new lattice has only $n' - 1$ sites. The concurrence is still given by $C = (2/n)|y|$, where

$$(3.6) \qquad y = \sum_{q=1}^{p} \sum_{k_1 < \cdots < k_p} \sum_{k'_1 < \cdots < k'_p} \left[\bar{b}_{k_1,\ldots,k_p} b_{k'_1,\ldots,k'_p} \delta_{k'_q, k_q+1} \prod_{r \neq q} \delta_{k'_r, k_r} \right].$$

We can express y more simply by introducing creation and annihilation operators for each site. We associate with site k the operators

$$(3.7) \qquad c_k = \begin{pmatrix} 0 & 1 \\ 0 & 0 \end{pmatrix} \text{ and } c_k^\dagger = \begin{pmatrix} 0 & 0 \\ 1 & 0 \end{pmatrix},$$

which are represented here in the basis $\{|0\rangle, |1\rangle\}$. In terms of these operators, we can write y as

$$(3.8) \qquad y = \langle \zeta | \sum_{k=1}^{n'-2} c_k^\dagger c_{k+1} | \zeta \rangle.$$

Our problem is beginning to resemble the nearest-neighbor tight-binding model for electrons in a one-dimensional lattice. The Hamiltonian for the latter problem—assuming that the spins of the electrons are all in the same state and can therefore be ignored—can be written as[3]

$$(3.9) \qquad H = -\sum_{k=1}^{n'-2} (c_k^\dagger c_{k+1} + c_{k+1}^\dagger c_k),$$

where we have taken the lattice length to be the same as in our problem, namely, $n' - 1$. From Eqs. (3.8) and (3.9) we see that $\langle \zeta | H | \zeta \rangle = -2 \operatorname{Re}(y)$. This expectation value is not quite what we need for the concurrence: the concurrence is proportional to the absolute value of y, not its real part. However, as in our earlier example, for the purpose of maximizing C there is no advantage in straying from real, nonnegative values of $b_{k_1,\ldots,k_p}$. If we restrict our attention to such values, then the absolute value of y is the same as its real part, and we can write the concurrence as

$$(3.10) \qquad C = -\frac{1}{n} \langle \zeta | H | \zeta \rangle.$$

Thus, maximizing the concurrence amounts to minimizing the expectation value of H, that is, finding the ground state energy of the tight-binding model, as long as the ground state involves only real and non-negative values of $b_{k_1,\ldots,k_p}$.

The one-dimensional tight-binding model is in fact easy to solve [16, 17]. Its ground state is the discrete analogue of the ground state of a collection of p non-interacting fermions in a one-dimensional box. In our case the "walls" of the box, where the wavefunction goes to zero, are at $k = 0$ and $k = n'$, and the ground state $|\zeta_0\rangle$ is given by the following antisymmetrized product of sine waves:

$$(3.11) \qquad b_{k_1,\ldots,k_p} \propto \mathcal{A}\left[\sin\left(\frac{\pi k_p}{n'}\right) \sin\left(\frac{2\pi k_{p-1}}{n'}\right) \cdots \sin\left(\frac{p\pi k_1}{n'}\right) \right].$$

---

[3]In Eq. (3.9) the operators c and $c^\dagger$ are fermionic, whereas those defined in Eq. (3.7) are not, because they do not anticommute when they are associated with different sites. We could, however, use our c's to define genuinely fermionic operators in terms of which the extremization problem has exactly the same form [16].

Here $\mathcal{A}$ indicates the operation of antisymmetrizing over the indices $k_1, \ldots, k_p$. In the range of values we are allowing for these indices, that is, $0 < k_1 < k_2 < \cdots < k_p < n'$, the coefficients $b_{k_1,\ldots,k_p}$ are indeed non-negative, so that Eq. (3.10) is valid.

The ground state energy, from which we can find the concurrence, is simply the sum of the first p single-particle eigenvalues of H. There are exactly $n' - 1$ such eigenvalues, one for each dimension of the single-particle subspace; they are given by

$$(3.12) \qquad E_m = -2\cos\left(\frac{m\pi}{n'}\right), \quad m = 1, \ldots, n' - 1.$$

Thus the concurrence is

$$(3.13) \qquad C = -\frac{1}{n}\langle \zeta_0 | H | \zeta_0 \rangle = \frac{2}{n}\sum_{m=1}^{p} \cos\left(\frac{m\pi}{n'}\right).$$

Doing the sum is straightforward, with the following result:

$$(3.14) \qquad C = \frac{1}{n}\left[\frac{\cos(p\pi/n') - \cos((p+1)\pi/n') + \cos(\pi/n') - 1}{1 - \cos(\pi/n')}\right].$$

Recall that $n' = n - p + 1$. Eq. (3.14) gives the largest value of C consistent with our conditions, for fixed values of n and p. Note, for example, that when $n = 5$ and $p = 2$, Eq. (3.14) gives $C = \sqrt{2}/5$, just as we found before for this case.

We still need to optimize over n and p. It is best to make the block size n very large—any state w that is possible with block size n is also allowed by block size $2n$—so we take the limit as n goes to infinity. Let α be the density of occupied sites—that is, $\alpha = p/n$—and let n approach infinity with α held fixed. In this limit, the concurrence becomes

$$(3.15) \qquad C_{\text{lim}} = \frac{2}{\pi}(1 - \alpha)\sin\left(\frac{\alpha\pi}{1-\alpha}\right).$$

Taking the derivative, one finds that C_{lim} is maximized when

$$(3.16) \qquad \tan\left(\frac{\alpha\pi}{1-\alpha}\right) = \frac{\pi}{1-\alpha},$$

which happens at $\alpha = 0.300844$, where $C_{\text{lim}} = 0.434467$. This is the highest value of concurrence that is consistent with our method of constructing the state of the chain and with our three conditions on $|\xi\rangle$. Note that it is considerably larger than what we got in our first example, in which a string of singlets was mixed with a shifted version of the same string—one might call this earlier construction the "bicycle chain" state. Unlike the bicycle chain state, our best state breaks the symmetry between the basis states $|0\rangle$ and $|1\rangle$: the fraction of qubits in the state $|1\rangle$ is about 30% rather than 50%. Of course the entanglement would be just as large if the roles of $|1\rangle$ and $|0\rangle$ were reversed.

It is interesting to ask what value of entanglement of formation the above value of concurrence corresponds to. As a function of the concurrence, the entanglement of formation is given by

$$(3.17) \qquad E_f = h\left(\frac{1 + \sqrt{1 - C^2}}{2}\right),$$

where h is the binary entropy function $h(x) = -[x\log_2 x + (1-x)\log_2(1-x)]$. For the above value of concurrence, one finds that the entanglement of formation is $E_f = 0.284934$ ebits. (For the bicycle chain state, the entanglement of formation between neighboring pairs is only 0.118 ebits.)

If one can prove that this value is optimal, then it can serve as a reference point for interpreting entanglement values obtained for real physical systems. A string of spin-1/2 particles interacting via the antiferromagnetic Heisenberg interaction, for example, has eigenstates that typically have some non-zero nearest-neighbor entanglement. It would be interesting to find out how the entanglements appearing in these states compare to the maximum possible entanglement for a string of qubits.[4]

Clearly the problem we have analyzed here can be generalized. One can consider a two or three-dimensional lattice of qubits and ask how entangled the neighboring qubits can be. If we were to analyze these cases using assumptions similar to those we have made in the one-dimensional case, we would again find the problem reducing to a many-body problem, but with less tractable interactions. Assuming that pairwise entanglement tends to diminish as the total entanglement is shared among more particles, one expects the optimal values of C and E_f to shrink as the dimension of the lattice increases.

I would like to thank Kevin O'Connor for many valuable discussions on distributed entanglement.

References

[1] See, for example, D. P. DiVincenzo, Science **270** (1995), 255.
[2] C. H. Bennett, H. J. Bernstein, S. Popescu, and B. Schumacher, Phys. Rev. A **53** (1996), 2046.
[3] V. Vedral, M. B. Plenio, M. A. Rippin, and P. L. Knight, Phys. Rev. Lett. **78** (1997), 2275; V. Vedral, M. B. Plenio, K. Jacobs, and P. L. Knight, Phys. Rev. A **56** (1997), 4452; V. Vedral and M. B. Plenio, Phys. Rev. A **57** (1998), 1619.
[4] C. H. Bennett, D. P. DiVincenzo, J. Smolin, and W. K. Wootters, Phys. Rev. A **54** (1996), 3824.
[5] S. Hill and W. K. Wootters, Phys. Rev. Lett. **78** (1997), 5022; W. K. Wootters, Phys. Rev. Lett. **80** (1998), 2245.
[6] V. Bužek, V. Vedral, M. B. Plenio, P. L. Knight, and M. H. Hillery, Phys. Rev. A **55** (1997), 3327.
[7] N. Cerf, J. Mod. Opt. **47** (2000), 187.
[8] D. Bruß, D. P. DiVincenzo, A. Ekert, C. A. Fuchs, C. Macchiavello, and J. A. Smolin, Phys. Rev. A **57** (1998), 2368.
[9] A. Karlsson and M. Bourennane, Phys. Rev. A **58** (1998), 4394.
[10] M. Murao, D. Jonathan, M. B. Plenio, and V. Vedral, Phys. Rev. A **59** (1999), 156.
[11] D. Bruß, Phys. Rev. A **60** (1999), 4344.
[12] V. Coffman, J. Kundu, and W. K. Wootters, Phys. Rev. A **61** (2000), 052306.
[13] I. Cirac and P. Zoller, Phys. Rev. Lett. **74** (1995),4091.
[14] A. Uhlmann, Phys. Rev. A **62** (2000), 032307.
[15] P. M. Hayden, M. Horodecki, and B. M. Terhal, quant-ph/0008134.
[16] E. Lieb, T. Schultz, and D. Mattis, Annals of Phys. **16** (1961), 407.
[17] See, for example, G. D. Mahan, *Many-Particle Physics*, 2nd ed. (Plenum, New York, 1990), pp. 25-27, 51.
[18] K. M. O'Connor and W. K. Wootters, Phys. Rev. A **63** (2001), 052302.

---

[4]Since the original version of this paper was written, the question about the antiferromagnetic Heisenberg chain has been answered for the ground state [**18**]: though the nearest-neighbor concurrence of the ground state is high ($C = 0.386$), it is not optimal.

(William K. Wootters) DEPARTMENT OF PHYSICS, WILLIAMS COLLEGE, WILLIAMSTOWN, MA 01267, USA

E-mail address, William K. Wootters: `William.K.Wootters@williams.edu`

URL: `http://www.williams.edu:803/Physics/wwootters`

Titles in This Series

305 **Samuel J. Lomonaco, Jr. and Howard E. Brandt, Editors,** Quantum computation and information, 2002

304 **Jorge Alberto Calvo, Kenneth C. Millett, and Eric J. Rawdon, Editors,** Physical knots: Knotting, linking, and folding geometric objects in $\mathbb{R}^3$, 2002

303 **William Cherry and Chung-Chun Yang, Editors,** Value distribution theory and complex dynamics, 2002

302 **Yi Zhang, Editor,** Logic and algebra, 2002

301 **Jerry Bona, Roy Choudhury, and David Kaup, Editors,** The legacy of the inverse scattering transform in applied mathematics, 2002

300 **Sergei Vostokov and Yuri Zarhin, Editors,** Algebraic number theory and algebraic geometry: Papers dedicated to A. N. Parshin on the occasion of his sixtieth birthday, 2002

299 **George Kamberov, Peter Norman, Franz Pedit, and Ulrich Pinkall,** Quaternions, spinors, and surfaces, 2002

298 **Robert Gilman, Alexei G. Myasnikov, and Vladimir Shpilrain, Editors,** Computational and statistical group theory, 2002

297 **Stephen Berman, Paul Fendley, Yi-Zhi Huang, Kailash Misra, and Brian Parshall, Editors,** Recent developments in infinite-dimensional Lie algebras and conformal field theory, 2002

296 **Sean Cleary, Robert Gilman, Alexei G. Myasnikov, and Vladimir Shpilrain, Editors,** Combinatorial and geometric group theory, 2002

295 **Zhangxin Chen and Richard E. Ewing, Editors,** Fluid flow and transport in porous media: Mathematical and numerical treatment, 2002

294 **Robert Coquereaux, Ariel García, and Roberto Trinchero, Editors,** Quantum symmetries in theoretical physics and mathematics, 2002

293 **Donald M. Davis, Jack Morava, Goro Nishida, W. Stephen Wilson, and Nobuaki Yagita, Editors,** Recent progress in homotopy theory, 2002

292 **A. Chenciner, R. Cushman, C. Robinson, and Z. Xia, Editors,** Celestial Mechanics, 2002

291 **Bruce C. Berndt and Ken Ono, Editors,** q-series with applications to combinatorics, number theory, and physics, 2001

290 **Michel L. Lapidus and Machiel van Frankenhuysen, Editors,** Dynamical, spectral, and arithmetic zeta functions, 2001

289 **Salvador Pérez-Esteva and Carlos Villegas-Blas, Editors,** Second summer school in analysis and mathematical physics: Topics in analysis: Harmonic, complex, nonlinear and quantization, 2001

288 **Marisa Fernández and Joseph A. Wolf, Editors,** Global differential geometry: The mathematical legacy of Alfred Gray, 2001

287 **Marlos A. G. Viana and Donald St. P. Richards, Editors,** Algebraic methods in statistics and probability, 2001

286 **Edward L. Green, Serkan Hoşten, Reinhard C. Laubenbacher, and Victoria Ann Powers, Editors,** Symbolic computation: Solving equations in algebra, geometry, and engineering, 2001

285 **Joshua A. Leslie and Thierry P. Robart, Editors,** The geometrical study of differential equations, 2001

284 **Gaston M. N'Guérékata and Asamoah Nkwanta, Editors,** Council for African American researchers in the mathematical sciences: Volume IV, 2001

For a complete list of titles in this series, visit the
AMS Bookstore at **www.ams.org/bookstore/**.